# A HISTORY OF WORLD AGRICULTURE

*From the Neolithic Age to the Current Crisis*

MARCEL MAZOYER *and* LAURENCE ROUDART

*Translated by* JAMES H. MEMBREZ

## EARTHSCAN

*London · Sterling, VA*

First published by Earthscan in the UK in 2006
Copyright © Marcel Mazoyer and Laurence Roudart, 2006

ISBN-10:   1-84407-399-8      paperback
           1-84407-400-5      hardback
ISBN-13:   978-1-84407-399-3  paperback
           978-1-84407-400-6  hardback

For a full list of publications please contact:

Earthscan
8-12 Camden High Street
London, NW1 0JH, UK
Tel:  +44 (0)20 7387 8558
Fax: +44 (0)20 7387 8998
Email:  earthinfo@earthscan.co.uk
Web:  www.earthscan.co.uk

Printed in Canada

22883 Quicksilver Drive, Sterling, VA 20166-2012, USA

*Earthscan publishes in association with the International Institute
for Environment and Development*

A catalogue record for this book is available from the British Library

Cet ouvrage, publié dans le cadre d'un programme d'aide à la publication,
bénéficie du soutien du Ministere des Affaires étrangères et du Service Culturel
de l'Ambassade de France aux Etats-Unis.

This work, published as part of a program of aid for publication, received
support from the French Ministry of Foreign Affairs and Cultural Services
of the French Embassy in the United States.

# Contents

# Acknowledgments

Each of this book's chapters was first written by one of us, then rewritten by the other and finally gone over again, corrected and polished together. In fact, this work on world agricultural systems would not have seen the light of day without the intense and continual questioning born of our rural origins and our training, without our engagement in research and development activities in many countries, and without the numerous profound and thoughtful exchanges that we have had with peasants from the four corners of the world who alone possess the original and intimate knowledge of their own practices. This work is part of the tradition associated with the chair of Comparative Agriculture of the National Agronomic Institute, Paris-Grignon (INA P-G), so ardently represented by Professor René Dumont.

This book took shape slowly and could not have been completed without the knowledge accumulated over the last few decades by historians, geographers, anthropologists, sociologists, economists, and agronomists. It would not have been written without the pressing demand from numerous auditors, professionals, students, and friends nor without the support of mesdames Francine Bassu, Isabelle Lemercier, Laure Vilcosqui, Mary Hermand, and of messieurs Hervé Bichat, Daniel Reitzer, Paul Vialle, Gilles Bazin, Philippe Guérin, and Christian Férault. We warmly thank all of them.

Finally, our deepest gratitude goes to those of our colleagues and close relations who, by their patience and confidence, supported us throughout the course of this work.

—M. MAZOYER AND L. ROUDART

# Preface

Five years after the first edition of this book, a number of new facts have invited us to review the recent evolution of world agricultural systems and human nutrition, and their possible futures.

## PEASANT POVERTY AND UNDERNOURISHMENT

At the beginning of the twenty-first century, around half of the planet's approximately six billion human beings live in poverty, with a purchasing power equivalent to less than two U.S. dollars per day. Close to two billion suffer from serious deficiencies of iron, iodine, vitamin A, and other vitamins or minerals.[1] More than one billion people have no access to potable water, and around 840 million are victims of undernourishment, which means that they do not have continuous access to a food ration sufficient to cover their basic energy needs.[2] In other words, they are hungry almost everyday.

Famines, which occur in various places on the occasion of some drought, flood, storm, illness of plants, animals or humans, or even war, are no less the ultimate consequence of poverty and undernourishment. In fact, these climatic, biological, or political accidents lead to famine only in regions of the world where large parts of the population already suffer from such a high level of poverty and food insecurity that they do not have the means to fight effectively against these calamities and their consequences.

This tragic situation is not new nor is it improving. Certainly, the *portion* of the total world population that is undernourished has diminished over the last three decades of the twentieth century, but the *number* of underfed persons in the world has hardly gone down. This is why the more than 180 heads of state and government gathered in Rome in 1996 for the World Food Summit committed themselves "to deploy a constant effort to eradicate hunger in every country and, for the time being, to reduce by half the number of underfed people by the year 2015 at the latest." That assumes that the world will have around 400 million underfed people in 2015. But the means mobilized in this effort have been neither as substantial nor as efficacious as predicted. Five years later, in 2001, it is necessary to recognize that the world will still count 600 to 700 million undernourished in 2015. At this rate, it will be more than a century before this curse can be made to disappear.

Thus the conventional means for struggling against hunger, even though strengthened, have once again turned out to be incapable of defeating it within a sufficiently short time frame to be morally acceptable, socially supportable, and politically tenable. In order to reduce the extreme poverty that leads to hunger and sometimes even to famine and death, it is not sufficient to attend to the most shocking symptoms of these evils. Rather, it is necessary to attack their profound causes, and to do that it is necessary to have recourse to other analyses and other means of struggle.

First, it is important to take into account that *close to three-quarters of the undernourished individuals in the world live in rural areas.* Among the rural poor, one finds that the large majority are peasants, who are especially poorly equipped, poorly situated, and bad off, or agricultural workers, artisans, and merchants who live on the basis of their relationship to the peasants and are as poor as they are. As for the other undernourished people, most are recent rural residents pushed by poverty toward refugee camps or underequipped and underindustrialized urban peripheries, where they still cannot find sufficient means of existence. Since the number of poor and hungry in the countryside has hardly diminished, even though each year several tens of millions of people leave the rural areas, one can only conclude that an almost equal number of new poor and hungry are formed each year in the countryside.

Most of the world's hungry people are not urban consumers and purchasers of food but peasant producers and sellers of agricultural products. Further, their high number is not a simple heritage from the past but the result of an ongoing process leading to extreme poverty for hundreds of millions of deprived peasants.

In order to explain this process, we will discuss the following questions: How large are the inequalities between different agricultures in the world? How has the contemporary agricultural revolution, carried out by a minority of farmers in the developed countries and in some developing countries, greatly increased these inequalities? Why has the Green Revolution, pursued by close to two-thirds of the farmers in the developing countries, only partially reduced these inequalities? How has the tendency for real agricultural prices to fall, which results from these agricultural revolutions, blocked development and led to extreme poverty for more than one-third of the planet's peasants?

## VERY UNEQUAL AGRICULTURES

One can measure the raw productivity of agricultural labor by looking at the production in quintals (one quintal = 100 kilograms) of cereals or cereal-equivalents per agricultural worker per year.[3] In a little more than a half century, the difference in productivity between the least efficient agriculture in the world, practiced exclusively with manual implements (hoe, spade, digging stick, machete, harvest knife, sickle) and the best-equipped and most efficient agriculture has increased dramatically. The gap has widened from 1 to 10 in the interwar period to 1 to 2000 at the end of the twentieth century.

## The Contemporary Agricultural Revolution

In the course of the second half of the twentieth century, the contemporary agricultural revolution (large-scale motorization and mechanization, selection of plant varieties and of animal species with a strong potential for high yields, widespread use of fertilizers, concentrated feed for livestock, and pesticides for plants and domestic animals) has greatly progressed in the developed countries and in some limited sectors of the developing countries.

In the developed countries, farmers who were already relatively productive have benefited from policies of support for agricultural development, as well as from real agricultural prices, which, at the beginning of the period under consideration, were much higher than today, thereby enabling maximum opportunities for investment and progress. But, ultimately, fewer than 10 percent of the farms have succeeded in going through every stage of this revolution. Today, the best equipped, the best proportioned, the best situated among them attain a raw productivity on the order of 20,000 quintals of cereal-equivalents per worker per year (200 hectares per worker multiplied by 100 quintals per hectare = 20,000 quintals or 2,000,000 kilograms per worker). The gains in agricultural productivity thus obtained have been so rapid and so high that they have exceeded those of industry and services. As a result, there has been a strong reduction in real agricultural prices. Depending upon the product, these prices have been reduced by 2, 3, or 4 times in the course of the second half of the twentieth century. Consequently, during this time more than 90 percent of the farms, the least advantaged, have been blocked in their development and impoverished by the lowering of prices to the extent that, one after another, they disappeared and thereby provided a labor force for expanding industry and services.

In the developing countries, the immense majority of peasants have not had the means to attain the costly large-scale motorization and mechanization of agriculture. In a few regions, however (Latin America, the Middle East, South Africa), some large agricultural entrepreneurs, having at their disposal thousands of hectares of land and poorly paid day laborers, have profited from inflation and the relatively high international agricultural prices from the first half of the 1970s, as well as advantageous credit terms, to equip themselves in turn. Today, the most successful of these large farms have labor productivity as high as the best-equipped North American or western European farms, but with a much smaller labor cost.

## The Green Revolution

Beginning in the 1960s, the Green Revolution, a variant of the contemporary agricultural revolution but without the large-scale motorization and mechanization, developed widely in the developing countries. Based on the selection of varieties of rice, maize, wheat, soya, a heavy utilization of synthetic fertilizers and pesticides and, if necessary, on firm control over water for irrigation and over drainage, the Green Revolution was adopted by farmers capable of acquiring these new means of production in regions where it was possible to realize a return on their investment. We emphasize that in several countries, the public powers have greatly favored the diffusion of this revolution by adopting policies of agricultural price supports, subsidies for inputs, preferential interest rates for borrowing, and investments in the infrastructures for irrigation, drainage, and transport. Today a farmer fully utilizing the means of the Green Revolution can attain a raw labor productivity on the order of 100 quintals of cereal equivalent if that farmer has only manual tools (1 hectare/worker  100 quintals/hectare), on the order of 500 quintals if they have equipment which uses animal power (5 hectares/worker  100 quintals/hectare), and even more if the farmer can make several harvests per year.

## Orphan Agricultures

A very large number of peasants in the developing countries have never had access to the means of production for either one of these agricultural revolutions. Motorization and mechanization are practically absent, and specially selected seeds, fertilizers, and pesticides are used only a little or not at all in large cultivated areas watered by rain or insufficient irrigation in the intertropical forests, savannas, and steppes of Africa, Asia, and Latin America. And even in regions fully penetrated by one or the other of these revolutions, numerous peasants have never been in a position to acquire the new means of production and make progress in both profitability and productivity. They too have been impoverished by the lowering of real agricultural prices and, moreover, have sometimes suffered from the ill consequences resulting from these two revolutions (for example, diverse pollutions, lowering of the ground water table, increasing salinity of irrigated and poorly drained soils).

Consequently, hundreds of millions of peasants continue to work with strictly manual tools, without fertilizers or pesticides, and with plant varieties that have not been the object of research and systematic selection (tef, finger millet, fonio, millet, quinoa, sweet potato, oca, taro, yam, plantain, manioc). The yields obtained in these conditions are less than 10 quintals of cereal-equivalents per hectare (for example, the average yield of millet in the world today is hardly 8 quintals per hectare). Since a manual tool set hardly allows a single worker to cultivate more than one hectare, the raw productivity does not surpass 10 quintals of cereal-equivalents per active worker per year (1 hectare/worker  10 quintals/hectare).

## Majority of the World's Farmers Involved in Manual Agriculture

In the last analysis, for an active agricultural population in the world of 1.3 billion people, or half of the total working population, there are only 28 million tractors,  or some 2 percent of those working in agriculture![4] Note that the total agricultural population of the world (working and not) is around 3 billion people, or half of humanity.

In addition, it is estimated that two-thirds of the working population in agriculture have benefited from the Green Revolution. Approximately half of them use animal power, while the others still work with manual tools. Consequently, one-third of the world's peasants, or more than 400 million of the working population (which corresponds to more than a billion persons to feed), work not only with strictly manual tools but without fertilizer, feed for livestock, pesticides, and specially selected varieties of plants and breeds of animals.

## Indefensible Inequality of Access to Land

What is more, in numerous ex-colonial countries (Latin America, South Africa, Zimbabwe) or ex-communist countries (Ukraine, Russia) which have not experienced recent agrarian reforms, the majority of these poorly equipped peasants are more or less deprived of land because of the existence of large estates of several thousand or tens of thousand hectares, estates which are private, public, or in the process of privatization. The size of the land area these  peasants have access to is much smaller than the one they could cultivate with their meager tools, and smaller than the minimum size necessary to cover the subsistence needs of their families. These peasants on "minifundia" are thus obliged to look for hand-to-mouth work on the large estates, the "latifundia," for $1 to $2 per day.

## Current Reasons for the Extreme Impoverishment of Hundreds of Millions of Peasants

The increases in productivity and production resulting from the contemporary agricultural revolution and from the Green Revolution have not only provoked a sharp drop in real agricultural prices in the countries concerned, they have also allowed certain countries to unload exportable surpluses at low prices. International trade in basic agricultural products involves only a small fraction of world agricultural production and consumption (some 12 percent of cereals, for example). The corresponding markets are residual markets, which consist of surpluses difficult to sell except at particularly low prices. At such prices, even the producers who have benefited from the agricultural revolution or the Green Revolution can win part of the market, or simply preserve their position, only if they have additional competitive advantages. Such is the case for certain South American, South African, Zimbabwean, and now Ukrainian and Russian, latifundia-based agro-exporters, who are not only well equipped but have access to huge areas of land at low cost and to some of the lowest paid workers in the world. Today, on this type of latifundia, an agricultural worker making less than $1,000 per year can produce more that 10,000 quintals of cereals (1,000,000 kilograms), which reduces the labor cost per kilo of cereal to less than a thousandth of a dollar (1,000 dollars/worker/year divided by 1,000,000 kilograms/worker/year). Consequently, the price of a quintal of exportable cereal from these regions is less than $10.

At this price, a number of American or European farmers would have a nil or negative income. Consequently, they would not be able to win a share of the market nor withstand imports nor  persist in their business if they did not live in high-income, developed countries concerned about their food sovereignty and where, as a result, they benefit from important public assistance.

Finally, in certain developing countries, notably in Southeast Asia (Thailand, Vietnam, Indonesia), the increase in production due to the Green Revolution is combined with local levels of income and wages so low that these countries have become exporters of rice, while undernourishment is rife in the countryside.

But for the immense majority of the world's peasants, the international prices of basic food products are far too low for them to support themselves and renew their means of production, much less allow them to invest and grow. As a result of the lowering of transport costs and the growing liberalization of international agricultural exchanges, the always renewed strata of underequipped peasantry, who are poorly situated, poorly endowed with land, and not very productive, are confronted with competition from commodities sold at very low prices in international markets. This competition blocks their development, leading ultimately to extreme poverty and hunger.

In order to understand this process better, consider a Sudanese, Andean, or Himalayan cultivator of cereal, using manual tools and producing 1,000 kilograms net of grain (seeds deducted), without fertilizer or pesticides. Around fifty years ago, such a farmer received the equivalent of $50 in 2001 dollars for 100 kilograms of grain. The farmer had to sell 200 kilograms in order to renew tools, clothes, etc., retaining 800 kilograms to feed four persons modestly. By cutting back on consumption a little, the farmer could even sell 100 kilograms more in order to buy some new and more effective tool. Twenty years ago, the same farmer received no more than the equivalent of $20 in 2001 dollars for 100 kilograms. Thus the farmer had to sell 400 kilograms in order to renew tools and retained only 600 kilograms for food, this time insufficient for four persons. It was no longer possible to buy new tools. Finally, today, if this farmer receives no more than $10 for 100 kilograms of grain, more than 600 kilograms must be sold in order to renew the equipment, which is well nigh impossible since one cannot feed four persons with 400 kilograms of grain. In fact, at this price, the farmer can neither completely renew the tools, however pitiable, nor eat properly and renew his/her labor power. This farmer is condemned to indebtedness and an exodus toward underequipped and underindustrialized shantytowns where unemployment and low wages hold sway.

In these conditions, it is easy to understand why development policies that consist of pushing the contemporary agricultural revolution and Green Revolution further in the advantaged regions and food policies that are designed to provide cities and towns with food products at always lower prices are particularly contraindicated in the fight against hunger. In fact, these policies impoverish the most destitute peasants even more, who, we have seen, form the majority of the world's undernourished people.

## AGRICULTURAL AND FOOD PERSPECTIVES
### FOR THE YEAR 2050

In 2050, our planet will have some 9 billion inhabitants (between 8 and 11 billion) according to estimates published by the United Nations in 2001. In order to feed such a population properly, without undernourishment or shortages, the quantity of vegetable products designated as food for humans and domestic animals will have to more than double for the whole world. It will almost have to triple in the developing countries, more than quintuple in Africa, and increase more than ten times in several African countries.[3] In order to obtain such an enormous increase in vegetable production, agricultural activity will have to be extended and intensified on a long-term basis in every region of the world where that is possible.

## LIMITED POSSIBILITIES FOR PROGRESS

Many people think of the progress made as a result of the contemporary agricultural revolution and Green Revolution as a way to obtain the necessary increase in agricultural output. But in the regions where these revolutions are already very advanced, it would be difficult to continue augmenting outputs through increased use of conventional means of production. In many places, abuses have occurred while using these conventional means, leading to many negative conse-quences, indeed to a change for the worse in the ecological, or social order—diverse pollutions, threats to the quality and safety of food, excessive concentra-tion of production and accompanying abandonment of entire regions, degrada-tion of soil, harm to the environment. In these conditions, in order to reestablish the quality of the environment or products, it will undoubtedly be necessary to impose restrictions on using these means of production, which will, in turn, imply no new increase in outputs.

On the other hand, the regions where the contemporary agricultural revolu-tion and the Green Revolution have already penetrated, even though they are not fully developed, undoubtedly contain a real potential for growth in production. But the actualization of this potential by an increased use of fertilizers and pesti-cides will come up against the same negative consequences as mentioned above. As for large-scale motorization and mechanization, these are not, in themselves, means to increase outputs and production significantly. What is more, the cost is so high that it is forever out of the reach of the great majority of the peasants in the developing countries, while its adoption by the large estates that employ wage labor will reduce by 90 percent the need for that labor power, thereby increasing rural poverty, migration, and unemployment in the same proportion.

As far as genetically modified organisms (GMOs) are concerned, the latest avatar of the two agricultural revolutions, these are not likely to miraculously restore such a disastrous worldwide agricultural and food situation. Their use-fulness presumes, in fact, that GMOs are not essentially a means of appropriating the genetic inheritance of plants and animals, that the environmental and health risks entailed are exaggerated or nonexistent, that the hopes and ambitions they sustain outweigh the reactions of fear and rejection they arouse, and that the development of GMOs resistant to agricultural pests and tolerant of climatic extremes and sterile soil is more rapid than the local selection of species and varieties appropriate to the needs and possibilities of the peasantry of specific areas. In addition, the use of GMOs is very expensive and that preventative con-trol over their possible ecological and nutritional harm is even more expensive. It is so expensive, in fact, that research is essentially oriented toward the needs of the most solvent producers and consumers, so expensive that GMO seeds and the means of production necessary to valorize them will not be any more accessi-ble to poor peasants than those of the Green Revolution.[5]

In the final analysis, neither GMO nor seeds specially selected in the classic manner or other technical means associated with them can eradicate the extreme poverty and hunger of poorly equipped peasants. At the prices paid for agricultural products today, these peasants are less than ever able to buy and profit from such means.

## The Necessary Reorganization of International
## Agricultural Trade

In order to allow all the world's peasants to construct and exploit cultivated ecosystems capable of producing a maximum of secure and high-quality commodities over the long term, without harming the environment, it is absolutely necessary to stop the international agricultural price war. It is necessary to break with the liberalization of trade that tends to align prices everywhere with the lowest ones offered by the exporters of surplus produce. We have seen that such prices impoverish and starve hundreds of millions of rural inhabitants, which increases the rural exodus, unemployment, and urban poverty, thereby reducing solvent demand well below needs. Moreover, by excluding entire regions and millions of peasants from production, and by discouraging production by those who remain, these prices limit agricultural production well below what would be possible with familiar and sustainable techniques of production. Such prices, which engender both nutritional underconsumption and under-utilization of agricultural resources, are doubly Malthusian. What is more, they place a negative burden on the environment and the safety and quality of produce. Agricultural and food products are not commodities like others. Their price is that of life, and, below a certain threshold, it is that of death.

To promote sustainable peasant agriculture everywhere possible without harm, one that is capable of ensuring both quantitatively and qualitatively the food security of 6 and soon 9 billion humans, it is above all necessary to guarantee sufficiently high and stable prices to peasants so they can live with dignity from their labor. It is the price of our future. To this end, it is necessary to create an organization of international agricultural trade much more equitable and effective than the current one, a new organization with the following principles: establish large common agricultural markets on a regional basis, grouping regions with similar agricultural productivities (West Africa, South Asia, western Europe, eastern Europe, North Africa, and the Middle East, etc.); protect these regional markets against all imports of low-priced agricultural surpluses by variable customs duty, guarantee to poor peasants of disadvantaged regions sufficient and stable enough prices to allow them to live from their labor and also to invest and expand; negotiate international agreements, product by product, that equitably establish an average purchase price for the product on international markets, as well as an agreed-upon quantity and export price for each of these large markets and, if necessary, for each country.

Moreover, in countries where the land is monopolized by a minority of large landowners (latifundia owners), it will still be necessary to implement true agrarian reform and legislation guaranteeing access to the land and security of tenure to the greatest number of people.

In the interior of these large markets, income inequalities between more or less favored agricultural zones could be corrected by a differential land tax, and income inequalities between farms that are more or less well equipped in means of production could be corrected by an income tax.

Finally, it will be important to support publicly based agricultural research, both national and international, and to focus such research in a way that gives priority to the needs of poor peasants in disadvantaged regions, with care for the ecological viability of cultivated ecosystems (renewing fertility) as well as their economic and social viability (the increase and equitable sharing of material well-being).

Some of the analyses and propositions in this book run counter to dominant economic and political thinking. However, they are widely shared and continue to gain support. Things are happening quickly, alas, that tend to corroborate our statements that global change in agricultural and food policies appears more urgent each day. Such policy changes also appear more attainable if one can judge by the numerous discussions we are called upon to lead with representatives of farmer unions, nongovernmental, governmental, and international organizations, universities, and centers of research, in France and many other countries.

PARIS, DECEMBER 2001
MARCEL MAZOYER, LAURENCE ROUDART

# Introduction

To want everything, absolutely everything, in a landscape, a region, a civilization, to belong to a rigid unified system, is this not a dream of a centralizing philosopher? Is it not better to accept that this landscape, this region, this civilization are made, after long historical accretions, of elements which possibly have relations of causality or interdependence, or not, and are juxtaposed to one another, sometimes at the price of mutual confusion? [...] Should not geographers and others see the world as full of questions, and not as a system to which they pretend to have the key?

—PIERRE GOUROU, *Riz et Civilisation*

If humanity were to allow every cultivated ecosystem of the planet to lie fallow, each would quickly return to a state of nature close to that in which it existed 10,000 years ago. Wild flora and fauna far stronger than those existing today would overwhelm cultivated plants and domesticated animals. Nine-tenths of the human population would perish because, in this Garden of Eden, simple predation (hunting, fishing, gathering) would certainly not feed more than 500 million people. If such an "ecological disaster" were to occur, industry would be of little assistance, since it is not yet in a position to synthesize food for humanity on a large scale, and will not be able to do so quickly. There is no other way to feed 20 billion people or 5 billion people than to continue to cultivate the planet by increasing domestic plants and animals while controlling wild ones.

But if the return to nature is only a pleasant utopian notion, and industrialized production of nutritional products an evanescent chimera, then the commonly accepted idea that the best means to respond to the growing needs of humanity would be to extend to the entire planet the type of motorized-mechanized agriculture that has been developed in the industrialized countries for a half century, and which is such a large consumer of mineral fertilizers, is also a mistaken idea. To give only one-fourth of the farmers in the developing countries such costly means of production, it would be necessary to invest billions of dollars, i.e., several times the annual income of these countries, which is obviously unrealizable in a short period of time. Moreover, by replacing people with machines, this response would throw three-fourths of the world's agricultural labor force onto the labor market, thereby at least doubling the number of unemployed in the world. At a time when no one dares pretend any longer that the development of industry can ever reduce the already existing unemployment, one can easily assess the disastrous economic, social, and political consequences of such a tidal wave change.

## 1. HUMANITY'S AGRARIAN HERITAGE

"Modern" agriculture, utilizing much capital and little labor power, has triumphed in the developed countries. Despite the billions spent in promotion, it has only penetrated limited sectors in the developing countries. The great majority of peasants in these countries are too poor to be able to afford the huge machines and large quantities of fertilizers. Around 80 percent of the farmers of Africa, and 40 to 60 percent in Latin America and Asia, continue to work exclusively with manual tools, while 15 to 30 percent of them use animal traction, and less than 5 percent use motorized traction. Modern agriculture is thus far from having conquered the world. Other forms of agriculture remain predominant and these continue to employ the majority of the active population in the developing countries.

Certainly, the most disadvantaged and least productive among these farmers are continually marginalized, plunged into crisis and eliminated by competition from stronger farmers. But those who have the means to maintain themselves and advance offer proof of an unsuspected wealth of inventiveness and continue to develop in their own ways. It is an error to consider agricultures in the developing countries as traditional and unchanging. They are continually in transformation and continually participate in the creation of modernity. It is another error to conceive of agricultural development as the pure and simple replacement of these agricultures with the only one that is supposedly modern, i.e., the motorized and mechanized one. Undoubtedly, this modern agriculture will be expanded further and will be of immense service. But it is difficult to conclude that it can be both generalized to the whole world and renewable in the long term, if only because of the probable exhaustion of phosphate reserves, which it uses in great quantities.

Considering the role that all of the world's agricultural systems should play in the construction of a livable future for humanity, it is disturbing to note how far both common and educated opinion are from agricultural realities, and to what extent that those who are in charge of agriculture are unaware of the wealth of humanity's agrarian heritage. Certainly, works of historians, geographers, anthropologists, agronomists, economists, and sociologists that study agriculture are not lacking. But, despite their richness and their value, they consistently lack, it seems to us, a body of synthetic knowledge that explains the origins, the transformations, and the role of agriculture in the evolution of humanity and of life, in different time periods and in different parts of the world. They are missing a body of knowledge that can simultaneously be integrated with general knowledge and form a conceptual, theoretical, and methodological foundation for all those who desire to intervene in agricultural, economic, and social development.

⌐ Projects and policies of agricultural development should respond to the needs of the populations in question, ensuring their agreement and encouraging their participation, otherwise these interventions will be neither effective nor legitimate. But they should also be based on a real competency. Just as a doctor would not legitimately listen to a person's heart, make a diagnosis and prescribe a treatment without preexisting knowledge of anatomy, physiology, reproduction, and human growth and aging, so one is not able to analyze a given agricultural system, formulate a diagnostic, and propose projects and policies of development without being grounded in a systematic knowledge of the organization, functioning, and dynamics of different sorts of agricultural systems.

This book attempts precisely to build this type of knowledge, under the synthesized form of a *theory of historical transformations and geographical differentiations of agrarian systems.* This aims to be a theory based on numerous direct observations, without which nothing original could be conceived, but also on observations reported by others and on a sum of historical, geographical, agronomic, economic, and anthropological knowledge that has been considerably enriched over the last several decades. This theory is necessary in order to apprehend agriculture in its complexity, diversity, and movement.

## 2. A THEORY OF AGRARIAN SYSTEMS

Every form of agriculture practiced in a given place and time appears first of all as a complex ecological and economic object, composed of several categories of production units that exploit different types of terrains and diverse species of cultivated plants and animals. Furthermore, observable forms of agriculture vary according to place, such that, from one region of the world to another, it is possible to classify them into very different genres (e.g., aquatic rice growing, pastoral animal breeding, cultivation involving rotation, arboriculture). Finally, over time, every agricultural system is transformed, and in a given region of the world different species of agriculture can succeed one another, forming the stages of an "evolutionary series" characteristic of the history of this region. (In Europe, for example, these forms succeeded each other: slash-and-burn cultivation in prehistoric times, cereal cultivation using an ard in antiquity, cereal cultivation using a plow in the Middle Ages, polyculture and animal breeding without fallowing in the modern period, motorized and mechanized based cultivation today.)[1] As we will see later (chapter 1, point 3), the theory of agrarian systems proposed in this work has been conceived precisely as an intellectual tool enabling one to apprehend complexity and construct a general outline of the historical transformations and geographical diversity of the world's agricultural systems.

In order to sketch this theory, we take into consideration first of all that the earliest systems of cultivation and animal breeding appeared in the Neolithic epoch, at least 10,000 years ago, in a few relatively small regions of the world. They developed out of the self-transformation of some of the quite varied systems of predation that predominated in all of the inhabited areas of the world. These first forms of agriculture were most probably practiced in the areas around dwellings and on alluvial deposits resulting from receding flood waters, that is, on ground already fertilized and requiring hardly any clearing.

From there, Neolithic agriculture expanded across the world in two principal forms: systems of pastoral animal breeding and systems of slash-and-burn cultivation. Systems of pastoral animal breeding were extended into grassy areas that could be used directly as pasturage. They have been maintained up to the present in the steppes and savannas of diverse regions, in northern Eurasia, central Asia, the Middle East, the Sahara, the Sahel, the high Andes, etc. Systems of slash-and-burn cultivation progressively conquered most of the temperate and tropical forests where they have survived for centuries, even millennia, and still exist in certain forests of Africa, Asia, and Latin America. Since this pioneer epoch, population growth led to deforestation and even, in some cases, to desertification in most of the originally forested regions. Systems of slash-and-burn cultivation gave way to numerous post-forest agrarian systems, differentiated according to climate, which are the origin of distinct and relatively independent evolutionary series.

Hence, in arid regions, hydraulic agrarian systems, based either on annual floods or irrigation, were formed at the end of the Neolithic epoch in Mesopotamia, in the valleys of the Nile and Indus, and in the oases and valleys of the Inca Empire. In humid tropical regions (China, India, Vietnam, Thailand, Indonesia, Madagascar, Guinean coast of Africa, etc.), different hydraulic systems based on aquatic rice-growing were developed in successive stages, first by utilizing well-watered and well-drained places (piedmont and interfluve areas), then in hilly areas (high valleys) or areas difficult to protect and drain (lower valleys and deltas), or even in places requiring irrigation. Parallel to this development, implements were improved and the number of possible harvests each year was increased.

In the intertropical regions with average precipitation, deforestation led to the formation of varied savanna systems, such as temporary cultivation with the hoe and no animal herding, as found in systems on the Congolese plateaus; cultivation with pasturage and the accompanying animal herding, as found in systems in the high altitude regions of East Africa and in diverse Sahelian areas; and cultivation and arboriculture with animal herding, as found in Sahelian systems associated with plantings of the *Acacia albida* tree.

After deforestation in the temperate regions of Europe, a whole series of post-forest systems succeeded one another that, from agricultural revolution to agricultural revolution, led to current systems. The agricultural revolution of antiquity gave birth to systems of rainfed cereal cultivation with fallowing, pasturage, and animal herding, in which manual tools such as the spade and hoe were used, and an implement of *animal-drawn cultivation*, the ard plow. Centuries later in the northern half of Europe, the agricultural revolution of the central Middle Ages gave birth to systems of fallowing and animal-drawn cultivation, using wagons and plows. Then, from the sixteenth to the nineteenth centuries, the first agricultural revolution of modern times engendered systems of cereal and feed grain cultivation without fallowing.

After the voyages of discovery, European agrarian systems were enriched with new plants from America (potatoes, maize, etc.), and these systems were extended into settler colonies in the temperate regions of the Americas, South Africa, Australia, and New Zealand. At the same time, in the tropical regions, agro-exporting plantations were developed within preexisting systems, sometimes to the point that the latter were replaced by the former, which then gave birth to new, specialized systems (sugarcane, cotton, coffee, cacao, palm oil, bananas, etc.).

Finally, the second agricultural revolution of modern times, last in the evolutionary series of the agrarian systems of the developed temperate regions, produced the motorized, mechanized, specialized systems of today, with their reliance on synthetic chemicals.

Millennia of separate but occasionally intersecting evolutions have produced a whole range of fundamentally different and unequally productive agrarian systems that occupy various exploitable areas of the planet.

## 3. AGRARIAN CRISIS AND GENERAL CRISIS

From the end of the nineteenth century, with the revolution in transport, all of these agrarian systems progressively confronted each other in the same increasingly unified world market that daily revealed all kinds of inherited inequalities and their resulting disparities in productivity and income. Then, in the twentieth century, productivity gains from the second agricultural revolution (motorization, mechanization, mineral fertilizers, special selection of seeds, crop specialization) were so enormous that they entailed a significant lowering of real prices (deductions made for inflation) for most agricultural commodities. Furthermore, the ratio between the gross productivity of labor in the least productive manual agriculture and that of the most productive motorized and mechanized agriculture has increased by several dozen times, going from 1 to 10 at the beginning of the twentieth century to 1 to more than 100 today.

Confronted with such harsh competition and hit by lower prices, the least well-equipped and productive farmers saw their incomes collapse. Having become incapable of investing and developing, they were condemned to regression and elimination. In this way, tens of millions of small and medium farms in the developed countries have disappeared since the beginning of the century. For the past several decades, the same causes producing the same effects have seen hundreds of millions of underequipped peasant farms in the developing countries plunged into crisis and eliminated, adding to the growing rural exodus, unemployment, and rural and urban poverty.

This immense wave of planetary unemployment and poverty limits the growth of solvent world demand that is already, on a global level, insufficient for strong industrial and agricultural development. Even if the archipelago of prosperity formed by the large industrial centers and their satellites continues to develop and expand, it finds itself stifled by the lack of outlets and overwhelmed and threatened by poverty growing a little more each day.

Our diagnosis is the following: the contemporary general crisis is rooted in the massive and ever-growing crisis of the least well-off peasant farmers, a crisis that essentially results from competition with the most productive agricultural enterprises. The greatest peril of our epoch is that the reduction in agricultural employment will continue to prevail over the creation of employment in other sectors of the economy and, as a result, unemployment and poverty will spread on a global level much faster than employment and material well-being.

There is no doubt that the world's rapid population growth considerably exacerbates the consequences of this phenomenon. But, paradoxically, population growth is itself encouraged by the lowering of agricultural prices over the past few decades because that, in turn, contributes to lowering the cost of the dietary needs of human life.

If the essential problem of the world economy today truly lies in the destructive confrontation between the very different and unequally productive agricultures that form the agrarian heritage of humanity, then the solution to the contemporary general crisis necessarily lies in a coordinated policy on the world scale that would allow poor farmers to support themselves and develop. This policy must be one that would finally make it possible to end the rural exodus, growing unemployment, and poverty. Moreover, it must restore solvent demand in the poor countries on a large scale, which alone is capable of giving a boost to productive investments and the world economy.

To give or return to all types of agrarian systems inherited from the past the possibility of participating in the construction of a viable future for humanity is the true way to resolve the general crisis of the contemporary world economy.

## 4. THE PLAN OF THIS BOOK

The book's first objective is to establish a methodical knowledge of the genealogy and characteristics of humanity's great agrarian systems. From there, it aims to explain the role that the agrarian crisis in the developing countries plays in the formation of the general crisis. Further, we attempt to show how the safeguarding and development of the ill-equipped and relatively unproductive agriculture of the poor, which is by far the most widespread in the world today, can contribute to resolving the contemporary crisis.

The first of the eleven chapters situates agriculture in the evolution of life and history of humanity, while the second chapter recounts the origins of agriculture in the Neolithic epoch.

The eight chapters that follow are devoted to the study of the principal agrarian systems that form humanity's agrarian heritage:

— Systems of slash-and-burn cultivation in forested areas and the consequences of deforestation (chapter 3); tropical savanna systems and systems of aquatic rice-growing in the humid tropical regions are briefly presented in the same chapter
— Hydraulic agrarian systems in arid regions, with the example of the Nile Valley (chapter 4)
— The Inca agrarian system, an example of a terraced mountain system (chapter 5)
— Systems of animal-drawn cultivation based on the ard, fallowing, and accompanying animal herding in the temperate regions of Europe: the agricultural revolution of Antiquity (chapter 6)
— Systems of animal-drawn cultivation based on the plow, fallowing, and accompanying animal herding in the cold temperate regions: the agricultural revolution of the Middle Ages (chapter 7)
— Systems of animal-drawn cultivation using the plow and without fallowing, resulting from the first agricultural revolution of modern times in the temperate regions (chapter 8)
— Mechanization of animal traction and transportation and the first world crisis of agricultural overproduction (chapter 9)
— Motorized, mechanized, specialized systems using mineral fertilizers resulting from the second agricultural revolution (chapter 10).

Finally, the agrarian crisis of the developing countries and its relationship with the general crisis is treated in chapter 11.

Each of the great agrarian systems is first defined and situated in time and space. Then, we try to comprehend its origin and explain its genesis. We analyze its organization (cultivated ecosystem, social productive system), its functioning (clearing, renewing fertility, management of cultivation and breeding) and the more or less long-lasting results that follow from all of that, as well as its dynamics and its geographical and historical limits. Finally, for each of these systems, we attempt to apprehend the conditions and the demographic, economic, social, and political consequences of its development.

Even if each chapter can be read independently of the others, the order in which they appear is not unimportant. Each chapter, in its place, contributes to the construction of an organized knowledge of agriculture and the comprehension of today's agrarian problems.

# 1

# Evolution, Agriculture, History

We have given to thee, Adam, no fixed seat, no form of thy very own, no gift pecu-
liarly thine, that thou mayest feel as thine own, have as thine own, possess as thine
own the seat, the form, the gifts which thou thyself shalt desire. A limited nature in
other creatures is confined within the laws written down by Us. In conformity with
thy free judgment, in whose hands I have placed thee, thou art confined by no
bounds; and thou wilt fix limits of nature for thyself.

—PICO DELLA MIRANDOLA, *On the Dignity of Man*

Life began to develop around 3.5 billion years ago in a solar system and on a
planet formed 4.6 billion years ago in a universe whose origin is unknown but
whose oldest light rays reach us from such a distance that we are led to conclude it has been expanding for 15 billion years. Since then, evolution has produced hundreds of millions of living species, many of which have disappeared
in the course of time. The first to appear were plants, of which there are more
than 500,000 species still living today, and then animals, which number nearly
a million species. Not all living species have yet been identified, and every year
new ones are discovered. The totality of individuals of a species living in a particular place at a given moment in time form a *population* of this species. The
totality of plant and animal populations living in this place form a *biocenosis*.
The biocenosis and the inanimate environment, or *biotope* (geology, morphology, climate) that it inhabits, form an *ecosystem*. All the ecosystems of the planet form the *ecosphere*.

   All living beings, be they plants or animals, are formed from organic matter,
water, and other minerals. Organic matter is formed from complex molecules
(sugars, fats, proteins, nucleic acids) which, besides forming living beings, are
also the source of the energy necessary for life and reproduction. Plants are
*autotrophs*: they are capable of synthesizing, by means of solar energy, their own
organic substance from water, carbonic gas, and other elements that they find in
the atmosphere and in the soil. By contrast, humans and animals do not have
this ability: they are *heterotrophs*. They live upon organic matter provided

directly by plants which have produced it or provided indirectly by animals which have first consumed and assimilated it.

The *biomass* of an ecosystem is the total mass of organic matter that it contains, including waste products and excrement. Only plant species are productive of biomass; humans and animals do not produce it. They only feed upon it and transform it. These are *exploitive* species. That is why the fertility of an ecosystem, that is, its capacity to produce biomass, is ultimately measured by its capacity to produce plant biomass.

Most animals are simple predators that are content with obtaining their food by force from the wild species of plants and animals that they exploit. Some among them, however, provide a service for the exploited species. The bee, for example, transports the pollen it gathers from the flower, thereby facilitating its fertilization. But curiously, millions of years before the present, evolution produced several species of ants and termites that cultivate fungi or raise aphids. These are domestic fungi and aphids that the ants and termites exploit intensely through the constant work of managing the environment by multiplying the populations and promoting their development.

Humans are a much more recent species and not born to be farmers or stock-breeders, unlike these ants and termites. They became so after hundreds of millions of years of hominization, that is, biological, technical, and cultural evolution. It was only in the Neolithic era, less than 10,000 years ago, that humans began to cultivate plants and breed animals that they themselves had domesticated. Subsequently, they introduced these plants and animals into all sorts of environments, where they endeavored to propagate them. In this manner, the original natural ecosystems were transformed into cultivated ecosystems, fabricated and exploited by human care and attention. Since then, human agriculture has conquered the world: it has become the principal factor in the transformation of the ecosphere, and its gains in production and productivity have respectively influenced the increase in the number of people and the development of social groups which do not produce their own food.

Our intention in this chapter is to situate agriculture in the evolution of life and in the history of humanity. More exactly, we aim to respond to three essential questions:

— What is agriculture as a particular relation between living species?
— At what moment in the process of hominization did humans become farmers and why?
— Since then, what is the role played by agriculture in the historical development of humanity?

Compared to our own views on the concept of agrarian system and on the relation between agriculture and history, the rudiments of ecology, paleontology, soil science, and history presented here do not claim to teach anything to the specialists of each of these sciences. May they pardon us for having so outrageously diminished their knowledge. Our intention is simply to present in as concise and intelligible a manner as possible the essence of what one should know in order to respond to the questions we just posed and understand the rest of this book.

## 1. LIFE, EVOLUTION, AND AGRICULTURE

It is useful to present briefly some concepts from ecology in order to understand the nature of agriculture as a relationship between an exploiting species and one or several exploited species existing in a cultivated, human-made ecosystem.

## Limiting Factor and Ecological Valence

All living beings find in the environment the resources necessary for their material existence: space, habitat, food, and the possibility to throw away waste materials derived from their life functions. Resources in any given environment are limited. Thus, there necessarily appears at one time or other conflict between the growing needs of a species that is multiplying within a given environment and the limited resources of this environment. When the population density becomes too great, when the quantities of water, minerals, pastures, or prey available at a particular critical period are totally consumed or become too scarce to remain easily accessible, then the growth of this population is blocked. The same thing happens when the waste material thrown away by a particular species encumbers the sites they occupy, diminishing or polluting its sources of provisions. The element of the environment that determines the maximum density the population of a species can attain over the long term at a given site is called the limiting factor. Of course, limiting factors vary from one species to another and vary from one environment to another for the same species.

In certain environments, a particular limiting factor for the development of a species (temperature, rainfall, food) can be found below a threshold of minimum tolerance or above a threshold of maximum tolerance, on the basis of which the development of this species becomes impossible. The level of this threshold varies according to species and their tolerance with respect to characteristics of the environment. The higher animals, humans, and certain domestic animals in particular, are very tolerant in relation to their environment. Their capacity to populate varied environments, that is, their *ecological valence*, is higher and their area of geographical extension is vast. On the other hand, some species demand very narrowly defined and rarely realized environmental conditions. Hence they are not widely dispersed and their ecological valence is weak.

The term "ecological valence" will be used here in a larger sense. It will designate not only the ability of a species to occupy varied environments, but also its capacity to populate them more or less densely. In this sense, the ecological valence of a species designates its potential for development: it is measured not only by the geographical extent of the species' distribution but also by the maximum population density it can attain at the peak of its development.

## Competition, Exploitation, Symbiosis

Often, two or more species struggle over the same resources. The opposition between the population of each species and the limitations of the environment is coupled with an opposition between the populations of each species in competition for the same resources. This competition, whether or not it involves an open struggle between competing populations, leads to their coexistence, within certain parameters, or to the elimination of one or several species.

One species can also exploit another, which acts as support, pasture, or prey for it. This exploitation can harm the development of the exploited species but, conversely, the development of the exploiting species can be conditioned by that of the exploited species. Such is the case when the latter forms an irreplaceable resource for the former. For example, a population of pandas is limited by the population of bamboos upon which it feeds exclusively.

Sometimes there exists between two species a reciprocal and necessary relation of exploitation, a relation that can be considered mutually beneficial to those species. Such a situation is called mutualism or symbiosis. For example, the nitrogen binding bacteria lodged in the bulges (or nodules) of the roots of leguminous plants contribute to supplying those plants with nitrogen. Ruminants and horses harbor bacteria in their intestines that facilitate the digestion of the cellulose materials essential in their dietary regime. Certain plants can only be pollinated by insects that gather the pollen.

## Labor, Fabrication of the Environment,
## Agriculture, and Breeding

Some species transform the environment where they live to make it more accommodating and increase the available resources for their own use. They thereby increase their own ecological valence. Numerous animals build nests, shelters, and even an artificial environment (e.g., the collective urbanism of beavers, bees, termites, ants) that is necessary for their development. This trans-formation, this fabrication of the environment, is the product of a labor that is not, as is sometimes said, unique to the human species.

Moreover, some animal species go beyond the exploitation of other species by simple predation. They are devoted to transforming the environ-ment in such a way as to create fabricated conditions of life that favor the development of the species they exploit. These exploited species, which could hardly develop without the support of the exploiting species, are called domestic. Some species of ants and termites cultivate mushrooms, which they eat. Other species of ants raise aphids whose honeydew they consume. In order better to understand the nature of the relations between cultivating or breeding species and domestic species, a quick analysis of the manner in which some ants manage their environment and organize the life of the species they exploit is not without interest.

The origin of ants goes back some 180 million years and evolution has pro-duced around 18,000 species with different anatomies and modes of life. The oldest forms are generally insectivores, the forms from the middle period of evo-lution are omnivores, and the later forms practice specialized dietary regimes. By forcing the analysis a little, one could say that after the hunter nomadism of the early forms, a sedentary mode of life with the gathering of food appeared. Developing this metaphor, one could say that about a hundred of these species practice agriculture and breeding.[1]

## The Cultivator Ants

Several species of tropical American ants live in association with a particular species of domestic fungus. These ants manage the environment by con-structing nests, galleries, and caves for the fungi. Among some species, the galleries go down several meters in depth and emerge into rooms with flat floors and vaulted roofs, sometimes as long as 1 meter and as wide as 30 cen-timeters, where the mushroom gardens are set up. In the heart of this layout, the immense central nest is sometimes linked up with several dozen small satellite nests within a radius of 200 meters. These ants also build a trans-portation infrastructure, a radiating network of trails made of built-up earth, several dozen meters long, 1 to 2 centimeters wide and set up for double cir-culation: one column of ants leaves for the harvest, while another returns to the nest with its cargo.

In order to multiply the mushrooms they eat, these ants methodically prac-tice a whole series of cultivation processes. They prepare a bed for cultivation by collecting diverse organic debris (pieces of leaves, wood, roots, or tubers) from the outside which they tear up, grind, and fashion into mushroom beds. They plant fragments of cultivated mushrooms in these beds and systematically eliminate any other species of mushroom that begins to develop. Finally they regularly cut the filaments of mycelium, which prevents the fructification of the mushrooms and causes the formation of bulges, the mycotetes, which is what they exclusively eat. The social division of labor is well defined. The largest individuals guard the entrances to the nest and rarely leave those positions. The midsized individuals go outside the nest to harvest the plant debris, which they break up and mix into pellets. The smallest individuals maintain the mushroom gardens, feed the young larva and leave the nest only at the end of their lives. But this apparently well-regulated division of labor does not prevent some individuals from being undisciplined or even lazy. In exchange for all of the work involved in fabricating the environment and caring for the mushrooms in order to facilitate their multiplication, the ants receive abundant food, which can support the needs of hundreds of millions of them.

## The Breeder Ants

Other species of ants live in association with a species of aphid, or mealybug. This partnership is a true form of breeding. In order to protect the aphids that they exploit, the breeder ants dig caves and lay out shelters in the ground or in a sort of carton, which are eventually linked up by galleries. The individuals in charge of guarding the shelters ward off the aphids' predators and tear the wings of those that attempt to escape.

Among some species, the breeding is done by permanent underground stabling. The aphids are placed in chambers dug out around the roots of plants, where they can directly take the sap they feed on. Among other species, the breeding is done in the open air and the ants organize the food for the aphids by transporting them to better pasturage, namely to still-growing, young shoots. The reproduction of the aphids is carried out in good conditions, because the reproducing females are kept in underground chambers where the eggs are sheltered during the winter. The ants eat the aphids' honeydew, their excrement, which is rich in sugars and other organic molecules derived from the sap of the plants they have ingested. To accomplish this, the ants rub the abdomens of the aphids with their antennae, stimulating them to excrete their honeydew.

The species of aphids raised by the ants are different from wild species. These are true domestic species whose wild ancestors are unknown. But one can assume that each species of domestic aphid is the result of a coevolution that simultaneously produced the species of breeder ant with which it is associated.

## Agriculture and Breeding

The relationship between these ants and the mushrooms or aphids is not a pure and simple one of exploitation. The ants act upon the environment and on the mode of life of the domestic species they eat. They work to favor their development and protect them. They thus increase the ecological valence of the species they exploit and, as a result, extend the nutritional limits of their own development.

Increasing the ecological valence of the exploited species in order to increase that of the exploiting species is the basic logic governing the particular relations between species that characterize agriculture and breeding. Cultivating or breeding a species, far from marking the end of its exploitation, is only, on the contrary, the extension and intensification of this exploitation by other means. Agriculture and breeding are thus elaborated forms of mutualism, but a dissymmetric mutualism in which the development of the exploited species is controlled by the labor of the exploiting species and the development of the exploiting species is, in turn, conditioned by that of the exploited species.

## 2. HOMINIZATION AND AGRICULTURE

Homo sapiens sapiens, current or modern humans, *thinking* and *knowing* humans, is a very recent species among the thousands that evolution has produced in 3.5 billion years. This species appeared on earth only some 50,000 or 200,000 years ago, according to different authors. It then rapidly spread to all the continents and, since about 10,000 years ago, it has practiced agriculture and animal breeding, thereby completely changing most of the planet's ecosystems.

However, humanity as a product of evolution is not endowed with specialized anatomical tools nor a genetically programmed mode of life that would, from the start, enable it to exercise a strong effect on the outside environment. Deprived of pincers, hooks, stingers, fangs, tusks, serrated teeth, hooves, or claws, a human being instead has hands which, even if they are the most flexible and versatile of tool-holders, are in themselves only a weak tool and a feeble weapon. Slow moving, bad climbers, poorly protected, essential and fragile parts of their anatomy exposed because of an upright posture, endowed with or rather afflicted with a weak capacity for reproduction and a belated maturity, humans are naked and defenseless beings who had at the outset a much poorer ecological valence than usually thought. They could barely survive by collecting plant products or capturing the most accessible animals in environments that were either benign or protected. Knowing little, poor in instincts,

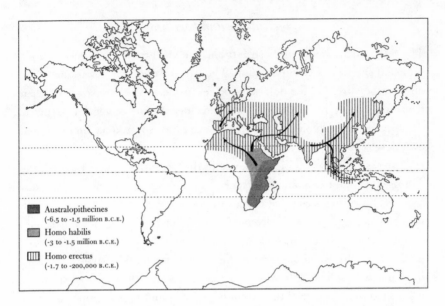

Figure 1.1 *The Spread of the Australopithecines, Homo Habilis, and Homo Erectus*

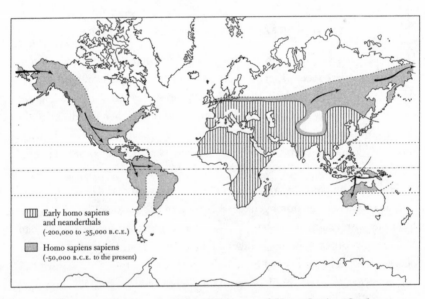

Figure 1.2 *The Spread of Homo Sapiens and Homo Sapiens Sapiens
up to 10,000 B.C.E.*

but immensely educable, their principal asset resided in the variety of dietary regimes and modes of life that could suit them. Humans are eclectic, omnivorous, and adaptable; such are their primary advantages.

According to the most commonly accepted theory, the current human species is the unique and latest representative of the evolutionary branch of hominids, which separated from other primates 6 to 7 million years ago. This branch successively engendered the Australopithecines, then *Homo habilis*, *Homo erectus*, and finally *Homo sapiens*. But opinions diverge concerning the more precise origin of *Homo sapiens*. According to one hypothesis, *Homo sapiens* appeared in Europe a hundred thousand years ago, in the form of a subspecies, *Homo sapiens neandertalensis*, the Neanderthals. Another human subspecies then appeared in the Middle East around 50,000 years ago, *Homo sapiens sapiens*, current or modern humanity.[2] According to a more recent hypothesis, *Homo sapiens* and *Homo neandertalensis* are two distinct species directly descending from *Homo erectus*. *Homo sapiens* appeared in southern Africa 200,000 years ago and then spread to the Middle East around 50,000 years ago. There, *Homo sapiens* encountered *Homo neandertalensis*, another species which itself had appeared in Europe 100,000 years ago and which disappeared 35,000 years ago for unknown reasons.[3]

One has to wonder how the different hominid populations that succeeded one another, from the Australopithecines to *Homo sapiens*, could increase their ecological valence to the point of conquering the entire earth and multiplying their numbers to millions then billions of individuals.

## The Australopithecines

The Australopithecines occupied East Africa from 6.5 to 1.5 million years before the present. But these "apes of the south"—such is the etymology of the word *Australopithecine*—were very different from modern humans. Of medium height, their cranial capacity was around 500 cubic centimeters, one-third that of a modern human, and they were imperfectly bipedal. Several species of Australopithecines were vegetarian. Other Australopithecines, such as *Australopithecus afarensis* (the species represented by the famous Lucy), one of the supposed ancestors to the genus *Homo*, were omnivores. They lived by gathering and supplemented their diet in the dry season by hunting small mammals, reptiles, insects, etc. They possibly used stones and sticks for this purpose.

Should we attribute to them the very first stones intentionally broken to give them a cutting edge? Some researchers think so.[4] In any case, it seems that the Australopithecines remained animals without any true technical and cultural history throughout the 5 million years of their existence.

## Homo habilis and Homo erectus

Contrary to the Australopithecines, who do not belong to the genus *Homo*, there are two old and long-vanished species which do belong to this genus. However, they, too, are very different from current humans.

The first of these species, though it is sometimes contested, is *Homo habilis*, "clever human," whose traces, found in eastern and southern Africa, date back 3 million years. The cranial capacity varies between 500 and 800 cubic centimeters and the remains of teeth attest to an omnivorous dietary regime. The first intentionally fabricated tools, i.e., purposely worked stones, have been attributed to this species. These stones, chosen for their size and form, were transformed in the most elementary way possible: they were broken by percussion in order to form a sharp edge for use in fracturing, cutting, and scraping. When these stones were cut, the resulting fragments of various types were also used as knives, scrapers, etc.[5]

The second of these species, *Homo erectus*, "upright human," is attested between 1.7 million and 200,000 years before the present. It is probable, however, that this species lasted much later. The *Homo erectus* species was not human as we understand it today. Cranial capacity was around 1,000 cubic centimeters, double that of the Australopithecines, but only two-thirds that of current humans. Skull bones show evidence of very few convolutions in the brain, and thus a relatively reduced amount of gray matter. The vocal apparatus, similar to that of a newborn human today, would undoubtedly not have enabled them to make use of a truly articulated language. Besides, members of this species are sometimes called Pithecanthropes, i.e., "ape-men."

*Homo erectus* appeared in East Africa 1.7 million years ago. Then, beginning about 1.5 million years ago, they occupied most of Africa and colonized large parts of Europe and Asia. Adapted to hot and temperate climates, they could not venture into the cold plains in the northern part of the old world and thus could not reach America from the eastern extremity of Siberia. Even though they occupied Indonesia, then connected to the continent, they could not reach Australia nor Oceania, because they had no knowledge of navigation. However, their colonization, despite its limits and slowness, extended much further than that of the Australopithecines and *Homo habilis*.

In Eurasia, *Homo erectus* was confronted with very long periods of glaciation (from 1.2 million to 700,000 years, from 600,000 to 300,000 years and from 250,000 to 120,000 years before the present) and lived in grottos and caverns. Undoubtedly, they began to use fire, the first traces of which date back nearly 500,000 years, but its use was not widespread during this epoch. It is assumed that the fire they used was of natural origin (fires, lightning, swamp fires) and that, even if they knew how to preserve it, they did not know how to produce it.

From the very beginning, *Homo erectus* fractured stones and nodules of flint by removing fragments from one face in order to fabricate tools with one or two sharp edges (one at each end). These were simple or double monofaced chopping knives. One million years later, around 700,000 years before the present, the first tools cut on two faces appeared in Europe and Africa: the *bifaces*. In Southeast Asia, superficially worked stones and the monofaced tools lasted a long time, and it was thought for many years that this part of the world did not have bifaced tools. But recent discoveries, though still few in number, show that this region also had such tools, undoubtedly later than elsewhere.[6] At the beginning, the bifaces were crude and their profile sinuous. The stone that served as the primary material for the tool was not cut on every surface. But beginning 250,000 years before the present, the bifaces were more finely cut, thanks to more and more elaborate cutting techniques. Perhaps Homo erectus even invented the effective method of cutting the stone called Levalloisian debitage. Until then, the final form of a tool was obtained by successive touching up of a stone chosen for that purpose, while with Levalloisian debitage, a type of rough biface was cut first, then well-defined fragments were cut into various shapes. Each of these fragments, in turn, was fabricated into a particular tool: point, scraper, knife, chisel, leather knife, etc. Yet it could be that these sophisticated industries should be attributed to the precursors of Homo sapiens (pre-Neanderthalian or pre-sapiens).

Little is known about the social organization of *Homo erectus*. It seems, however, that beginning 400,000 years before the present, maybe even before, the hunt for large, isolated mammals (elephant, bear, rhinoceros) led them to organize hunting groups of five or six, each group corresponding to a community of a few dozen individuals. These generally mobile groups established more or less lasting encampments, and perhaps built some rudimentary shelters.

Thus, contrary to the Australopithecines, *Homo habilis* and *Homo erectus* had a true technical and cultural history, which led from simple stone-chopping tools to specialized bifaces, from simple predation to organized hunts of large game, from nomadism to the occupation of grottos and the establishment of homes. Moreover, it is assumed that they developed a minimal language for communication. Their technical history corresponds to the Old Stone Age, which is the longest period of prehistory.

## Homo sapiens neandertalensis

The oldest Neanderthal fossils, discovered in Israel, date back some 120,000 years and the most recent date to around 35,000 years. Over thousands of years, these humans lived as nomads and hunted in the forests and on the tundras of

Eurasia. Although their cranial capacity was the same size as that of modern humans, which varies between 1,000 and 2,000 cubic centimeters, Neanderthals have clearly distinct morphological characteristics: a muzzle face, prominent forehead, receding chin, and a high larynx, which would have prevented them from pronouncing certain articulated sounds.

Middle Stone Age techniques are generally attributed to the Neanderthals, but recent discoveries give rise to the thought that certain techniques long considered as characteristic of *Homo sapiens sapiens* (Cro-Magnon man) were also known to the Neanderthals. Stone tools formed by percussion remained predominant during this whole period, but were differentiated and specialized thanks to the practice of Levalloisian debitage, a cutting technique which, as we have seen above, perhaps began at the time of *Homo erectus* and is the basis for the subsequent evolution in the methods for fabricating flaked stone. Bone work remained crude, as in the Old Stone Age, but the use of fire was generalized, which tends to prove that its production had been mastered. Large organized, collective hunts to drive entire herds toward natural traps seem to have begun in this epoch.

The discovery of traces and fragments of coloring give rise to the thought that the Neanderthals were familiar with artistic preoccupations, as suggested, in addition, by the discovery of ornaments made from collections of teeth, shells, and precious stones. However, no obvious artistic production is known to exist. On the other hand, the first individual or collective sepulchers in organized funerary sites should be attributed to them.

### *Homo sapiens sapiens*

By comparison, *Homo sapiens sapiens* made very rapid and varied technical progress. The first period of its history, the later Paleolithic, which extended from 40,000 to 11,000 years before the present, saw a profusion of new inventions. Hard stone tools were more and more finely flaked, by percussion but also by pressure and sometimes even after preliminary heating. They were also increasingly varied and specialized. Different types of chisels, drills, scrapers, knives, leather knives, axes, oil lamps, etc. were made. Thanks to progress in flaking techniques, the yield from these manufactures grew: up to 17 meters of useable cutting edges per kilogram of stone were obtained, as opposed to only 4 meters from Neanderthal techniques and 0.6 meters for all the first bifaces of *Homo erectus*.[7] To the tools and weapons intended for immediate use were added specialized tools intended for the manufacture of other tools. In addition to simple tools and weapons, there were tools and weapons composed of two or more parts made of stone, bone or wood. Bone and ivory work, rudimentary up until then, developed rapidly and supplied harpoons, awls, forks, throwing

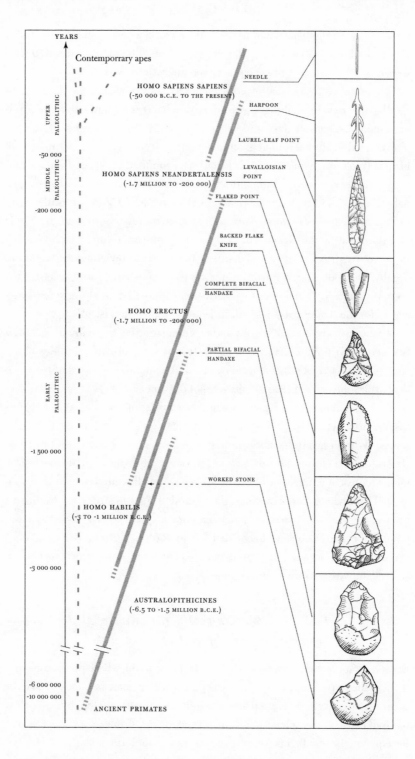

*Figure 1.3 Diagram of the Biological and Technological Evolution of Hominids*

sticks, arrow straighteners, and needles. Woodwork, which has left fewer traces, was undoubtedly considerably developed as well. The manufactured objects were carefully finished and sometimes even finely decorated.

This new equipment allowed humans to hunt new species of large and small game, develop fishing, more effectively harvest certain vegetable products, build artificial shelters, and hence occupy and exploit new environments. The collective hunt assumed great importance. It brought together dozens of beaters (including women and children) who rounded up entire herds of animals (reindeer, horses, bison, aurochs) in order to push them either toward natural obstacles (cliffs, rocky dead ends) or artificial traps (stockades, trap doors, ambushes). *Homo sapiens sapiens* thus rapidly conquered the whole area occupied by the Neanderthals. Then they moved beyond that by settling in Japan, Australia, and some islands, due to their knowledge of navigation. Finally, they penetrated into America via eastern Siberia and Alaska. Thus, 20,000 years ago, humans were already present on all land above water level, except for the two icecaps, high altitude areas, and certain islands.

A fantastic profusion of objects and representations without immediate utility emerged with *Homo sapiens sapiens*. It is as if the creative faculties of the species surpassed its material needs and could respond to all sorts of transcendent aspirations, whether aesthetic, symbolic, or memorial. This surplus of creativity is apparent from the paintings and engravings on the walls of some grottos, the ornamentation of objects for everyday use such as weapons, tools, and diverse costumes, and the fabrication of art objects such as statuettes, small bas-reliefs, carved stones, and disks and sticks carved from bone or ivory. The motifs represent animals and hunting scenes above all, more rarely human subjects. The Chauvet cave paintings, discovered in 1994 in Ardèche in southern France, are truly so expressive, so contemporary, and in fact so obviously modern that it is hard to believe they date back 30,000 years. And if these paintings affect us so, it is because, through them, the *Homo sapiens sapiens* of that epoch so brilliantly express their affinity with us.

## *The End of the Paleolithic: Differentiation of Modes of Predation and Specialization of Tools*

Between 16,000 and 12,000 years before the present, the whole planet was once again in the grip of large ecological disruptions. The climate became warmer, the polar icecaps partially melted and the billions of cubic meters of water thus released caused the sea level to rise by several meters. From the polar icecaps to the equator, the continents were covered by new plant formations:

— tundra, taiga, mixed forests of conifers and broad-leaved trees
in the cold regions
— forests of trees that lose their leaves in winter in the cold temperate regions
— oceanic moors and continental prairies
— evergreen forests in the hot temperate and Mediterranean regions
— sparse forests, wooded savannas and steppes in the Sahara region
(where desertification dates back less than 10,000 years)
— tropical forests of trees that lose their leaves in the dry season
— dense and evergreen equatorial forests.

Humans adapted themselves to these new ecological conditions by implementing new forms of predation. The means for hunting large animals were again
improved thanks to weapons, traps, and large collective beats. In fact, the effectiveness of hunting intensified to such a point that some species were considerably reduced, such as the horse and bison in Europe, or even vanished, such as
the mammoth in the north and the rhinoceros in the south. Towards 12,000
years before the present, humans began hunting non-herd game, middle-sized
game (elk, stags, roe deer, gazelles, wild boars, donkeys) and small game (rabbits, birds), as well as fishing and gathering mollusks (snails, oysters, limpets),
which left enormous piles of shells in some places. In zones rich in wild cereals
and legumes, the consumption of grains took on real importance.

These new modes of predation were remarkably differentiated between
regions. To each mode a whole set of specific tools and weapons corresponded, which made possible the exploitation of the resources belonging to a given
environment. Most often, hunters, fishers, and gatherers moved from encampment to encampment, after having exhausted the resources in the vicinity.
Sometimes, in particularly privileged places, rich in conservable plant products
(seeds, dried fruits) or in continually replenished animal products (necessary
crossing points for migratory birds or other game, shores of seas, lakes, and
rivers rich in fish), resources were so abundant that it was possible for large
groups to settle for a whole season, and even to settle permanently, thanks to the
progress made in conservation processes (drying, smoking, cold, silos).

This relatively short period at the end of the flaked stone era is called the
Mesolithic. Systems of predation were differentiated and specialized tools
abounded. Compound tools multiplied, some formed from a wooden or bone
support into which very small flaked stones called "microliths" were inserted.
With the microliths, *Homo sapiens sapiens* obtained 100 meters of useable cutting
edge per 1 kilogram of stone.[8] Humans had thus nearly attained the limits of their
present living area, which extends from the southern point of the South American
continent, where the now vanished Frigian people lived, to the Arctic polar

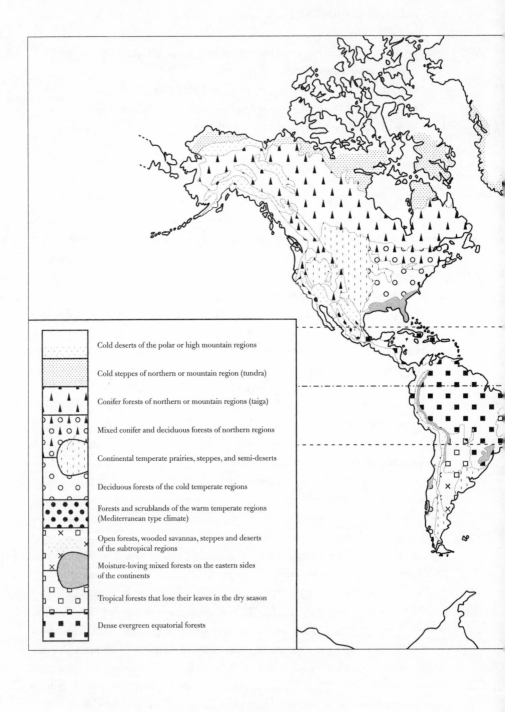

Cold deserts of the polar or high mountain regions

Cold steppes of northern or mountain region (tundra)

Conifer forests of northern or mountain regions (taiga)

Mixed conifer and deciduous forests of northern regions

Continental temperate prairies, steppes, and semi-deserts

Deciduous forests of the cold temperate regions

Forests and scrublands of the warm temperate regions
(Mediterranean type climate)

Open forests, wooded savannas, steppes and deserts
of the subtropical regions

Moisture-loving mixed forests on the eastern sides
of the continents

Tropical forests that lose their leaves in the dry season

Dense evergreen equatorial forests

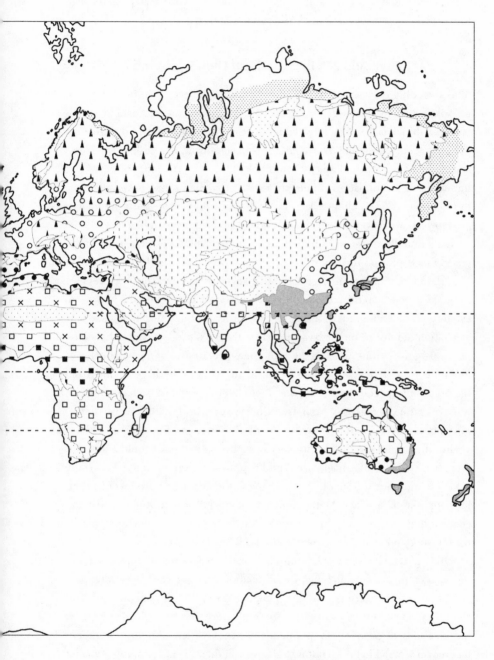

Figure 1.4 Schematic map of "original" plant formations 10,000 years ago

regions where the Eskimos live and up to 5,000 meters above sea level in the high valleys of the mountains of Central Asia and the Andes.

## Hominization: A Biological and Cultural Evolution

The preceding analysis shows that hominization, the evolution from the Australopithecines to *Homo sapiens sapiens*, was a complex transformation, simultaneously biological and cultural, which accelerated as it progressed. While the progress accomplished by *Homo habilis* is insignificant, that realized by *Homo erectus* in 1.5 million years is more notable. However, the latter appears less important in comparison with that achieved by *Homo sapiens neandertalensis* in 100,000 years. But ultimately it is with *Homo sapiens sapiens* during the last 40,000 years that a veritable technical and cultural explosion took place.

Undoubtedly, the growth in the volume and convolutions of the brain influenced this profusion in creativity, just as the development of the larynx and of articulated language facilitated technical and cultural exchanges. Inversely, progress in tool making and in cultivation certainly influenced the biological evolution of the hominids. Throughout the process of hominization, each new generation developed on the basis of technical and cultural ground enriched by previous generations, so that the biological precursors of a new species of hominids were dependent on the technical and cultural heritage coming from the preceding species. Of course, one could hypothesize that each new species was created independently from preceding species, and that it was capable of reproducing all at once the entire history of technique produced by the latter.

But if one sticks to the evolutionary hypothesis, then one should admit that there is no biological, social, or cultural rupture between one species and the next. The movement from one human type to the next is the product of a double cooperation: a sexual cooperation, which ensures the diffusion of advantageous mutations, and a technical and cultural cooperation, which guarantees the transmission of acquired knowledge and know-how.

One can then conceive of hominization as the process of the emergence and replacement of one human type by a succeeding type, the later one always possessing more efficacious biological capacities and technical and cultural resources. The latest, *Homo sapiens sapiens*, clearly has an ecological valence, i.e., an ability to conquer and populate the world, which is superior to that of its predecessors. That does not mean, however, that it is necessary to consider modern humans as the victors in some sort of struggle for life, understood as an incessant fight among unequally evolved populations in which the most advanced would eliminate at each moment the most backward. Rather, hominization appears as the result of labor: generation after generation, hominid populations multiplied efforts to provide themselves with the means to exploit different environments more intensely and broadly. Some of them succeeded in conquering larger amounts of territory and increasing their numbers more than others, such that, after a certain period of time, they could absorb the "backward" minorities biologically and culturally. In this sense, very different from the "struggle for life" mentioned above, hominization can be considered as the result of an incessant activity of the species for survival, expansion, and multiplication, i.e., to increase its ecological valence. That being said, it certainly does not exclude the possibility that the biological and technical advantages acquired at a given moment by a more numerous and better-equipped population could have been used to suppress and diminish the less evolved populations gradually.

Hominization is thus simultaneously an evolution and a history. The biological progress of a species conditions its later technical and cultural advances, though, in turn, the technical and cultural heritage of a species forms a humanized environment, historically constituted, that conditions its future biological evolution. Thus from one species of hominid to the next, the growth of its population and enrichment of its technical and cultural baggage multiply the possibilities for innovations, which appear more and more quickly and are concentrated toward the end of each species' period of existence.

## The Neolithic and the Appearance of Agriculture and Animal Breeding

Around 12,000 years before the present, a new technique for making tools began to develop: polishing stone. This new technique opened the last period of prehistory, the Neolithic, which lasted until the appearance of writing and metallurgy. In addition to the axes and adzes that could be made by polishing all sorts of hard stones and sharpening them several times, this epoch is marked by other innovations, such as the construction of long-lasting dwellings, terra-cotta pottery, and the first developments of agriculture and animal breeding.

Between 10,000 and 5,000 years before the present, some of these Neolithic societies had begun to sow plants and to keep animals in captivity for the purpose of increasing their numbers and using products derived from them. Consequently, after some time, these specially chosen and exploited plants and animals were domesticated. By doing this, these societies of predators transformed themselves gradually into societies of cultivators and breeders. After that, these societies developed the domesticated species and introduced them into most of the planet's ecosystems, thereby transforming the latter, through their labor, into cultivated, human-made ecosystems, increasingly different from the original natural ones. This change from predation to agriculture, in other words, the *Neolithic agricultural revolution*, was undoubtedly, as V. G. Childe emphasized, "the first revolution which transformed the human economy."[9]

From the beginning, human agriculture is thus very different from that of ants or termites. Each species of cultivating or breeding ant or termite is associated with only one domestic species, which they raise or cultivate always in the same manner with the aid of anatomical tools (mandibles and forelegs) and according to a nearly immutable social organization. These cultivating or breeding species are directly produced by evolution, while humans are not born farmers. When they first appeared, humans of the *Homo sapiens sapiens* species were hunter-gatherers. When they began to practice cultivation and animal breeding, they did not find any species already domesticated in nature, nevertheless they domesticated a large number. They did not possess anatomical tools adapted to agricultural work, but they fabricated all sorts of increasingly powerful tools. Finally, no innate or revealed knowledge laid out for humans the art and manner of practicing agriculture, thanks to which they could freely develop extraordinarily varied systems of agriculture and animal breeding, adapted to different environments of the planet and changing with their needs and equipment.

Each form of cultivation and breeding practiced by ants and termites rests on the exploitation of one species according to a unique mode of organization and functioning, while human forms of agriculture rest on the combined exploitation of several species according to diverse modes of organization and functioning. This diversity is due to the fact that, contrary to ant or termite societies, human societies of farmers and animal breeders are not the relatively stable product of evolution of the species but are the changing product, over time and according to place, of a never-ending history.

## 3. THE CONCEPT OF AGRARIAN SYSTEMS

As we indicated at the beginning of this book, the theory of agrarian systems is an intellectual tool that enables us to comprehend the complexity of each form of agriculture and to explain in broad terms the historical transformations and geographical differentiation of human agricultures. In order to understand what an agrarian system is, it is necessary first to distinguish, on the one hand, agriculture as it is effectively practiced and can be observed, which constitutes a *real object of knowledge*, and on the other hand, what the observer thinks of this real object and says about it, which constitutes an ensemble of abstract knowledges that can be methodically elaborated in order to construct a true conceptualized object, or *theoretical object of knowledge* and reflection.

## Complexity and Variety of the Observable Forms of Agriculture

Agriculture as observed in a given place and time appears first as a complex ecological and economic object, composed of a cultivated environment and of a group of related agricultural production units (or farms) that maintain and exploit the fertility of this environment. Looking further, one can also observe that the forms of agriculture practiced at one given moment vary from one locality to another. And if one were to observe a given place over a long period of time, one would note that the form of agriculture practiced there changes from one epoch to another.

In other words, agriculture appears as an ensemble of local forms, variable in space and time, as diverse as the observations themselves. Sometimes, despite this diversity, one also observes that local forms of agriculture practiced in a given region and epoch are enough alike to be compared and classed in the same category. But if one extends the observations even further and for a longer period of time, one discerns very different forms of agriculture that should be classed in other categories. Thus by degrees, one discovers that the presently observable multiple forms of agriculture and earlier identifiable multiple forms of agriculture can be classed in a finite number of categories, each category occupying a determinate place in time and space, in the same way that one classes other changing objects such as living beings, soils, plant populations, etc.

The observable forms of agriculture appear, as we have said, as complex objects that can nevertheless be analyzed and conceived in terms of a system. To analyze and conceive a complex object in terms of a system, it is first necessary to delimit it, i.e., trace a virtual frontier between this object and the rest of the world, and consider it as a whole, composed of hierarchical and subsystems. For example, the anatomy of a higher living being is conceived as a system (or organism) composed of skeletal, muscular, circulatory, respiratory subsystems. Each of these subsystems can be broken down into organs, each organ into tissues, each tissue into cells, etc.

To analyze and conceive of a complex and living object in terms of a system, it is also necessary to consider its functioning as a combination of interdependent and complementary functions, which ensure both internal circulation and external exchanges of matter, energy, and, if it is a question of an economic object, value. For example, the functioning of a higher living being is conceived as an ensemble of digestive, circulatory, respiratory, reproductive functions, etc., which contribute to the renewal of the organism. Thus, to analyze and conceive of agriculture practiced at a given time and place in terms of an agrarian system consists of breaking it down into two principal subsystems, *cultivated ecosystem* and *social productive system*, studying the organization and the functioning of each of these subsystems, and studying their interrelations.

## The Cultivated Ecosystem and Its Renewal

The cultivated ecosystem has a structure composed of several complementary and proportionate subsystems. These include gardens, plowable lands, meadows for mowing, pastures, and forests. Each of these subsystems is organized, maintained, and exploited in a particular manner, and contributes its part in satisfying the needs of domestic animals and humans. Each subsystem in turn can be split up into parts. Plowable lands, for example, are composed of several areas established on different terrains. Each area is composed of several plots (fallow lands, winter wheat, spring wheat), themselves composed of parcels. The system of animal breeding is composed of herds of different species (bovines, ovines, porcines, etc.). Each herd can be organized into units managed separately (milk cows, breeding of calves, bull-calves, heifers, etc.).

A cultivated ecosystem is also renewed. This activity can itself be broken down into several functions. These include the function of clearing and suppressing wild vegetation (slash-and-burn, plowing, either manually or with a plow, hoeing, weed-killer treatment) and the function of renewing the fertility (allowing the land to lie fallow for a long period of time, adding animal excrement, manure, mineral fertilizers). Also included are the management of cultivation (rotations, technical methods, farming operations) and the management of herds (reproduction, fodder schedules).

These functions, which ensure the internal circulation of matter and energy in the cultivated ecosystem, also open the latter to more or less important external exchanges with near or distant ecosystems: water supply and drainage, erosion and deposition of sediments by streams and rivers, transfers of fodder and fertility, and transfers, voluntary or not, of wild or domestic species. Through these exchanges, the transformations of a cultivated ecosystem can influence remote ecosystems. For example, the deforestation of a watershed basin can give rise to floods and sediment deposition in a lower valley. Conversely, an hydraulic installation in an upper valley can deprive the lower valley of water. Deforestation of vast continental spaces can bring about the drying up of the climate in sometimes very remote peripheral regions. An agrarian system cannot be studied in isolation from these distant exchanges and influences.

## The Social Productive System and Its Renewal

The social productive system (or technical, economic, and social system) is composed of *human resources* (labor power, knowledge, and know-how), *inert resources* (productive implements and equipment), and *living resources* (cultivated plants and domestic animals). The agricultural population uses these resources to create and expand the activities involved in renewing and exploiting the fertility of the cultivated ecosystem, in order to satisfy its own needs directly (by consumption) or indirectly (by exchanges).

These means of production and productive activities are organized into units of production. The latter, in turn, are characterized by the type of production system they practice and by the social category to which they belong. The production system of a farm is defined by the combination (the nature and proportions) of its productive activities and its means of production. The social category of a farm is defined by the social status of its labor (familial, wage, cooperative, slave, serf), by the status of the farmer, by its mode of access to the land (free access to common lands, manorial reserve, serf tenure, quit-rent tenure, cultivation by the owner, tenant farming, sharecropping) and by the size of the farm.

In a given agrarian system, farms can practice similar production systems and belong to the same social category. But they can also be very different and complement one another. For example, in many agrarian systems, farms specializing in animal breeding and others specializing in cultivation complement one another by exploiting different parts of the ecosystem. They then exchange the resulting products: the former, manure and animal products, and the latter, grains and other vegetable produce. In systems combining latifundia and minifundia, wage labor for the large farms is provided by a large number of peasant farms too small to employ fully their own familial labor and fulfill their own needs. In a similar manner, in medieval Europe the forced labor used on the manorial estates was provided by the subjugated serfs.

Certainly, one could break down the productive system into as many subsystems as there are units of production or reduce the diversity of units of production to a misleading average or divide these units according to an unsystematic classification, not to say a stupid one (for example, classifying them by surface area, defined in a purely numerical manner, such as 5 by 5 or 10 by 10 hectares). By grouping and classifying the farms according to the production system they practice, then classifying the farms practicing the same production system by social category, the social productive organization of any agrarian system appears as a particular combination of a limited number of types of farms, defined technically, economically, and socially.

The social production system renews its means of production and its particular productive activities year by year. To ensure this renewal, each unit of production (or, simplifying, each type of production unit) can produce its seeds, animals, fodder, and some of the tools and other equipment (*self-supplying*). It can equally produce a portion of the goods consumed on the farm by the producers and their families (*self-consumption*). But it can also sell all or part of its products in order to buy most of the consumer and producer goods necessary for its reproduction.

The total production of each farm must cover all of its expenses in producer goods (current expenses and amortization) and consumer goods, whether this be by self-supplying and self-consumption or by selling its products. Moreover, the farm's income must also be used to pay financial obligations of various kinds: tribute, quitrent, farm rent, taxes, interest on capital, etc. These monies can be partially reinvested by their beneficiaries in the productive system itself and thus contribute to its development. But they can also be purely and simply transferred to the profit of other social spheres and contribute to the impoverishment of the agricultural system.

## Dynamics of Agrarian Systems

The development of an agrarian system results from the dynamics of its production units. We say that there is *general development* when every type of farm makes progress, by acquiring new means of production, developing their operations, and increasing their economic size and income. The development is *unequal* when some units grow much quicker than others. It is *contradictory* when some units progress while others are in crisis and regress. The crisis of an agrarian system is general when every type of production unit regresses and tends to disappear.

In some cases, the farms that progress are able to adopt new means of production, develop new practices and new systems of cultivation and animal breeding, and thereby engender a new cultivated ecosystem. In that way, a new agrarian system emerges. Such a change in an agrarian system is called an *agricultural revolution*. In the course of time, agrarian systems in a given region of the world can be born, develop, decline, and succeed one another in an evolutionary series characteristic of this region. For example, there are the evolutionary series of hydraulic agrarian systems in the Nile Valley (systems based on winter crops grown in flood-water basins, systems of irrigated cultivation at different seasons; see chapter 4), the evolutionary series of the agrarian systems of the temperate regions of Europe (systems of slash-and-burn cultivation; systems based on animal-drawn cultivation with an ard, fallowing, and accompanying animal breeding; systems based on animal-drawn cultivation with a plow, fallowing, and accompanying animal breeding; systems based on animal-drawn cultivation with a plow and accompanying animal breeding, but without fallowing; motorized, mechanized and specialized systems; see chapters 3 and 6 to 10) or the evolutionary series of hydroagrarian systems of the humid tropical regions (chapter 3).

The analysis of the dynamics of agrarian systems in different parts of the world and at different times allows us to apprehend agriculture's general movement of transformation in time and differentiation in space and to express it under the form of a theory of the evolution and differentiation of agrarian systems. Analyses and theorizations of the same type have been developed in response to the need to study other complex, varied, living and evolving objects. The systematic classification and theory of evolution of living species (Linnaeus, Darwin), classification and theory of the formation and differentiation of the main types of soil in different regions (Dokoutchaev), and classification and theory of the relation among languages (Saussure) are all examples.

Thus conceived, each agrarian system is the theoretical expression of a historically constituted and geographically localized type of agriculture, composed of a characteristic cultivated ecosystem and a specific social production system. The latter makes the long-term exploitation of the fertility of the corresponding cultivated ecosystem possible. The production system is characterized by the types of tools and energy used to prepare the soil of the ecosystem in order to renew and exploit its fertility. The types of tools and energy used are themselves conditioned by the division of labor dominant in a society of a particular epoch.

An agrarian system cannot then be analyzed independently of the upstream activities that provide it with the means of production, any more than it can be analyzed independently of the utilization of its products by downstream activities and by consumers. Nor can it be analyzed independently of other agrarian systems that contribute to satisfying a society's needs.

## Why a Theory?

In the final analysis, the concept of agrarian system is an intellectual tool that makes it possible to comprehend the complexity of each form of agriculture by the methodical analysis of its structure and its functioning. This concept also makes it possible to classify the innumerable agricultural forms identifiable in the past or observable today in a limited number of systems, each characterized by a type of organization and functioning. The theory of the evolution of agrarian systems is a tool that makes it possible to represent the continual transformations of agriculture in a region of the world as a succession of distinct systems, forming a definite historical series. Finally, the theory of the differentiation of agrarian systems makes it possible to apprehend and explain in broad outlines the geographical diversity of agriculture in a given epoch.

These intellectual tools have a heuristic function: they make it possible to apprehend, analyze, understand, and explain an infinitely complex, extremely diversified and constantly changing reality. As R. Thom writes in *La Rencontre théorie-expérience*, "In order that the nonmathematical verbal description of a spatio-temporal form be widely accepted, it is necessary that this form be conceptually classified and stabilized. This last condition is essential. If we do not have the concept corresponding to a form, we are incapable of recognizing this form, even of perceiving it. ... Does not the construction of a taxonomy demand a theory which can make it possible to recognize if two forms are or are not to be placed under the same concept?"[10]

But, on the other hand, by methodically explaining the structure and functioning of an agrarian system, a sort of archetype is formed that necessarily provides a coherent and harmonious image to the corresponding species of agriculture. This archetype, which clarifies the rationality of a particular species of agriculture in space and time, that is, its reasons for being, expanding and surviving through adaptation, is necessary in order to identify and classify the observable forms of agriculture belonging to this species and recognize their particularities and possible failures. However, the conception of a typical agrarian system should not lapse into idealization and even less into apology. It should also entail analysis of the system's spatial and temporal limits.

## Why Concrete Analyses?

The theory of agrarian systems does not exhaust the richness of agrarian history and geography, and it does not pretend to do so. It is not the sum of the multitude of accumulated knowledges in these domains. It gives an account of the most widespread and longest-lasting forms of agriculture. It does not give an account of the particularities of form nor the fate of each singular agriculture. These particularities can be known and understood only through observation and concrete analysis of each agriculture, for which the theory offers a method and a proven system of reference, but certainly not a preconceived knowledge of reality that can act as a substitute for observation and analysis. The theory is not dogma.

No more than human anatomy and physiology can exempt a doctor from the art of examining his/her patient, the theory of agrarian systems does not allow the analyst to dispense with the observation, investigation, and analysis of each particular agriculture. Conversely, if the theory of the human body is necessary to give a meaning to the auscultation of a patient and reasonably justify a diagnosis and treatment, a theory is necessary to give meaning to the study of an agriculture and justify proposals for appropriate intervention in reference to it, i.e., projects and policies.

### 4. BIOMASS, SOIL, AND FERTILITY

Since humans became farmers, they have fed themselves less and less from organic matter taken from wild species and more and more from organic matter coming from domestic species propagated through human care and attention within all kinds of cultivated ecosystems. But all organic matter thus produced is not consumable. Important parts of the organic matter coming from domestic plants and animals are by-products that are sometimes difficult to use or eliminate (residue from cultivation, animal excrement). Moreover, a cultivated ecosystem is still made up of many wild plants and animals, sometimes useful but often not, indeed, sometimes even harmful.

The overall fertility of a cultivated ecosystem, which measures its capacity to produce plant biomass, is much greater than its *useful fertility*, i.e., its capacity to produce over a long period of time vegetal organic matter useful to humans or domestic animals—in other words, harvests. Let's see how the biomass and fertility of an ecosystem are formed and renewed.

## Production and Destruction of the Biomass

Organic matter, essential constituent of living beings, is first produced by plants, which then feed, directly or indirectly, all the animals. It originates as a combination of water, drawn from the soil by the roots, and carbon dioxide from the air, absorbed by the leaves. This combination is made in the green parts of the plants, thanks to the radiant energy coming from the solar rays that is captured by the chlorophyll. This is called photosynthesis and complies with the following equation:

$$\text{carbon dioxide + water + photons} \longrightarrow \text{sugar + oxygen}$$
$$\text{in the presence of chlorophyll}$$

or

$$CO_2 + H_2O \ (+ \text{ light + chlorophyll}) \longrightarrow (HCHO) + O_2$$

Thus photosynthesis produces sugars, or glucides, composed of carbon, hydrogen, and oxygen. These sugars, which are present in various forms (glucose, saccharose, amidon, cellulose), serve as the raw material for the fabrication of most other organic substances (lipids, proteins, nucleic acids). The latter also are principally composed of carbon, hydrogen, and oxygen, but some of them contain nitrogen, phosphorus, and sulfur. Moreover, four metallic elements (sodium, potassium, calcium, magnesium), which fulfill various functions indispensable to life, are frequently associated with organic substances. Finally, twenty other elements (iron, chlorine, fluorine, boron, bromine, iodine, silicon, aluminum, copper, manganese, zinc, molybdenum, arsenic, vanadium, etc.), present in very small quantities in living beings, are activators of various biochemical reactions.

Thus plants, which live principally on water and carbon dioxide, also live on various minerals that they absorb through their roots in the form of salt solutions from the water in the soil. Water represents about 80 percent of the weight of plants. It envelops and transports all the other organic and mineral substances that form the *dry matter* or biomass in the strict sense of the term. The latter represents only around 20 percent of the weight of plants. Part of this water is used in different reactions of biological synthesis (such as photosynthesis) and another much larger part is discharged into the atmosphere, in the form of water vapor, through *transpiration.*

Humans and animals, whether they consume plants (primary consumers) or animals (secondary or tertiary consumers), form their own organic substance from the organic matter initially provided by plants.

Part of the organic matter derived from photosynthesis provides plants themselves, as well as animals, with the energy necessary for their subsistence and reproduction. The origin of this energy is found in the inverse reaction from photosynthesis, called *respiration*, which is represented by the following equation:

$$\text{sugar} + \text{oxygen} \longrightarrow \text{carbon dioxide} + \text{water} + \text{energy}$$

or

$$(HCHO) + O_2 \longrightarrow CO_2 + H_2O + \text{energy}$$

As this equation shows, respiration is in fact an oxidation, or combustion, of sugars. All living beings breathe, and in so doing, they absorb oxygen, burn sugars, and discharge carbon dioxide and water.

Organic substances also serve as matter for plants and animals to build their own bodies and at death these substances are found in the form of dead organic matter, or *litter*, more or less dispersed in the soil. This litter contains above all carbon, hydrogen, and oxygen, but also contains all the other elements that nourished the plants and were retained for a time in the living biomass, be it plant or animal. The litter decomposes by using the oxygen and releasing the water, carbon dioxide, and mineral salts.

When an ecosystem is in equilibrium, i.e., when the quantity of organic matter produced each year by photosynthesis is equal to the quantity of organic matter destroyed by respiration and decomposition of the litter, then the quantities of carbon dioxide, water, nitrogen, and various mineral salts, which are absorbed and stabilized in organic matter, are in principal equal to those released by respiration and decomposition. In the same way, the quantities of oxygen released by photosynthesis are compensated by those used by respiration and decomposition. A stable ecosystem neither "creates" nor "loses" anything; it recycles everything.

It is different when a part of the dead biomass accumulates without decomposing, as in the tundra or peat bogs, or when the living biomass increases. Then, the ecosystem fixes water, carbon dioxide, nitrogen, and other mineral elements and releases oxygen. Conversely, when the biomass is destroyed, its decomposition or combustion returns the water, mineral salts, nitrogen, and carbon dioxide to the soil or the air by using oxygen.

## Fertility

The overall fertility of an ecosystem is its capacity for long-term production of plant biomass. The biomass thus produced acts, on the one hand, as compensation for the losses due to respiration and, on the other hand, as food for animals and humans. If need be, it also helps to increase the total biomass.

The fertility of an ecosystem depends, in the first place, on temperature and hours of sunshine, which must be sufficient for the water of the soil to be absorbed by the roots of the plants, for the sap to rise, and for photosynthesis and transpiration to take place. Fertility depends in particular on the length of the vegetative periods, during which these conditions are found together. Beyond these requirements, fertility depends on the quantity of nutritive matter (carbon dioxide, water, mineral salts) that the environment can supply to the plants. Carbon dioxide from the air is not generally lacking, so the growth of plants during the vegetative periods is essentially conditioned by the presence of water in the soil and by the richness of the nutritive mineral salts dissolved in this water (the soil solution).

In a given climatic zone, the conditions of temperature, hours of sunshine, and amount of rain are nearly equal. The possibilities for the plants to feed on water and mineral salts, hence the fertility of a local ecosystem, vary according to the physical, chemical, and topographic characteristics of its geological bedrock. This bedrock largely conditions the volume and the circulation of the soil solution, just as it conditions its mineral richness during vegetative periods, depending upon whether or not it is easily alterable, rich or poor in soluble nutritive minerals, and more or less permeable and uneven. Therefore it conditions the fertility of the place in question. But if the fertility of the soil is indeed conditioned by the climate and the geomorphology (the *biotope*) of a particular place, it is also conditioned, as we will see, by the living population (the *biocenosis*) that develops there. Fertility depends on the age, the size, the composition and the functioning of this population. We are going to see how a soil is made, when life develops in it, and how its fertility is formed and renewed.

## The Formation of the Soil

The soil, the superficial part of Earth's crust, is formed from the decomposition of its rocky geological bedrock, the parent rock, and from the decomposition of the litter, the dead organic matter stemming from the living population that develops in the soil.

## The Decomposition of the Parent Rock

Under the effect of the action of climatic, chemical, and biological agents (variations in temperature, water, oxygen, carbon dioxide, soil acids, micro-organisms, roots, earthworms, etc.) decomposition is effected first, for compact rocks, by breaking up into mineral particles. These are classified in terms of their size into stones, gravels, sands, alluvia (particles that vary in diameter between 0.20 and 0.002 millimeters) and clays (particles with a diameter smaller than 0.002 millimeters). The proportions between these different classes of particles determines the granulometric composition, or texture, of the soil. This is quite variable. There are coarse soils and fine soils, soils that are predominantly pebbly, sandy, alluvial, or clay, and all sorts of mixed soils. The physical properties, the possibilities for agricultural use and fertility are very different for each type of soil. The texture of a soil evolves slowly. Therefore it constitutes a rather stable sort of granulometric heritage, which nevertheless is subjected to forms of decomposition consisting of physical transformations (hydration, swelling), physico-chemical transformations (conversion of micas into clays), or chemical transformations.

The most important result of all these transformations is, in the end, the solubility of the parent rock, which gradually releases the mineral salts it contains, in a form that is soluble in the soil's water and absorbable by the roots. Hence, most of the mineral salts absorbed and incorporated into the biomass of a plant population originally come from the solubility of the parent rock, with the notable exception of nitrogenous salts, which are formed from atmospheric nitrogen. The mineral fertility of a soil is a function of the nature of the parent rock, which is more or less rich in nutritive elements, and of the extent of its decomposition.

## Fixation of Atmospheric Nitrogen

Atmospheric nitrogen is introduced into the soil in different ways. Electrical discharges produced by thunderstorms synthesize nitrogen oxide from oxygen and nitrogen in the atmosphere. Rainwater carries it into the soil, which is consequently enriched with several kilograms of nitrogen, in either nitrous or nitric form, per hectare per year.

In addition, some bacteria living in the soil, such as those of the genus *Azotobacter*, synthesize nitrogen compounds directly from atmospheric nitrogen. After the death of these bacteria, their bodies quickly decompose and mineralize, enriching the soil with twenty to thirty kilograms of mineral nitrogen, assimilable by plants, per hectare per year in temperate environments. The same phenomenon is produced with photosynthetic organisms like the cyanophyceae (blue-green algae), which live in partnership with *Azolla water fern*. In hot and humid tropical environments, combined cyanophyceae and *Azolla* actively fix nitrogen to such an extent that continual rice growing is made possible.

Finally, nitrogen-fixing microorganisms live in symbiosis with certain plants, which provide them with organic matter. In return, the microorganisms supply the plants with nitrogen compounds. The latter return to the soil after the death of the host plant. Moreover, the soil solution is directly enriched by these nitrogen compounds in the vicinity of the roots. In this manner, bacteria of the genus *Rhizobium* penetrate into the roots of leguminous plants, where they cause the formation of bulges or nodules. These very efficient bacteria can fix more than 100 kilograms of nitrogen per hectare per year. Other microorganisms, associated with trees such as alders and casuarinaceae (ironwood trees), also fix atmospheric nitrogen.

### *The Decomposition of the Litter and Formation of Humus*

Before a soil is formed, the parent rock is bare, deprived of all biomass and directly exposed to the action of climatic agents, which begin to break it down. The parent rock then constitutes a not very fertile substratum, which can only be colonized by nitrogen-fixing bacteria and by mosses and lichens that require few mineral elements. Thanks to these first occupants, a litter begins to be formed which, while decomposing, helps to feed the soil-in-formation with nutritive mineral salts. New species of plants, taking root more deeply and requiring more mineral elements, gradually develop until they constitute, at the end of several decades or centuries, a fully developed, relatively stable plant formation called *climax* and an evolved climactic soil whose litter is regularly fed with organic matter from the bodies and debris of plants and animals.

The decomposition of the litter is a process that occurs very rapidly. Dead organic matter is transformed into humus under the action of some microorganisms. Then, under the action of other microorganisms, the humus is oxidized and decomposes, releasing the water, carbon dioxide, and mineral salts it contains. In short, it is mineralized. By doing this, it restores to the soil solution the minerals that had been absorbed and fixed, for a time, in the biomass.

Humus contains humic acids that accelerate the decomposition of the parent rock and are combined with fine particles of clay to form a clay-humus complex. This complex, which has a great capacity for "adsorption" of basic ions from water and mineral salts, forms a vast reservoir of nutritive elements that can be exchanged with those in the soil solution. Moreover, the clay-humus complex serves as a link, as mortar between soil particles (sands, alluvia). It binds them into aggregates and lines the interstices (or lacunae) of the soil, thereby facilitating the circulation of water and air. In brief, it provides a lighter, softer structure to the soil, which is more favorable to the penetration of roots. Finally, humus favors the life of microorganisms in the soil, which accelerates the solubility of the parent rock.

## Migration of the Fine Ingredients

Beyond the decomposition of the parent rock and of dead organic matter, a third process contributes to the formation of the soil. This process is the migration of soluble salts, some oxides and acids, and fine particles of clay that is caused by the circulation of water in the soil. Salts, notably nutritive salts, are carried toward the bottom by the seepage of water from rainfall and overhead irrigation, drained completely down to groundwater level, and often lost forever to the local ecosystem. Fine clays suspended in water are leached, i.e., carried down several dozen centimeters in depth, where they settle and accumulate. In rainy weather, this draining and leaching deplete the upper levels of the soil of salts and other fine components. In dry weather, however, water rises by capillary action to the upper levels of the soil through evaporation and enriches them.

In regions where at certain periods of the year precipitation is much greater than evaporation from the soil and transpiration (evaporation) from the leaves surface, the upper levels of the soil are particularly leached and exhausted. This is the case for the podzols of the cold regions (taiga), humid temperate regions (the Atlantic moors), and some very wet equatorial regions. In moderately watered temperate and tropical regions, the soils are more or less leached. In arid regions, on the contrary, the evaporation and rising of ground water by capillary action can largely prevail over seepage and drainage. The upper levels of the soil are enriched with salts which, above a certain level of concentration, can become toxic for vegetation and even crystallize to form soil with a superficial saline, gypsum, or chalky crust. Finally, in continental-type temperate regions, the evaporation and rising of water by capillary action during the summer, particularly hot and dry summers, offsets seepage and drainage during the rest of the year. The soils of these regions, neither leached nor saline, retain all their mineral richness. Such is the case with the black, or chernozem, soils of central Europe and the Ukraine.[11]

## The Recycling of Mineral Elements

Once occupied by a plant and animal population, a soil is then doubly supplied with mineral fertilizers. This happens through the decomposition of the parent rock and the fixation of atmospheric nitrogen or by the decomposition of the litter, which restores to the soil the minerals previously absorbed by the vegetation and fixed for a time in the biomass. However, even the minerals thus recycled once or several times originally came from the decomposition of the parent rock or the fixation of atmospheric nitrogen.

But if a soil is constantly supplied with minerals, it is also subjected to losses of minerals. In the humid season, as we have seen, part of the salts are carried by rain and drainage deeply into the groundwater. Or, denitrifying bacteria break down nitrogenous salts and return the nitrogen to the atmosphere. Finally, in certain circumstances, soluble salts are "retrograded"; i.e., they are recrystallized to form insoluble compounds that no longer take part in recycling.

All things considered, over the course of a given period, the fluctuations in the inflow and outflow of minerals in the soil solution are equilibrated according to a sort of balance sheet. On one side are the additions of minerals from several sources (solubility of the parent rock, fixation of atmospheric nitrogen, decomposition of the humus and organic manure, additions of chemical fertilizers) to which it is necessary to add the stock of preexisting minerals. On the other side are the losses of minerals during the period under consideration (drainage, denitrification, recrystallization, removal of minerals through harvests of plant and animal products, and, if need be, the gathering up of animal excrement) and the residual mineral stock.

Note that the mineral materials that are absorbed and incorporated into the biomass during a given vegetative period are, consequently, removed by losses due to drainage, denitrification, and recrystallization. If these mineral materials had not been stored in the biomass, most of them would have been well and truly lost. Consequently, a portion of the mineral materials restored to the soil during the decomposition of the litter is a net addition (or more precisely, a non-loss), which is added to the supplies coming from the solubilization of the parent rock and the fixation of atmospheric nitrogen. The soil solution is thus enriched and the plant populations that subsequently develop benefit from this increased fertility. The quantity of mineral matter recycled increases from season to season, at least until reaching a climactic maximum. In an analogous manner, the humus content of the soil can vary over the course of time. This positive or negative variation results from the balance between the quantity of humus that the soil receives or which is formed by the decomposition of dead organic matter of diverse origins (litter and organic manure) and the quantity of humus that it loses through mineralization.

Thus if the humic and mineral fertility of a cultivated soil is indeed conditioned from the beginning by the climate, parent rock, and original population, this fertility is not given once and for all. It can be maintained at a constant level on condition that this soil receive exactly sufficient quantities of organic and mineral matter to compensate for both the losses of humus through mineralization and the losses of minerals through drainage, denitrification, and harvests.

The fertility can be reduced if these contributions are insufficient or increased in the opposite case. In fact, from the moment a soil is cultivated, its fertility becomes a historical variable, largely influenced by successive agrarian systems.

## Modes of Renewing the Fertility of Cultivated Soils

An agrarian system can develop and perpetuate itself only if the fertility of the cultivated soil is maintained at a level sufficient to ensure the harvests necessary to the population over the long term. There are only a few soils, certain chernozems or certain slightly leached loessial and alluvial soils, in which the mineralization of the parent rock and the fixation of atmospheric nitrogen allow the indefinite production of sufficient harvests each year to support the needs of the population. To each (lasting and widespread) agrarian system there necessarily corresponds an effective method of renewing the fertility.

The first of these methods consists of allowing the wild vegetation, after clearing a wooded terrain and cultivating it for some time, to reconstitute itself and restore to the soil sufficient quantities of organic and mineral matter to make up for the losses caused by its cultivation, after which one can again clear and cultivate this ground. Such is, as we will see, the mode of renewing fertility in systems in which temporary slash-and-burn cultivation alternates with allowing the land to lie idled in its original wooded condition[12] for a long period of time and is also the mode in systems in which cultivation with a hoe alternates with allowing the land to lie idled in its original grassy condition[13] for a period of medium duration (see chapter 3).

The second method consists of concentrating the cultivation on the best lands and using the other lands as natural pasturage. Grazing all day on these pastures, the animals are penned up at night on the fallow land[14] where they leave their excrement, thus transferring part of the biomass that they had grazed on from the uncultivated land (the *saltus*) to the cultivated lands (the *ager*). This mode of renewing fertility is used in systems of manual cultivation or animal-drawn cultivation with an ard, fallowing, pasturage, and associated animal breeding (chapter 6).

The third method consists of mowing part of the pasturage in order to feed the livestock in the stable and produce manure, which will be buried when the fallow land is plowed. This method is used in systems based on animal-drawn cultivation with a plow, fallowing, mowing of meadows, and associated animal breeding.

A fourth method consists of replacing the fallow land by a cultivation that produces a great quantity of biomass and fixes a maximum of mineral matter, thus saving it from drainage and denitrification. Then this organic and mineral matter is restored to the cultivated soil, either by burying it directly as "green manure" or by making it available for consumption by animals, whose excrement is gathered and then buried. This mode of renewing fertility is practiced in systems that do not involve fallowing (chapter 8).

Another method is maintaining a number of large trees above the cultivated ground whose roots draw out the mineral elements from deep in the soil and are then restored, to the cultivated soil, either directly through falling leaves and other types of dead organic matter or indirectly through manure from animals that have consumed leaves and small branches. This mode of renewing fertility is used in agrarian systems that unite arboriculture and annual cultivation (chapter 3, point 4 and chapter 7, point 2). Moreover, the basic principle of these systems, based on cultivating tiered rows of associated plants has been known for a long time. The elder Pliny (first century c.e.) described the agriculture of the south Tunisian oases in these terms in his *Natural History*: "The olive grows in the shadow of the noble palm tree, the fig under the olive, the pomegranate under the fig, the grape under the pomegranate, the wheat under the grape, then the legumes, finally the lettuce. Every plant lives and grows in the shadow of another during the same year."

In many of the hydraulic agrarian systems, flood and irrigation waters, full of alluvial deposits and soluble minerals coming from the side basins which feed them, also contribute to renewing the fertility of cultivated soil (chapters 3 and 4). Moreover, in aquatic rice-growing in the tropical regions, the blue-green algae associated with *Azotobacter* greatly contribute to supplying the rice fields with nitrogen.

We add that, in all these systems, the presence of legumes, whether grown for fodder or not, whether herbaceous or arboreal, whether cultivated in rotation or in association with other plants, can also contribute to enriching the soil with nitrogen. Lastly, in some older systems and in many contemporary systems, organic or mineral fertilizers are gathered outside the cultivated ecosystem and transported by humans to the cultivated areas. Thus, already in pharaonic Egypt, organic sediments, from plant, human, and animal origins, deposited during thousands of years in ancient village sites in the Nile Valley, were exploited and used as fertilizers. In the same way, during pre-Inca and Inca periods in Peru, guano deposits from the Pacific Coast, containing phosphates and nitrates produced by the decomposition of the excrement and carcasses of millions of marine birds, were exploited (chapters 4 and 5). The use of mineral fertilizers is thus very old, but due to poor means of exploitation and transport, remained limited for a long time. By contrast, in the twentieth century, the extraction, transformation, synthesis, long-distance transport, and use of synthetic fertilizers and various soil amendments were used on a wide scale in the agriculture of the developed countries and in some sectors of agriculture in the developing countries (chapter 10).

## Fertilizers and Amendments

Fertilizers are, in the strict sense, mineral or organic materials that are incorporated into soil for the purpose of providing the plants with nutritive minerals, and possibly other substances such as growth hormones, which they need. Fertilizers are distinguished from *amendments*, which in principle are organic or mineral materials that are incorporated into the soil for the purpose of improving its composition and its physical and chemical properties. Clays and marls are used to correct the lightness of a soil and its lack of absorbability. Calcium and magnesium amendments correct excess acidity; leaching with water corrects its salinity. Organic amendments elevate its humic content, increase its capacity to store water, improve its structure, and contribute to maintaining a thriving and diverse population of soil organisms.

Some amendments also contribute to restoring or elevating the soil's nutritive mineral reserves usable by plants; in other words, they act as fertilizers. Such is particularly the case with organic matter produced on the farm or collected in its vicinity (animal manures, composts, green manures, or algae) and with all types of plant and animal by-products that are dried, ground, transformed, and packaged in various ways (dried bird droppings, dried blood, fish and meat meal, bone powders, grape marc, urban refuse and sludge, etc.). While decomposing, these amendments and organic fertilizers also provide nutritive minerals to plants.

Mineral or synthetic fertilizers are extracted from volcanic, sedimentary, or saline layers of rock, which are then mechanically and chemically transformed. Nitrogen fertilizers can also be synthesized from atmospheric nitrogen. Mineral fertilizers are for the most part soluble (nitrogen, superphosphate, potassium fertilizers), that is, after spreading they are quickly found in the form of ions in solution in the water of the soil, absorbable by the roots. Others are called "insoluble" (natural phosphates, slag from dephosphorization, various ground rocks), but they are in fact slowly solubilized, like a finely crushed rock would be under the effect of chemical (soil acids) and biological (microorganisms and roots) agents.

In many ways, organic fertilizers (manure, compost, green manure, animal excrement, etc.) are more effective than mineral fertilizers. One fertilizer unit of nitrogen, phosphoric acid, or potassium contained in an organic fertilizer may result in a more substantial increase in the harvest than the same unit coming from a mineral fertilizer. In fact, minerals of organic origin are progressively released and absorbed to the extent that the plants need them during the summer season, while in the winter they are reserved in their organic form. Thus they are less subject to drainage. Moreover, they feed the soil solution in a more complete and balanced manner than mineral fertilizers, because they contain, beyond the main elements (nitrogen, phosphorus, potassium, calcium magnesium sulfur) trace elements (iron, manganese, zinc, copper, boron, molybdenum, chlorine). They also support the existence of microorganisms in the soil and supply them with various substances that stimulate the growth of plants (hormones). Finally, remember that the most important advantage of organic fertilizers is that they also act as soil amendments that increase the effectiveness of all mineral fertilizers, regardless of origin.

But do not forget that the fertility of a cultivated ecosystem does not only depend upon the mineral richness of the soil solution. It depends first on temperature, sunlight, and availability of water during the growing season. To increase the fertility of an ecosystem, one can thus also act on the temperature (possibly, heated greenhouses), on sunlight (provide shade), on the water supply and its

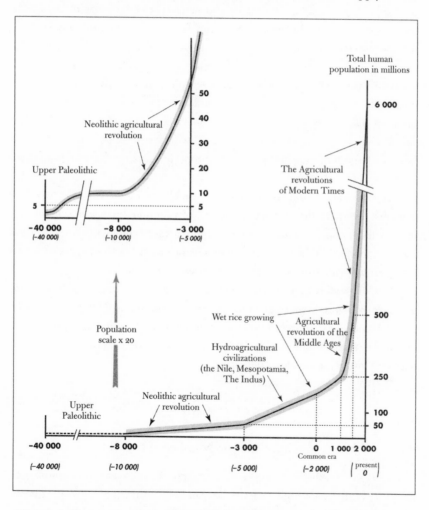

*Figure 1.5 The growth of Human Population in Connection with the Development of Agrarian Systems throughout the World*

organization (irrigation, drainage, windbreaks, soil covering that minimizes evaporation), and even on the carbon dioxide content of the air (greenhouses with an artificial atmosphere). But these large and costly facilities are not always necessary, feasible, or profitable. However, certain agrarian systems all over the world, particularly all the hydraulic agrarian systems, can only exist thanks to facilities of this type. Lastly, beyond these relatively stable characteristics of the environment, the useful fertility of an agricultural terrain over the course of a given period of time also depends, of course, on the nature of the exploited plant populations and on the way in which these are raised (natural pastures, cultivation of single crops or associated crops, rotations, farming work).

As you can see, in most agrarian systems, renewal of the fertility of the cultivated lands is provided by organic and mineral resources originating in the uncultivated parts of the ecosystem: fallow lands as part of a rotation system, whether forested or not, pasturage, mowing of meadows, side basins feeding irrigated land, etc. However, part of the territory must be reserved for other uses: forests, dwellings, roads, water reservoirs, etc. Finally, some lands are unsuitable for any use. That is, cultivated lands can only occupy a part, often quite small, of the ecosystem. In the last analysis, the useful fertility of a cultivated ecosystem, i.e., its capacity to produce harvests, does not result only from the fertility of the cultivated lands, properly speaking, but also from their relative size in the ecosystem.

The extent and fertility of actually cultivated lands are therefore the two variables that determine the production capacity of a cultivated ecosystem, and hence the maximum population density it can support. At each moment, these two variables are conditioned by the characteristics of the original ecosystem, more or less modified by prior successive agrarian systems, and governed by the mode of renewing the fertility of the current system. In other words, in each period of history the population level reached by humanity is conditioned by the nature and performance of the agrarian systems that develop in different parts of the world during the period in question, systems that themselves depend on preceding systems for most of their ecological heritage.

## 5. AGRICULTURE AND HISTORY

### Agriculture and the Number of Humans

The works of demographers give us an idea of the increase in the number of people over the last 50,000 years.[15] We can look at this increase in relation to the evolution of agrarian systems in different parts of the world, such as we present in this book.

On the eve of agriculture's appearance, the human population was rapidly expanding thanks to the development of increasingly diversified and effective modes of predation. However, even if in certain places humans had reached the limits of exploitability of some species, even so far as to cause them to diminish or even disappear, nothing justifies the contention that development of agriculture and animal breeding in the Neolithic responded to the need to overcome a sort of generalized crisis of systems of predation.

It is, however, undeniable that the tenfold increase in the human population, which grew from around 5 to 50 million inhabitants between 10,000 and 5,000 years before the present, is essentially due to the planetary development of Neolithic agriculture. The systems of slash-and-burn cultivation, which developed mainly in the cultivable forested environments of the planet, supported population densities of 10 to 30 inhabitants per square kilometer, densities which are much higher than those of systems of predation (see chapter 3).

Then, between 5,000 and 3,000 years before the present, i.e., between 3,000 and 1,000 B.C.E., the world population doubled, growing from around 50 to around 100 million individuals. This increase can be explained, to a certain extent, by the extension of slash-and-burn cultivation, but also by the development of large societies based on hydraulic agriculture in the valleys of the Indus, Mesopotamia, and the Nile. Certainly, agricultures organized around floodwaters and irrigation, which were organized in these privileged valleys, were limited in extent, but they could support impressive population densities of several hundred inhabitants per square kilometer (chapter 4).

In the course of the two millennia that followed, between 1000 B.C.E. and 1000 C.E., the world population more than doubled to around 250 million inhabitants, due to the development of hydraulic systems of aquatic rice growing in the valleys and deltas of China, India, and Southeast Asia and, to a lesser degree, to the development of systems of hydraulic agriculture (Olmecs, Mayas, Aztecs, pre-Inca societies) in America during this period (chapter 5). On the other hand, systems of cultivation based on rainwater and fallowing, which extended around the Mediterranean region and into Europe, contributed very little to this population growth, because they were not that much more productive than the slash-and-burn agriculture which they replaced (chapter 6).

The contribution of European agriculture to world population growth became noteworthy only with the agricultural revolution in the Middle Ages. From the eleventh to the thirteenth centuries, the development of agricultures based on fallowing and cultivation with an animal-drawn plow made it possible for the European population to triple or even quadruple (chapter 7). After having fallen during the great crisis of the fourteenth century, this population increased again in the sixteenth century. Then it doubled once more thanks to the agricultural revolution of the seventeenth, eighteenth, and nineteenth centuries, a revolution which gave rise to agrarian systems that did not use fallowing (chapter 8). But the growth in world population since the year 1000 was also due to the development of aquatic rice growing systems, particularly in Asia. In addition, from the sixteenth century, the population of European origin grew by extending its agriculture into the temperate regions of America, South Africa, Australia, and New Zealand, to the detriment of the indigenous populations.

Finally, even today, the world demographic explosion, whatever the other reasons may be, is only made possible by a gigantic increase in the production capacities of world agriculture. This increase essentially results from the expansion and improvement of aquatic rice growing from two to three harvests per year, principally in Asia, and from the development of motorized, mechanized, and chemicalized agriculture in the developed countries and in some limited sectors of the developing countries.

For all that, however, this immense increase in the number of people should not induce us to forget that famine, undernourishment, and the persistent difficulty in meeting the needs of humanity are all very real. At the end of the twentieth century, 800 million persons suffered from chronic undernourishment and more than 2 billion were nutritionally deficient in one or several nutrients (iron, iodine, vitamin A, proteins). And it is possible to believe, as do Professor René Dumont and Lester Brown, director of the Worldwatch Institute, that the growing needs of humanity are dangerously approaching, right now, the exploitable limits of water resources, planetary fertility, and even the possibility of using photosynthesis for food production.[16]

But if some regions are fully exploited and even sometimes dangerously overexploited, there are also many exploitable regions that are today unexploited or underexploited. More than two-thirds of the exploitable areas in the developing countries (China not included) are unexploited.[17] And even if half of this land is in fact difficult to exploit, the possibilities for expanding agriculture are still very great. Moreover, it is possible to believe that the twenty-first century will see the development of agrarian systems producing more basic provisions and able to support much larger population densities than the cereal-growing or pastoral systems predominant today. In fact, setting aside progress in irrigation, seed selection, and synthetic chemical agriculture, all kinds of highly productive and sustainable systems, closely combining annual cultivation, animal breeding, and arboriculture, are developing right now in the densely populated regions of the world in Southeast Asia, Central America, the Caribbean, and Great Lakes area of Africa. Systems of this type are labor intensive but not demanding in nonrenewable resources nor very polluting. They formerly existed in difficult and relatively populated regions of Europe (the chestnut groves of Corsica and the Cévennes, and various forms of *cultura promiscua* in the Mediterranean region). Finally, in the developed countries, many uncultivated regions today could, if necessary, again enter into production if the products fetched a higher price and agricultural work were better paid.

As we will see throughout this book, the overpopulation of an ecosystem is rarely absolute. It is generally relative to the capacities of the agrarian system at a given point in time. Thus, according to some people, on the eve of the Neolithic agricultural revolution, the planet, which only had several million inhabitants, was already overpopulated in relation to the means available from the system of predation. In the tenth century, with 10 million inhabitants, France was afflicted with famine. Three centuries later, after having adopted a cultivation system based on the animal-drawn plow, it fed nearly 20 million people. Then, after the horrible food crisis and the huge number of deaths in the fourteenth century, the population was restored. Up to the end of the eighteenth century, France again appeared "overpopulated" each time its population surpassed the level of 20 million inhabitants. At the end of the nineteenth century, however, thanks to the first agricultural revolution of modern times, France (within its current borders) fed nearly 40 million inhabitants. In the same way, for several decades, the rice-growing deltas of Asia, with only one harvest of rice per year, were considered overpopulated with 500 inhabitants per square kilometer. Today, there are well more than 1,000 inhabitants per square kilometer, thanks to the increase in output and the development of systems with two, three, or four harvests per year.

In truth, today no one knows how to estimate the planetary capacity for production of consumable biomass by humans and domestic animals without an enormous margin for error. According to the distinguished Department of Agricultural Research at the University of Wageningen, this production capacity could be 30 billion (at nearly 50 percent) to 72 billion tons of cereal-equivalent per year according to the type of agriculture practiced, with varying degrees of synthetic chemical use, or 7 to 18 times more than current production (which is around 4 billion tons of cereal-equivalent per year). That is enormous! Nevertheless, this estimate does not make it possible to know absolutely how many billions of people world agriculture will be able to feed at any particular future moment. The whole question is indeed to know what part of this potential will be effectively used in a particular time frame, who will benefit from it, up to what point, and who will be excluded from it.

## Agricultural Productivity, Social Differentiation, and Improvement in Diet

If, in any event, the volume of agricultural production strongly limits the number of people, the fact remains that an increase in agricultural production is not sufficient to bring about a population increase. For that to happen, other social and cultural conditions controlling natality and mortality must be fulfilled. But above all, in order for a given population to increase, or even simply renew itself, it is necessary that the production of an agricultural worker, that is, the productivity of agricultural labor, be at least equal to the sum of that worker's own needs and the needs of all those supported by that worker. Indeed, it is important to remember that in any society, the majority of individuals (elderly, children, disabled, people practicing occupations other than agriculture, etc.) do not produce their own food.

Thus, in an entirely agricultural society, without outside food supplies, in which there are four mouths to feed per working person (including the latter), agricultural productivity must be at least equal to four times the needs of an average individual (making no distinction for age or sex). In most developing countries today, the average food consumption does not surpass 200 kilograms of cereal-equivalent (the quantity of cereal having the same caloric value as the group of food products under consideration) per person per year, which corresponds to an average daily ration of 2,200 calories. Certainly, the caloric needs of a population vary according to its structure (age, sex, weight), its mode of life, and climate. Nevertheless, the fact remains that as a first approximation, one can consider an average ration of 2,200 calories per person per day as a minimum. In these conditions, in order to support just the needs of the agricultural population, labor productivity must be at least 4 x 200 kilograms = 800 kilograms of cereal-equivalent per agricultural worker. Below this minimum level of productivity, an agrarian system cannot reproduce itself.

What is more, in order to support the needs of nonagricultural social groups, agricultural productivity must be higher than this minimum level over a long period of time. Thus, under the same conditions as outlined above, in order to support the total needs of a population composed of a nonagricultural population as numerous as the agricultural population itself (which corresponds to 8 mouths to feed per agricultural worker), the average agricultural productivity must be at least doubled, 8 x 200 kilograms = 1,600 kilograms of cereal-equivalent per worker. Beyond the volume of production necessary to support the needs of the agricultural workers and their families, the increase in agricultural productivity makes it possible then to produce a surplus that conditions the possibilities for development of nonagricultural social strata (warriors, priests, administrators, artisans, merchants, workers, etc.). In the last analysis, an agricultural surplus determines the possibilities for social differentiation and urbanization.

But the increase in agricultural productivity can also be expressed by a quantitative and qualitative improvement in diet. Indeed, the consumption level considered above (200 kilograms of cereal-equivalent per person per year, or 2,200 calories per person per day) can be greatly surpassed. Thus today in the developed countries and in the well-off social strata of most developing countries, the average ration greatly exceeds 3,000 calories per person per day and an important part of it is made up of calories from animal products. Consider now an average ration of 3,200 calories per day, composed of some 2,200 calories from vegetables and 1,000 calories from animal products. As we have seen, in order to provide these 2,200 vegetable calories, it is already necessary to have available 200 kilograms of cereal-equivalent per person per year. Moreover, knowing that it takes around seven calories of vegetables to produce one calorie of animal products, it is then necessary to have around 7,000 vegetable calories (per person per day) to provide 1,000 calories of animal products, which corresponds to 640 kilograms of cereal-equivalent per person per year. In total, it is necessary then to have 200 + 640 = 840 kilograms of cereal-equivalent per person per year at one's disposal, or around four times more than the minimum considered earlier.

Given this greatly expanded dietary norm, in order to support just the needs of an agricultural population with four mouths to feed per worker, agricultural productivity can no longer be 4 x 200 = 800 kilograms of cereal-equivalent per worker, but 4 x 840 = 3,360 kilograms of cereal-equivalent, three-quarters of which are then consumed by animals. And, in these conditions, to support a population half of which is nonagricultural, with a total of eight mouths to feed per worker, average agricultural productivity must be 8 x 840 = 6270 kilograms of cereal-equivalent per worker.

Therefore if the output from the land (production per square kilometer) of an agrarian system determines the maximum population density that it can support, at the same time its productivity conditions the possibilities for social differentiation and for dietary improvements. The gross productivity of a system is the result of the output per hectare multiplied by the cultivated area per worker, an area that depends on the effectiveness of the tools and the power of the energy sources (human, animal, motomechanic) that this worker uses.

In systems of rain-fed cultivation with manual labor (slash-and-burn agriculture using the ax or machete in forested environments and the spade and hoe in deforested environments), the area cultivated by an active worker (with assistance from others) rarely exceeds one hectare. If the output per hectare is around 10 quintals of cereal-equivalents, productivity is barely sufficient to support the basic needs of the agricultural population itself. In these conditons, if there is no outside source of food provisions, social differentiation and the consumption level will necessarily remain low (see chapter 3).

On the other hand, in hydroagricultures using manual labor (aquatic rice growing, agriculture using receding flood waters or irrigation), even if the area cultivated by each worker is often less than one hectare, the higher net outputs generally make possible a much higher level of social differentiation (chapters 4 and 5).

In systems based on fallowing, animal-drawn cultivation using the ard and transport using the pack-saddle, the area cultivated per worker can reach three to four hectares, but as the mode of renewing fertility is not very effective, the outputs and thus the productivity remain low (chapter 6). On the other hand, in animal-drawn cultivation using the plow and wagon, the area cultivated per worker can reach four to five hectares, while, thanks to the possibilities of producing, transporting, and plowing in large amounts of manure, the outputs attain a much higher level (chapter 7). The development of agrarian systems based on fallowing and cultivation using an animal-drawn plow conditioned the demographic, artisanal, industrial, commercial, urban, and cultural progress of the Middle Ages in the West beginning in the year 1000. This progress was strengthened from the seventeenth to nineteenth centuries thanks to the development of agrarian systems built around cultivation using an animal-drawn plow but no fallowing (chapter 8).

From the end of the nineteenth century in the West, the mechanization of animal traction (Brabant plow, seed drill, mower, reaper) made possible the doubling of both the area cultivated by each worker and productivity (chapter 9). Finally, in the twentieth century, the motorization associated with large-scale mechanization made possible an increase in the area cultivated by each worker for cereal production to more than 100 hectares. This, combined with outputs that could go as high as 100 quintals per hectare, led to a gross productivity of 10,000 quintals per worker, or 1,000 times more than the productivity of a manual system of cultivation without the use of fertilizers (chapter 10). Today, the use of tractors and other powerful equipment makes it possible for each worker to cultivate in excess of 200 hectares. Thus, in North America and western Europe, an agricultural population that makes up less than 5 percent of the total population can feed everyone. It is important to note that remote-control or automatic machines, making it possible to multiply this productivity several times over, are currently in development and beginning to be used in some limited sectors of agriculture in the developed countries, while the vast majority of the peasants in the developing countries still use strictly manual tools.

We turn now to the Neolithic agricultural revolution.

# 2

# The Neolithic Agricultural Revolution

The *Sanskrit* language, whatever be its antiquity, is of a wonderful structure; more perfect than the *Greek*, more copious than the *Latin* and more exquisitely refined than either; yet bearing to both of them a stronger affinity, both in the roots of verbs, and in the forms of grammar, than could possibly have been produced by accident; so strong, indeed, that no philologer could examine them all three, without believing them to have sprung from some common source, which, peerhaps, no longeer exists. There is a similar reason, though not quite so forcible, for supposing that both the *Gothic* and the *Celtic*, though blended with a very different idiom, had the same origin with the *Sanskrit*; and the old *Persian* might be added to the same family, if this were the place for discussing any question concerning the antiquities of *Persia*.

—SIR WILLIAM JONES, "On the Hindus"

At the end of the Paleolithic (Old Stone Age), some 12,000 years ago, after hundreds of thousands of years of biological and cultural evolution, human societies were able to make increasingly varied, sophisticated, and specialized tools, thanks to which they developed differentiated modes of predation (hunting, fishing, gathering), adapted to the most diverse environments.[1] This specialization became more pronounced in the Neolithic (New Stone Age) and it is in the course of this last period of prehistory, beginning less than 10,000 years ago, that several of these societies, among the most advanced of the moment in question, began the evolution from predation to agriculture.

At the beginning of this change, the very first practices of cultivation and animal raising, which we will call *protocultivation* and *proto-animal raising*, were applied to populations of plants and animals which had not yet lost their wild characteristics. But, as a result of such practices, these populations acquired new characteristics, typical of domestic species, which are the origin of most of the species that are still cultivated or bred today.

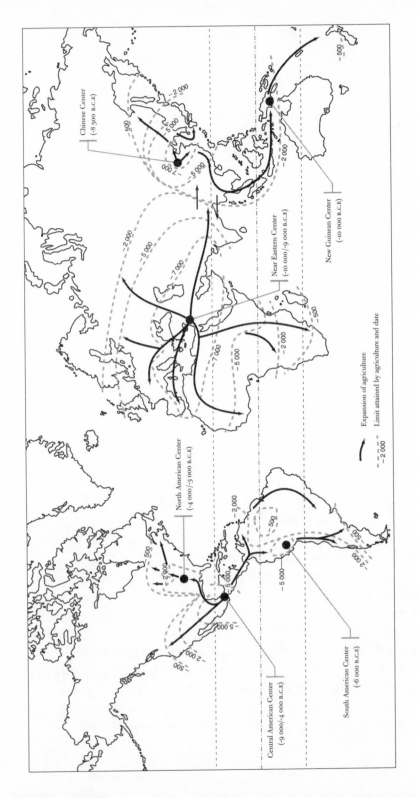

Figure 2.1 *Centers of Origin and Areas of Extension of the Neolithic Agricultural Revolution*

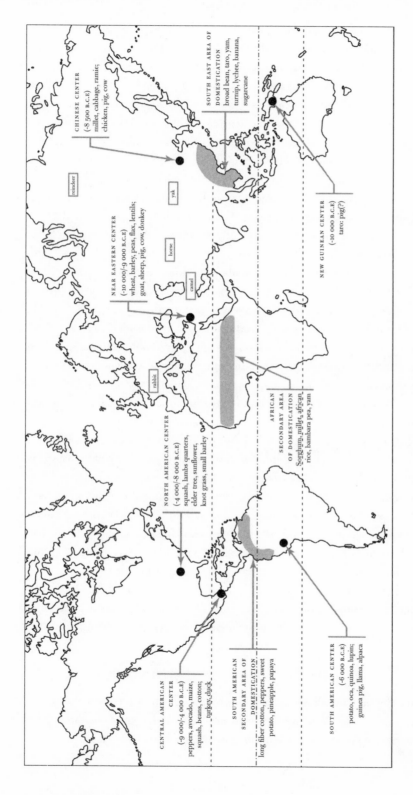

CHINESE CENTER
(~8 500 B.C.E)
millet, cabbage, ramie;
chicken, pig, cow

SOUTH EAST AREA OF
DOMESTICATION
broad bean, taro, yam,
turnip, lychee, banana,
sugarcane

reindeer

yak

NEAR EASTERN CENTER
(~10 000/~9 000 B.C.E)
wheat, barley, peas, flax, lentils;
goat, sheep, pig, cow, donkey

horse

camel

NEW GUINEAN CENTER
(~10 000 B.C.E)
taro: pig(?)

rabbit

NORTH AMERICAN CENTER
(~4 000/~8 000 B.C.E)
squash, lambs quarters,
elder tree, sunflower,
knot grass, small barley

AFRICAN
SECONDARY AREA
OF DOMESTICATION
Sorghum, millet, african
rice, bambara pea, yam

CENTRAL AMERICAN
CENTER
(~9 000/~4 000 B.C.E)
peppers, avocado, maize,
squash, beans, cotton;
turkey, duck

SOUTH AMERICAN
SECONDARY AREA OF
DOMESTICATION
long fiber cotton, peppers, sweet
potato, pineapple, papaya

SOUTH AMERICAN CENTER
(~6 000 B.C.E)
potato, oca, quinoa, lupin;
guinea pig, llama, alpaca

*Figure 2.1 (continued) Centers of Origin of the Neolithic Agricultural Revolution and Secondary Areas of Domestication*

The regions of the world in which human groups, living exclusively from predation on wild species, transformed themselves into societies living principally from the exploitation of domestic species are, in the end, not very numerous, not very large, and quite distant from one another. They form what we will call *centers of origin of the Neolithic agricultural revolution*, it being understood that the word center denotes an area, and not a point of origin. From some of these centers, which we will call *expanding centers*, agriculture spread to most of the other regions of the world. To each expanding center corresponds a specific area of extension, which encompasses every region taken over by the agriculture coming from that center. However, some centers did not give rise to a significant area of extension. Such centers expanded very little or not at all and were subsequently incorporated into one or another of the areas of extension mentioned above.

New species of plants and animals were domesticated in these areas of extension. Some areas supplied a great number of these new domestic species and, after the centers of origin, form true *secondary areas of domestication*. The societies of farmers and animal breeders that stemmed from the centers of origin generally spread their new mode of life by gradually colonizing diverse exploitable territories of the planet. In doing so, they also encountered more or less evolved, preexisting societies of hunter-gatherers who were sometimes themselves practicing a form of proto-agriculture. Some of the latter were, as a result of this contact, won over to practicing agriculture.

In both the centers of origin and the areas of extension, the first societies of farmers were primarily confronted with two main types of original ecosystems: nearly closed, forested ecosystems, in which they practiced diverse forms of slash-and-burn cultivation and some incidental animal raising; and grassy, open ecosystems where they mainly developed various types of pastoral stock-breeding, sometimes combined with cultivation. These societies also encountered various environments that were unexploitable by farming or animal breeding, which thus remained virgin or occupied by hunter-gatherers.

Where, when, and how did Neolithic agriculture appear? How did it expand across the world? What are the mechanisms of domestication? Such are, in short, the questions which we intend to answer in this chapter.

## 1. CENTERS OF ORIGIN OF NEOLITHIC AGRICULTURE

In the current state of research, six more or less well-attested centers of origin for the Neolithic agricultural revolution are generally cited. Four of them were largely expanding centers. The latter, which we will study in more detail later, are:

1. *Near Eastern center*, which was formed in Syria-Palestine, and perhaps more broadly in the whole of the Fertile Crescent, between 10,000 and 9,000 years before the present
2. *Central American center*, which was established in southern Mexico between 9,000 and 4,000 years before the present
3. *Chinese center*, which was first constructed in northern China 8,500 years ago on the loess terraces of the middle Yellow River, then was completed by expanding toward the northeast and southeast between 8,000 and 6,000 years before the present
4. *New Guinean center*, which perhaps had emerged in the center of Papua-New Guinea some 10,000 years ago.

Two other minimally or nonexpanding centers of origin were formed as well during the same time period:

5. *South American center*, which developed in the Peruvian or Ecuadorian Andes more than 6,000 years before the present
6. *North American center*, which appeared in the middle Mississippi basin between 4,000 and 1,800 years before the present.

For a long time, the emergence of Neolithic agriculture has been reduced to the invention and rapid generalization of a new productive technique made necessary by the insufficiency of wild resources. In this view, the insufficiency results either from the drying up of the climate (theory of the oasis) or from the scarcity of large game animals overexploited by a human population that had become too numerous. More recent archaeological studies focusing on the different centers of origin of Neolithic agriculture show that there is nothing to these theories.[1] The transformation of a society living by simple predation and making use of the necessary tools, social organization, and know-how to a society living principally from the products of cultivation and animal raising and making use of the corresponding material resources, social organization, and knowledge, appears as a complex chain of material, social, and cultural changes that condition one another over hundreds of years.

## The Important Expanding Centers

To begin with, let's look at the circumstances in which the four significant known expanding centers were formed, namely the Near Eastern, Central American, Chinese, and New Guinean centers.

## The Near Eastern Center

In the Near East, where one of the oldest and best-known centers of origin of Neolithic agriculture was formed, this slow transition from predation to agriculture lasted more than 1,000 years and it revolutionized all technical, economic, and cultural aspects of the human way of life.[2] In this region of the world, around 12,000 years before the present, the post-glacial warming up of the climate entailed the progressive shift from a cold steppe ecosystem, characterized by the dominance of artemisia to a savanna ecosystem characterized by the dominance of oaks and pistachios, rich in wild grains (barley, spelt, emmer wheat, etc.) and also other exploitable plant resources (lentils, peas, vetch, and other legumes), as well as various game animals (wild boars, deer, gazelles, aurochs, wild sheep, wild goats, rabbits, hares, birds, etc.) and in some places fish.

*Abundance of Resources and Sedentary Populations.* Cave dwellers abandoned the hunting of reindeer and other tundra game driven northward by the warming of the climate and gradually adopted new systems of predation centered on the exploitation of abundant wild grains capable of meeting the largest part of the caloric needs of the population. The protein complement of the dietary ration came from hunting, fishing, and gathering legumes. This mainly vegetarian diet was based on the exploitation of resources that were more abundant than they had ever been, so much so that it became possible for a numerous sedentary population to survive. The population grew, left the caves, and began to live in new, human-made houses grouped into small villages (0.2 to 0.3 hectare), composed of noncontiguous roundhouses with wooden superstructures, built over pits supported by low stone walls. The population gradually expanded over the whole of this privileged ecosystem.

*Specialization of Tools and Intensified Exploitation of the Environment.* The development of this new mode of sedentary life was conditioned by a whole series of innovations that made possible the much greater exploitation and use of the new resources. Sickles formed from a blade of flaked stone, whose characteristic lustering proves that they were used as knives for harvesting, and sickles composed of microlithic teeth inserted into a support of curved wood made it possible to harvest in several hours enough wild grain to feed an entire family.[3] The grinding stone, hollowed right out of the rock or in a large stone, on which a handful of grain was ground with the assistance of a roller (a type of large, flat stone), made it possible to produce farina, from which a paste was made as well as round, flat cakes that could be cooked under ashes or on large heated stones in large ovens. Other instruments to grind grain (mortars, pestles)

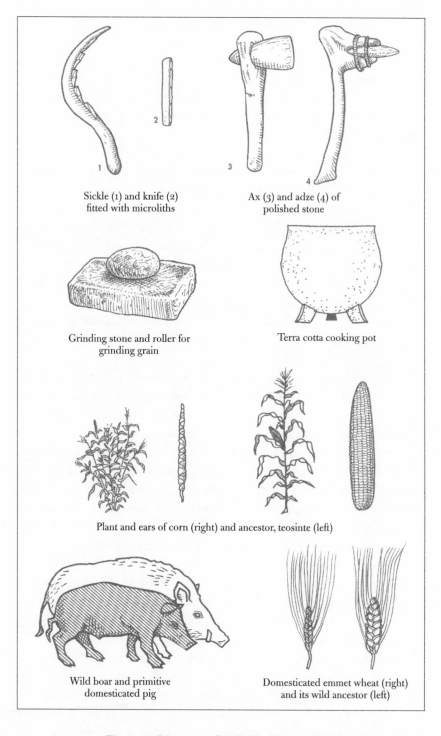

Sickle (1) and knife (2)
fitted with microliths

Ax (3) and adze (4) of
polished stone

Grinding stone and roller for
grinding grain

Terra cotta cooking pot

Plant and ears of corn (right) and ancestor, teosinte (left)

Wild boar and primitive
domesticated pig

Domesticated emmet wheat (right)
and its wild ancestor (left)

*Figure 2.2 Diagrams of Neolithic Tools and Wild
and Domesticated Plants and Animals*

are equally attested, as are silos allowing the grain harvested in summer to be put in reserve for the winter season.

The use of hearths built in pits lined with clay were the altogether fortuitous origin of the invention of ceramics, while the "discovery" of polished stone was linked to the use of grinding stones and rollers. Moreover, the first terra-cotta objects (figurines and very small containers) and polished stone objects (pendants and sticks) do not seem to have been of great utility. But subsequently, large terra-cotta pots, impermeable and fire resistant, were produced in great quantity, making it possible to cook porridges of grains and soups of peas and lentils. Likewise, axes and adzes of polished stone, used to cut and hew wood effectively, played a very important role in the construction of dwellings and, later, in the clearing of land for cultivation.

Sickles, grinding stones, rollers, mortars, pestles, axes and adzes, in brief all the materials that formed the tools of Neolithic farmers for millennia, existed for the most part prior to the development of agriculture. They had been developed, over the course of the preceding centuries, in the quite particular conditions of sedentary living and the increasingly intense exploitation of new resources, particularly wild grains.

*Proto-agriculture and Domestication.* In the Near East, the first traces of completely domesticated spelt (a species of wheat, *Triticum monococcum*) and emmer wheat (*Triticum dicoccum*) date from 9,500 years before the present. Barley, peas, lentils, chickpeas, two species of vetch and flax appear to have been domesticated toward 9,000 years before the present.[2] As far as animals are concerned, domestication goes back to 16,000 years before the present for the dog, 9,500 years for the goat, 9,200 years for the pig, 9,000 years for the sheep, 8,400 years for the cow, and 5,500 years for the donkey.[4] In order for these animals and plants to have been domesticated by these dates, protocultivation and proto-breeding must have begun dozens or even hundreds of years earlier.

It is generally thought that the first sowing took place in an accidental manner, close to dwellings, where the shelling and cooking of wild grains took place. Protocultivation would thus have developed on these same lands, already cleared and enriched by domestic wastes, and on the alluvial soil deposited by flooding rivers, which required neither clearing nor soil preparation.

However, these lands were limited. Cultivation had to extend onto forested lands, where axes of polished stone made it possible to clear the land easily enough by cutting down the trees and then burning them before planting. The practice of slash-and-burn cultivation has been attested quite early in the Near Eastern center, in the North American center, and undoubtedly also in the Chinese center.[5] In these conditions, the importance of polishing stones for the first developments of agriculture cannot be underestimated. The heavy cutting down of trees would have been difficult with axes of flaked stone, which are quickly cracked, worn out, and not easy to make. On the contrary, axes of polished stone were less fragile, they could be made with all types of hard stones, including stones that could not be flaked, and they could be sharpened at will.

*Other Changes in the Mode of Life.* Between 9,500 and 9,000 years before the present, the change from villages of small size (0.2 to 0.3 hectares) with roundhouses to villages of large size (2 to 3 hectares) with quadrangular houses, often joined together, can be observed. These changes are evidence of a population growth in the villages and a transformation in social organization. This epoch also coincides with the development of utilitarian fired pottery, rapid expansion in the production of axes and adzes of polished stone, proliferation of feminine statuettes and figurines, undoubtedly symbolizing fertility, and the preservation of clay-filled skulls with the face modeled over them.

It is difficult to establish relations of cause and effect among all these new things because they do not appear in a constant chronological order in the various excavated sites. However, it is clear that they are present at the same time across the entire Near Eastern center 9,000 years before the present, when domesticated plants and animals supplied humans with the bulk of their diet. In addition, we note that all these transformations in the mode of life were the common product of a large social space, coinciding in the Middle East with the distribution area of wild grains, particularly barley, rather than the result of the linear evolution of one or several closely related villages, whose new economic system would have expanded, already developed, into a larger area. This area includes enough common characteristics and, at the same time, enough variations and gaps so that the sharing of multiple experiences was both possible and enriching.

*Increase in the Time Required for Predation and the Transition to Agriculture.* If the conditions for the emergence of Neolithic agriculture in the Near Eastern center are increasingly better known, it remains to be seen why, toward the middle of the tenth millennium before the present, villages of sedentary huntergatherers, occasionally practicing proto-cultivation and proto-animal breeding, changed dramatically from an economy based essentially on predation to an economy resting on an extensive and enduring practice of cultivation and animal raising, sufficient to bring about the domestication of a whole series of plant and animal species.

To try to respond to this question, let's recall first of all that in this region, at the end of the tenth millennium, the size of villages of farmers and animal breeders was around ten times larger than the size of hunter-gatherer villages during the same time period. The products of gathering and hunting had a reduced role in these larger villages. Consequently, it seems logical to conclude that since the natural resources exploitable by simple predation on the territory belonging to each village were limited, the fast-growing population of these villages must have had increasingly greater recourse to the products of cultivation and animal breeding, when the products of predation became insufficient to feed them. But against this thesis, J. Cauvin emphasizes that there exists no proof of a crisis in predation during this epoch.[6] Moreover, some authors observe that hunter-gatherer societies hardly ever suffer shortages and generally spend less time obtaining their food than farmers.[7]

One can, however, object to these authors that if it is true of less numerous and mobile hunter-gatherer groups roaming over extensive territories, it may not be true for sedentary hunter-gatherers grouped into large villages, each making use of a territory limited either by the territory of neighboring villages or by the maximum range of the hunter-gatherers of each village. Undoubtedly, as these authors argue, an isolated individual needs only a few hours a day to harvest enough wild grains for a single family. But this individual would need much more time if he/she had to exploit a limited village territory at the same time as a hundred others and would not be able quite simply to "fill his/her basket" when in competition with hundreds of others. Mushroom collectors experience this harsh law of ecology every Sunday.

Since the question of the role played by population density in the transition from predation to agriculture is controversial, we will try to be a little more precise. It is clear that the volume of food which can be extracted on a long-term basis from a restricted village territory through simple predation is necessarily limited. In other words, a village territory, whatever it may be, has a limited exploitability through simple predation, which limits the maximum population density of the hunter-gatherers that this territory can support. In these conditions, when the population of a village of sedentary hunter-gatherers grows, the quantity of resources available for each individual hunter and/or gatherer diminishes. The appropriation of resources effected by some inevitably reduces the resources available for others. If the number of competing hunter-gatherers continues to grow, then inevitably a moment arrives when the time required for anyone of them to obtain the necessary food for him/herself and his/her dependents increases. Finally, when the population approaches its maximum (maximum corresponding to the level of exploitability by simple predation in the territory in question), then the necessary predation time for an individual hunter or gatherer increases exponentially.

Beyond this level, the overexploitation of the environment begins, which tends to reduce its production capacity and leads directly to famine for the population of the village under consideration. This will occur unless the population finds a way to stop its growth (limitation of births, etc.) or a way to obtain new resources, either by the displacement of all or part of the villagers to unoccupied or underexploited territories, by the conquest and colonization of already occupied territories, or by developing a new mode of exploiting the environment that is more productive than simple predation.

Thus, when the population of a village of sedentary hunter-gatherers grows, predation time increases and, beyond a certain level, becomes more time-consuming than the labor time necessary to satisfy the needs of the population through cultivation and animal husbandry. But the latter occurrence is not a sufficient condition for a hunter-gatherer society to be transformed into an agricultural society. Other necessary ecological and social conditions must still take place.

In the Near East, the inhabitants of villages whose size had increased tenfold over the course of the tenth millennium before the present were undoubtedly confronted with a situation of this type toward the end of this period. Since they had already made use of all the necessary tools and already occasionally practiced proto-cultivation and proto-animal breeding, it was possible for them to develop these practices when they became more advantageous than predation. Thus, since the technical conditions (tools, know-how) had already existed for a long time, when the demographic (population density) and economic (labor time) conditions required a change, the transition from predation to agriculture could occur quite rapidly. In our opinion, this is what would explain, in this place and at this time, the absence of an obvious crisis in predation.[8]

Having said that, it remains the case that a technical and economic change of such magnitude could not be produced without profound social and cultural transformations.

*Social and Cultural Conditions.* It is not difficult for a society to plant its grains of choice in already prepared ground or to capture, tame, and raise the animals easiest to master. Even hunter-gatherers know how to do that. What is difficult is to arrange a social organization and rules that make it possible for units (or groups) of producer-consumers to subtract from immediate consumption an important part of the annual harvest in order to save it as seed stock. What is also difficult is to exempt from slaughter enough reproductive and young animals to make it possible for the herd to reproduce itself. Again, what is difficult is to protect the fields planted by one group from the previously recognized right of other groups to "gather" in those areas and to protect the animals being raised from the right of those groups to "hunt" them. Lastly, what is difficult is to ensure the distribution of the fruits of agricultural work among the producer-consumers of each group, not only every day, but above all—and this is even more difficult—when the eldest die and when the group becomes too large and must be subdivided into several smaller groups.

The changes that occurred in housing (dimensions, subdivisions, arrangement, etc.), furniture, burials, and art bear witness to the importance of the transformations that took place in the social organization and culture of these societies during the time of their transition from predation to agriculture. Everything seems to indicate that domestic groups of production-consumption were formed that were capable of managing and maintaining agricultural activity and of redistributing its products. These family groups each possessed a home, a hearth, a silo, and, according to the season, seeds in reserve or in the ground, cultivated fields with immature grains or standing crops, as well as animals.

In these rather broad family units, the division of labor and responsibilities according to sex and age, the distribution of the products, where young men and women and certain goods went at the time of marriage, and even the handing down of responsibilities and goods upon the death of the elderly or at the time of the group's segmentation, necessarily obeyed a minimum of social rules enabling the balanced reproduction of the group and the cultivated plants and domestic animals upon which its survival depended. This does not mean that the prohibitions, morals, and obligations imposed by the familial or village authority were reduced to this function of economic regulation. Nor does it mean that these rules could not allow any contradictions, uneconomical arrangements, or exceptions. It means simply that among all the rules governing the life of the group, there existed a subset sufficient to enable this group to reproduce and renew its new means of existence. Moreover, one can assume that the nascent religion played a role in the establishment of these new rules for living.[9]

*The Neolithic Parent Languages.* Finally, it must be said that nothing of the new mode of life could have been understood, transmitted from one individual to another, retained from one generation to the next, and improved without the support of language. The latter had to be able to express the new material conditions, new productive practices, new organization and new social rules, as well as the corresponding ideas, representations, and beliefs. At the beginning of the new mode of life, there had to be the verb, that is, a new language.

The first articulated languages were formed in the Paleolithic, in the crucible of the organized hunt for large game.[10] According to some linguists, all the world's languages can be derived from a single common ancestor language. But current languages generally stem from several much more recent parent languages. The hypothesis that these parent languages were formed in the centers of origin of the Neolithic agricultural revolution and spread through differentiation at the same time as the first agrarian societies is increasingly accepted.[11]

In this hypothesis, the agriculture and language originating in each of the expanding centers expanded simultaneously, wandering across the continents, forming several large agrarian and linguistic areas of extension: an Indo-Afro-Asiatic area stemming from the Near Eastern center, an American area stemming from the Central American center, an Asiatic area stemming from the Chinese center, and maybe also much smaller areas in South and North America.

It is exactly this kind of hypothesis that scholars support in reference to the Indo-European languages.[12] Furthermore, the Arab, Hebrew, and Nilotic languages are related to the Indo-European languages, as are the languages of African farmers.[13] According to this hypothesis, the languages of the agricultural peoples of the Americas are related to one or another of the Central American, Andean, or North American parent languages, just as the languages of the agricultural peoples of the Far East are related among themselves.

If it should turn out that the languages of hunter-gatherers has a different structure from that of the languages of the agricultural peoples, one can also wonder if certain linguistic isolates are not the languages of hunter-gatherers won over to agriculture by contact with a wave of agrarian colonization, without being, for all that, entirely absorbed linguistically.

In brief, the Neolithic agricultural revolution, as other agricultural revolutions in history, was not only a vast change in the economic system prepared by a whole series of technical changes, but was also necessarily conditioned by a profound social and cultural revolution.

## *The Chinese Center*

The first sites of villages of sedentary Neolithic farmers in China belong to the Yang Shao civilization, characterized by its colored pottery. These sites are located at the heart of the Chinese Mesolithic system, on the upper terraces of poorly watered loess along the middle reaches of the Yellow (Huanghe) River. The oldest of these sites goes back to 8,500 years before the present and is found in the Henan, where the original north China center could have been situated. The latter then expanded toward the northeast in Shanxi (Yang Shao sites date from 7,000 years before the present), toward the west in Gansu (6,500 years before the present), and toward the southeast in Hebei (6,000 years before the present).[14] Birdseed, some legumes (cabbage, turnips), ramie (a type of nettle whose long fibers are used for textiles), as well as mulberry trees for raising silkworms contributed to the rather limited cultural complex of these regions of origin. The presence of domestic animal bones in ancient Chinese Neolithic sites is evidence for the development of animal raising. But even if one were to believe that some of these animals (hens, pigs, oxen) were domesticated in the Chinese center, it would appear to be improbable for others (sheep, horses).[15]

By expanding toward better-watered regions to the east, notably the lower valleys of the Yellow River and the Blue (Yangzi) River, this cultural complex was enriched with two very important cultivated plants: soya, coming from the Northeast, and rice, coming from the Southeast. It is in this southeastern zone of extension that the Long Shan culture emerged, in the seventh millennium before the present, characterized by its black pottery and the predominance of rice cultivation. Note, however, that the hypothesis that rice could have been domesticated independently in several regions of southeast Asia is supported by numerous researchers.[16]

## The Central American Center

The first center of origin for American Neolithic agriculture was gradually formed in the south of Mexico between 9,000 and 4,000 years before the present. According to J. R. Harlan, at the beginning of this period, small groups of nomadic hunter-gatherers could have begun to assemble during the wet season to carry out gathering activities and, undoubtedly quite incidentally, the cultivation of peppers and avocados. Much later, 7,000 years before the present, these temporary villages of seasonal farmers were already much more important, and spring and summer cultivation of early maize, squash, zucchini, and pumpkin were also practiced. Still much later, 5,000 years ago, the cultivation of beans was carried out. However, these populations continued to be nomads during the off-season and obtained a still important part of their subsistence by hunting and gathering. Around 3,500 years ago, cotton started to be cultivated and, secondarily, sapodilla and amaranth.

Starting in this period, American farmers then had a cereal, maize, and a food legume, the bean, which made it possible to cover their caloric and protein needs, as well as a textile plant, cotton. It is only then that agriculture became at least the clearly dominant, if not exclusive, mode of exploiting the environment and that populations settled in permanent villages in the Tehuacan valley and several other sites (Tamaulipas, Oaxaca, etc.). Note that the only domestic animals in Mexico were the turkey and the musk duck and that their domestication occurred very late, around 2,000 years ago.

Also note that in each of these three large expanding centers, the Near Eastern, Chinese, and Central American, a group of plants, sufficient to cover the essential needs of a population and adaptable to extended territories, was domesticated. This cultural complex in every case consisted of a grain that supplied glucides, a legume that supplied proteins, and a plant that provided textile fibers.

## The New Guinean Center

The cultivation of taro and other plants native to Southeast Asia and Oceania seems to have begun in the mountains of Papua-New Guinea close to 10,000 years ago. This date is, however, highly approximate because these plants have left practically no archaeological trace. Initially, these plants would have been protected and maybe even planted in their places of natural growth, dispersed within this densely forested region.

Then, 9,000 years before the present, these cultivated lands were grouped into gardens, which had been cleared and enclosed beforehand, perhaps, as some argue, to protect them from local domesticated pigs, but undoubtedly to protect them from wild boars, which did not need to be domesticated to rush toward a field of tubers! The domestic pig, which originated in Asia, arrived in New Guinea only 5,000 years ago and was then crossed with wild boars or boars in the process of being domesticated.[17] Around 7,000 years ago, taro gardens were extended into swampy zones and established in beds which had been cleared and drained beforehand.[18]

# The Minimally or Non-Expanding Centers

## The South American Center

In South America, archaeological research has not made it possible to clearly locate an original center for agriculture. However, the domestication of Lima beans, groundnuts, potatoes, oca (a small tuber), quinoa (a species of chenopodiaceae), lupin, as well as guinea pigs, llamas, and alpacas dates from some 6,000 years ago in the northern Andes. In these regions, domestication clearly began before agriculture from Central America could have arrived there, and thus certainly began independently. It is even likely that this South American agriculture had expanded into a sizable Andean space when it was engulfed, around 4,500 years before the present, by the wave of agriculture based on maize from the Central American center.

## The North American Center

In North America, recent research has revealed the existence of an original center situated between the Appalachians and the great continental prairie.[19] Between 4,000 and 3,000 years before the present, the swamp elder, squash, sunflower, and goosefoot (a type of potentilla, or false strawberry) were domesticated there. However, at that time these seasonal cultivations, practiced on the borders of lakes and rivers regularly cleared by the spring floods, played only a secondary role for populations that simultaneously exploited important resources from the aquatic environment and continued their nomadic way of life, hunting and gathering the rest of the year.

The conversion of these still predatory societies, which practiced agriculture secondarily, into societies of sedentary farmers occurred much later, between 250 B.C.E. and 200 C.E., with the domestication of three grains capable of ensuring the essential dietary ration: knotgrass, small barley, and a type of millet. As a result of this long transition, North American agriculture made use of seven cultivated plants, which provided around two-thirds of the diet for the settled farmers, who used axes, hoes, grinding stones for grains, pottery, and silos. Then maize from the Central American center arrived in this region and several centuries later it took first place among the cultivated plants in North America.

## A Doubtful Thai Center

In Thailand, less certain indications of slash-and-burn cultivation dating from the seventh millennium before the present have been discovered. As a result, some would like to see in this region a possible original center for Neolithic agriculture. But it appears rather that the first indisputable traces of agriculture (rice, pigs, cattle, poultry) in this region date from less than 5,000 years before the present, and that they quite simply come from the agriculture that expanded during this epoch to all of East Asia from northern and central China.[20]

But if there is no shortage of ill-founded claims of having discovered new origin centers for Neolithic agriculture, it is necessary to point out that some of the centers that are well-attested today were discovered by very recent research. For example, the North American center was still unknown a dozen years ago. It is thus not impossible that one day another center of origin, in West Africa or Southeast Asia, will be discovered. It is even probable that regions of the world in which human groups began, in the Neolithic epoch, to practice proto-agriculture are more numerous than they appear today. But it can be assumed that the beginnings of such a Neolithic revolution, overtaken and submerged by the speed of the agricultural wave stemming from one of the large expanding centers, did not have the time to succeed.

At this stage, regions of the world where human groups, alone and without external influence, embraced the new mode of subsistence appear to be neither numerous nor extensive.

## 2. AREAS OF EXTENSION

In the centers of origin, beyond the gardens and flood zones, which were certainly the first cultivated lands, Neolithic slash-and-burn cultivation was extended to forested terrain. The deforestation that followed, although it could in certain cases favor the development of raising herbivores, which are more at ease in open country, seems rather to have led to such a large decrease in arable land that the populations of these zones were forced to extend their clearings even farther.[21]

Outside the centers of origin, migrant farmers encountered two main types of nearly virgin plant formations: grassy formations, sometimes with trees or shrubs, but in all cases open (from north to south: tundra, high steppes, continental prairies, arid steppes, tropical savannas) and more or less dense and closed forested formations (from north to south: taiga, mixed forests of conifers and broad-leaved trees, temperate and Mediterranean broad-leaved forests, tropical forests that lose their leaves in the dry season, evergreen equatorial forests).

Equipped with their axes of polished stone, the farmers were able to extend the slash-and-burn cultivation they already practiced to forests that were the easiest to clear and the most fertile; however, since these environments remained forested and closed, animal breeding could have only a limited presence. Systems of slash-and-burn cultivation, which we will discuss in the next chapter, thus were mainly extended into temperate and tropical forests, leaving out the relatively unfertile taiga and the equatorial forest, which was too difficult to clear with the means available at the time.

Conversely, in the open grassy formations, easily penetrable and immediately exploitable by domestic herbivores, pastoral nomadic or seminomadic animal breeding could easily be extended. Since Neolithic farmers did not have the tools to work the soil and make it easy to clear the dense grassy ground cover of a prairie or savanna and, moreover, since the broken steppes are not very fertile, cultivation played only a secondary role in these systems. Thus were formed the pastoral cattle-raising societies of the Saharan savannas before their desertification, the yak breeding societies of the high steppes of central Asia, the horse breeding societies of the Eurasian prairies and steppes, the reindeer breeding societies of the tundra, the goat and sheepherding societies of the fields and grassy, bushy formations of the Mediterranean and Near Eastern highlands, the llama and alpaca breeding societies of the Andes, etc.

Beginning in the Neolithic, then, the first great geographic differentiation between societies of farmers and societies of animal breeders occurred. For all that, there was no absolute separation between cultivation and raising animals. Rare were the systems of cultivation that did not include some animal raising and the pastoral societies that did not practice some cultivation. Moreover, as we will see, subsequent agrarian systems often united cultivation and animal breeding more and more closely.

## Four Main Areas of Extension

Over several millennia, four main areas of extension of Neolithic agriculture developed from the four principal expanding centers. Neolithic agriculture from the Near Eastern center expanded step by step in every direction, starting 9,000 years ago. By the eighth millennium before the present, it had spread to the whole Near East and the eastern rivers of the Mediterranean. By the sixth and fifth millennia, it had spread to the western rivers of the Mediterranean and, via the Danube Valley, had penetrated into central Europe, then into northwest Europe. During the same time, it expanded toward the east as far as India and toward the south as far as central Africa, bypassing the large equatorial forest. By the fourth and third millennia, it had progressed toward the east, all along the thick band of broad-leaved forest that borders the south of the taiga, as far as the Far East where it came into contact with agriculture of Chinese origin. In Africa, it continued to expand toward the south, up until recent times.

By the ninth millennium before the present, agriculture of Chinese origin, with a base of millet, had hardly occupied more than the middle and lower valley of the Yellow River. By the eighth millennium, after having adopted the cultivation of rice, it extended as far as the Yangzi River, and 6,000 years ago it had spread to Manchuria, Korea, Japan, Central Asia, Southeast Asia, where it combined with agriculture of New Guinean origin, and South Asia (India), where it encountered agriculture of Near Eastern origin.[22]

Agriculture of Central American origin, based on maize, only began to expand outside of its center of origin in the sixth millennium before the present, reaching the South American and North American continents. Progressing toward the south, it reached the Andes and the Peruvian coast around 3,500 years before the present, and Chile around 2,000 years ago. In the process, it fused with the agriculture stemming from the South American center. American Neolithic agriculture was still expanding up to the first centuries of our era, toward the east and the southeast, bypassing the large Amazonian forest and without reaching the southern point of the continent. Advancing toward the north, it reached the south of California and the middle Mississippi at the beginning of our era, and it fused with the agriculture stemming from the North American center. From there, it continued to expand toward the north, moving up the valley of the Ohio River. Around the year 1000, it reached the shores of the Great Lakes and the Saint Lawrence, leaving aside the great central prairies and the west, as well as the Rockies and the far north.

As for agriculture of New Guinean origin based on taro, it was gradually dispersed into the Indonesian and Pacific islands until the beginnng of our era. In the process, it was enriched with plants (millet, yam, banana) and domestic animals of

Asian origin. Much later, the sweet potato of South American origin largely replaced taro and yam in most of these islands.[23] The expansion of Neolithic agriculture outside of its centers of origin thus appears as a slow phenomenon, lasting for millennia. Agriculture of Near Eastern origin took some 4,000 years to reach the rivers of the Atlantic and the Baltic, and more than 6,000 years to reach the Far East and the south of Africa. It advanced at an average speed of around one kilometer per year.

However, large regions of the globe were not affected by this first expanding wave of agriculture. In the fifteenth century, at the time of the Great Discoveries, these were Australia, the southern part of Africa and of South America, the northwestern part of North America and the far north of America and Eurasia. Furthermore, the large Amazonian, Central African, and Asian equatorial forests and the large prairies of the two American continents also remained outside of this extensive movement. Subsequently, colonies of white people and plantation economies occupied larges parts of these virgin territories. But, even today, agriculture is not universal. Difficult to penetrate or not very fertile environments, such as arid deserts, polar or high-altitude cold deserts, the taiga and a part of the equatorial forests are still uncultivated, and sometimes still exploited by hunter-gatherers (Bushmen of the Kalahari, Eskimos of Greenland, Pygmies of the Central African forests, Negritos of the Southeast Asian forests, Indians of the Amazon, etc.).

## Forms of Expansion of Neolithic Agriculture

There are two possible forms of expansion of Neolithic agriculture. In the first form, the agrarian societies of the expanding centers gradually colonized territories previously unoccupied or occupied by hunter-gatherers. In the second form, the expansion resulted from a gradual transfer of tools, domestic species, agricultural knowledge and know-how to preexisting hunter-gatherer societies, which then converted to an agricultural mode of life.

Most of the archaeological observations show that the areas of extension were generally colonized step by step by already-existing pioneer agrarian societies. For example, the so-called Danubian farmers spread through Europe from east to west by following the principal water courses, in particular the Danube and its tributaries. Equipped with tools of polished stone and domestic species of Near Eastern origin, they first colonized the river banks before venturing onto the plains and the less accessible plateaus. In these regions, as in most of the expansion areas, there is no trace of a gradual transformation of preexisting hunter-gatherer societies. Most often, all archaeological traces of a fully established Neolithic agricultural society are superimposed without transition on the earlier Paleolithic or Mesolithic levels. In support of this colonization thesis, it can be observed that the Pygmies, who have lived next to farmers and animal herders for millennia, have not been converted to agriculture as long as they have had large enough territories to exploit for hunting and gathering. On the other hand, when their territories were reduced in size due to clearing by advancing farmers, they gradually adopted an agricultural mode of life. In Rwanda and Burundi, for example, the Batwas Pygmies were won over to agriculture and craft industry when they no longer had enough territory to exploit for hunting.

In reality, agrarian colonization is not incompatible with adoption of agriculture by contact. In fact, over the centuries, immigrant farmers did not occupy all of the territories that they colonized  and thus lived next to hunter-gatherer peoples with whom they inevitably developed technical and cultural exchanges. Depending upon the case, less numerous hunter-gatherers were biologically and culturally assimilated or, in the end, converted to agriculture. Moreover, this cooperation could explain, in part, the modifications in the tools, housing, and pottery that occurred as agriculture further advanced into new territories.[24] In any case, the process of gradual conversion through contact is difficult to apprehend archaeologically. In the case of Japan, however, it indeed appears that hunter-gatherer peoples started to cultivate plants domesticated elsewhere. The Jomons of the west coast of Japan, settled as long as 12,000 years ago, used a diversified set of Neolithic tools and pottery and exploited a wide range of forest and marine resources by predation.[25] Around 5,500 years ago, they began to cultivate diverse types of cereals (millet, buckwheat), cucurbitaceous plants, and domestic peas originating in China. It was only much later, around 2,500 years ago, that farmers from China colonized the Japanese islands, spreading rice cultivation. But the case of Japan is most likely quite distinctive. The existence of a very advanced Neolithic society undoubtedly greatly favored its conversion to agriculture from the beginning. Its insular situation most probably delayed the arrival of Chinese agrarian colonization, giving the local populations time to adopt domestic species from the continent.

## Absorption of the Minimally Expanding Centers

As we have seen, the waves of agrarian colonization stemming from the large expanding centers encountered and submerged other centers in the process of their expansion. Little is known about how the New Guinean and South American centers were incorporated into the areas of expansion of Neolithic Chinese and Central American farmers. On the other hand, recent works show how the North American center was absorbed by the maize-based center emanating from

Central America. Around the year 200 of our era, a variety of maize with twelve rows, of Central American origin, was introduced into the region of the middle Mississippi, where seven locally domesticated plants were already cultivated. While the cultivation of these local plants developed more and more rapidly, there was only limited cultivation of this variety of Central American maize until the year 800, because it was poorly adapted to the region. After that year, a variety of maize with eight rows, with a short vegetative cycle and adapted to the colder local climate, made its appearance. Cultivation of that variety established itself rapidly and spread toward the north as far as the Great Lakes. Around the year 1000, a more adaptable and productive maize had largely supplanted the other domesticated species of North America.[26] By the sixteenth century, when the Europeans arrived, domestic species of Central American origin (maize, tobacco, beans, squash) were cultivated as far as the Saint Lawrence.

## Secondary Areas of Domestication

During their advance, Neolithic agrarian societies also encountered new, exploitable wild species which, in turn, could be domesticated. While some regions of the world provided only few domestic species (in Europe, for example, only rye and oats were domesticated), other regions provided many domestic species, so much so that they formed, after the centers of origin, genuine secondary areas of domestication. These regions were situated in:

— the north and west of the South American continent, where the groundnut, manioc, long-fiber cotton (*Gossypium barbadense*), peppers, lima bean, sweet potato, pineapple, etc. were domesticated
— tropical Africa north of the equator, where sorghum, small millet, African rice, the voandzu (Bambara pea), palm oil, okra, the African yam, etc., originated
— Southeast Asia, where the broad bean, taro, Chinese yam, turnip, lichee, banana, sugarcane, mandarin orange, etc., came from[27]

As far as animals are concerned, many species were domesticated rather early in the centers of origin: the ox, sheep, goat, pig, pigeon in the Near East; the hen, pig, ox, and maybe also the dog in China; the turkey and the musk duck in Central America; and the llama, alpaca, and guinea pig in South America. But many other species were domesticated in the areas of expansion: the zebu in Baluchistan (8,000 years ago); the horse on the extensive continental prairies of eastern Europe and the donkey in Egypt (5,500 years ago ); the dromedary in Arabia (5,000 years ago); the camel in Iran (4,500 years ago); the water buffalo in India

(2,750 years ago); the guinea fowl in the Mediterranean (2,500 years ago); the yak in Tibet; the gaur in Indochina and the reindeer in Siberia (2,000 years ago). As for the rabbit, it was domesticated in western Europe during the Middle Ages.[28]

## 3. DOMESTICATION AND DOMESTICABILITY

The origin of agriculture, and more specifically the origin of domestic plants and animals, has long been part of those mysterious phenomena surpassing human understanding. As a result, in order to explain these origins, there is often recourse to transcendent "causes," of a magical, miraculous, or divine nature, which abound in the founding myths of societies of farmers and animal herders. There is still a trace of this in modern scientific thought.[29] Archaeological and biological research of the last few decades clearly show that domestication is a process of biological transformation, which is an almost automatic consequence of proto-cultivation and proto-animal breeding when these processes are applied to certain wild species. Such a process of transformation can be explained by completely comprehensible genetic mechanisms.

The archaeological signs of the beginnings of cultivation and animal breeding are difficult to observe and interpret, because it takes time for the plants that humans begin to cultivate and the animals they raise to lose their original wild characteristics and acquire obvious domestic characteristics. In order to mark the beginnings of the cultivation of a wild plant species, one is reduced to measuring the increase in the number of its seeds in dwellings, the concentration of its pollen in certain terrains, from which one conjectures when the seeds were planted, or even to look for the presence of seeds or pollen from this species outside of its area of origin.[30] For animals, we can analogously measure the increase in bone remains near dwelling places, but this increase can also result from an intensification of hunting. A distribution of bones by age and sex conforming to what would be expected from the exploitation of a herd raised for the production of meat is more convincing, although it too can be the sign of more selective hunting. Morphological changes, such as reduced size and growth in variability, certain pathological manifestations (anomalies in dentition, fractures) and the presence of entire animal skeletons (whereas the skeletons of hunted animals are often incomplete because the useless parts were cut off the animal when it was slaughtered) are other indications of probable domestication. Finally, the presence of breeding implements (stones to fetter the animals, traces of animal pens), the presence of animals outside their area of origin, and clearly domesticated forms of their remains (reduction in size, bone deformations) are the only indubitable signs of animal domestication, above all when they are combined.[31]

## Select, Plant, and Breed

In order better to understand how domestic species are formed, recall that over millions of years, hominids were content with exploiting wild plant and animal populations by predation, chosen from among thousands of other species for their usefulness and their ease of exploitation. In the Neolithic, sedentary human groups began to change this way of doing things. They seized small collections of individuals belonging to one or another of these species in order to subject them to new, human-made conditions of growth and reproduction, resulting from the practices of proto-agriculture. From the moment they were cultivated or bred, these specially selected and exploited parts of a population, and their descendents, led a separate existence, different from that of their wild fellow creatures. After several generations, the descendents of some of the species subjected to proto-agriculture had lost some of their original wild genetic, morphological, and behavioral characteristics, which were not very compatible with their new mode of life, while they had acquired other characteristics which, if they were transmissible and advantageous, were from then on retained. From this moment, even if they continued to resemble in thousands of ways their ancestors and surviving wild populations, the new "domestic" forms thus obtained were distinguished by a small number of characteristics, forming what it is appropriate to call the syndrome of domestication. But let's see more exactly by what mechanisms this transformation could have taken place.

## The Domestication of Cereals

Any population of a wild species of cereal is heterogeneous. For example, some seeds that have fallen to the soil germinate during the first sufficiently hot and humid season, while others only come up two or three seasons later. This delay in germinating (dormancy) is conditioned by substances that inhibit germination, generally contained in the small scale-like bracts (the glumes and glumellas) enveloping the grains. As long as a population reproduces itself spontaneously, these tendencies, variable from one plant to another, contribute toward spreading the germination of seeds over several more or less favorable consecutive seasons, thus increasing the chances for reproduction and multiplication of the species. However, as soon as they are cultivated, that is, sown together during the first rainy season and harvested together during the following harvest, only the non-dormant seeds can be harvested and subsequently sown again. The common sowing and, subsequently, common harvesting of an initially wild cereal population thus tends to eliminate the descendents of dormant seeds surrounded by thick glumes and glumellas.

Moreover, the seeds germinating first and yielding the most vigorous seedlings gain the upper hand in the close competition among fellow plants sown in the same parcel and, as a result, have more numerous descendants than the others. The earliest and sturdiest seedlings generally come from the largest seeds in which the albumen, rich in sugars that can be rapidly mobilized, is relatively more developed than the germ, richer in proteins and fatty acids. Common sowing, then, tends to select the lineages with rapid germination and large seeds rich in sugars and relatively poor in proteins and fats.

But when the harvest is carried out only once, at the time the greatest number of grains have reached maturity, then those that mature later are eliminated from the lineage because grains harvested too early are infertile. Consequently, forms that include a great number of flowerings with a staggered maturity tend to be eliminated. Moreover, the common harvest tends to eliminate the forms whose ears or spikelets are carried by weak stalks or stemlets and whose easily removable grains fall to the ground too early and thus escape the harvest.

Thus a whole series of characteristics (dormancy, thick husks, small grains, numerous small flowerings, weak stalks and stemlets, easily shelled grains, etc.) favoring the reproduction and spontaneous diffusion of wild populations become counterproductive in conditions of reproduction imposed by human agricultural practices and tend, as a result, to be eliminated. The opposite characteristics (non-dormancy, thin husks, large grains rich in sugars and poor in proteins and fats, single or less numerous ears or flowerings of large size with abundant grains, strong stalks and stemlets, grains that are difficult to shell, etc.) multiply the chances for development of cultivated lineages that are harvested at maturity and reproduced by sowing.

This whole set of genetic, morphological, and advantageous behavioral characteristics, which forms the syndrome of domestication typical of most of the cultivated cereal populations, is therefore the product of a quasi-automatic mechanism of selection performed on lineages of cereals from the moment they are cultivated over several successive generations.

## A Small Number of Jointly Inheritable Genes

The natural potential of a cereal to be domesticated, a potential that we call *domesticability*, results from specific genetic and reproductive propensities.[32] For maize and millet, the genes that control the syndrome of domestication are less numerous, grouped on the same chromosome and therefore jointly inheritable, which greatly facilitates the transition from the wild form to the domestic form. In addition, since maize, sorghum, millet, wheat, oats, and rice are preferentially reproduced by self-fertilization (fertilization of each plant by its own pollen), the risks of hybridization with wild forms are reduced, while the isolation and preservation of acquired domestic characteristics are facilitated.

However, even if the selection of domestic characteristics is automatic, the observation, choice, and conscious action of the farmer can usefully be exercised to preserve and diffuse the obvious advantages acquired by selection. When a plant population subjected to proto-cultivation leads in several generations to the appearance of a syndrome of domestication, the farmer is able to choose the visibly most advantageous lineages, in order to sow them again preferentially and eliminate the wild members of the same species and the hybrids. If, however, the appearance of the syndrome of domestication is involuntary, the situation is quite different for the choice, preservation, and diffusion of the most advantageous species and lineages affected by this syndrome. Even today, in the Sahelian regions where millet was domesticated and where cultivated and wild millets coexist, the farmers continue to track down young hybrid plants. Moreover, it may appear amusing that the agronomists of an Indian research center charged with "preserving genetic resources" of cultivated millet simply "forgot" to eliminate the hybrids present in their collections and saw the latter rapidly deteriorate.[33] From this analysis, it is possible to conclude that domestication could not occur as long as most of the sown seeds came from gathering. In order for domestication to take place, the seeds resulting from proto-cultivation must become predominant and be sown for several subsequent generations. It is then highly improbable that domestication would have occurred in the centers of origin, so long as the easily gathered wild cereals were overabundant in relation to the needs of the population.

## The Domestication of Other Plants

### Other Plants with Seeds

Among seed plants other than cereals, the general pace of the process of domestication is rather similar. For example, while populations of wild legumes generally have pods that open easily at maturity, thereby facilitating the dissemination of the seeds, as well as dormant seeds with different germination periods, the domestic populations have lost these characteristics. Also, with domestication, there is a tendency toward making less numerous and larger flowerings, with numerous seeds and uniform maturation.

### Plants Characterized by Vegetative Reproduction

Among plants characterized by vegetative reproduction, which the farmers reproduce by taking a cutting from a fragment of the stem (manioc), by burying a tuber or part of a tuber (potato, yam) or by planting a shoot (banana), each cultivated plant inherits identical genetic characteristics from the parent plant, so that, it is commonly thought, the obvious qualities of a wild parent plant, selected for having given the best tubers, fruits or roots, will be handed down fully to its cultivated descendents.

However, it is not so simple: some plants do indeed yield a good harvest because of their own genetic characteristics and this quality is consequently transmissible. But other plants that do not possess these advantageous genetic characteristics occasionally yield results as good or even better when they develop in micro-local conditions, which are favorable in terms of soil, sunlight, humidity, or absence of competition. Conversely, genetically advantageous plants can be found in unfavorable conditions that prevent them from showing their intrinsic qualities. Therefore it requires time and attention to separate the genetically advantageous plants from the plants that are simply advantaged by their conditions of development.

## Favored Non-Domesticated Plants

It is also possible to favor a species without truly cultivating it. Some species, useful on several accounts, are simply protected. For example, the palm oil tree, which grows spontaneously on the edge of the equatorial forest, is saved during clearings; the baobab tree, whose fruits and leaves are consumed and whose bark supplies fiber, and the shea tree, whose fruit provides the butter of the same name, are protected from overexploitation. Other species, such as the *Acacia albida*, an off-season forage tree that contributes greatly to reproducing the fertility of much Sahelian agricultural land, is not only protected but also propagated outside of its natural area. However, every species favored in one way or another by humans does not acquire, for all that, any particular domestic characteristics.

## The Domestication of Animals

To take a wild animal population away from its natural mode of life in order to save it, protect it, and propagate it with a view to exploiting it more easily and intensely is exactly the principle of proto-animal breeding. From generation to generation, this population is subjected to conditions of life and reproduction different than those of the remaining wild populations. These new conditions tend to eliminate certain genetic, behavioral, and morphological characteristics and to select others, whether they be preexisting characteristics from the originally wild populations or characteristics that have appeared by mutation during the process of domestication. The mechanisms that govern this evolution are the same as those for plants, with this exception: among the animals, no totality of linked genes, which are able to be selected jointly and determine a "syndrome of domestication," has been discovered. That does not mean that there is no set of typical characteristics that distinguish most of the primitively domesticated animals from their wild fellows.

Thus, in the conditions of proto-animal breeding, the most timid animals that refuse to eat or reproduce in captivity have no descendents. The aggressive, violent, and dangerous animals are generally eliminated by the breeders, who also, out of preference, slaughter animals of worthwhile size for consumption. The management of large herds makes it possible for the most vulnerable animals, which are the best protected, to survive, while they would have been eliminated in small herds living in the wild. By castrating or taking some of the males away from the females during the rutting season, the breeders allow animals that are less forceful and bold to participate in reproduction. Breeding animals are often subjected to shortages and deficiencies to which smaller animals are more resistant than larger ones. Finally, from generation to generation, proto-breeding tends generally to select animals that are less sensitive, less nervous, less forceful, and of smaller size, all typical characteristics of primitive domestic animal species.[34] If domestic plants then appear at first as "improved" in relation to their wild ancestors (more numerous and larger grains, etc.), primitive domestic animals appear, on the contrary, as "debased." But whether they appear "improved" or "debased," domestic species are in all ways better adapted than their wild ancestors to their new conditions of life, and thus more advantageous for the farmer and breeder. Whatever these advantages may be, it remains the case that, as a whole, they were obtained involuntarily. A domestic species is in reality the final product, unknown and inconceivable at the beginning, of a process of selection governed by a whole series of cultivating and breeding activities, each of which leads in the short term to something entirely different than the absolutely unforeseen and distant final result.

We add that if indeed some plant species were proto-cultivated without ever being domesticated, it is also true that many animal species were captured and subjected to diverse breeding practices without, for all that, having been domesticated. These practices left hardly any traces, except in that time period. Thus, in ancient Egypt, for example, pelicans and herons were force-fed for a long time, and hyenas, gazelles, and oryxes (a species of antelope) were kept in captivity without resulting in their domestication. It must be said that not every animal species is able to be domesticated. Species that do not reproduce in captivity, those whose young require long years of care and the fragile, capricious, or violent species hardly lend themselves to domestication. Those species that are not very sociable, live in limited families, and mark their territory are not easily domesticated either.

## 4. CONCLUSION

As J. R. Harlan has written, "agriculture was neither discovered nor invented." In the current state of knowledge, it appears as the result of a long evolutionary process that affected several societies of *Homo sapiens sapiens* at the end of prehistory, in the Neolithic epoch. Societies of predators that were subsequently transformed into societies of farmers were among the most advanced of the epoch. They made use of sophisticated stone tools, exploited plant resources abundant enough to allow them to live together in settled villages, and undoubtedly practiced ancestor worship. Finally, if J. Cauvin, in *The Birth of the Gods and the Origins of Agriculture*, is to be believed, they already worshiped several divinities.

The particular technical, ecological, and cultural conditions in which the first agrarian societies emerged were brought together only recently and then only in some privileged regions of the planet. Thus we can understand why the Neolithic agricultural revolution could not have taken place at the time of *Homo erectus* or early *Homo sapiens* nor in every region of the world at once. Neolithic agriculture subsequently spread over an unequally developed world, most often by direct agrarian colonization or by the gradual conversion of hunter-gatherer societies, which were advanced enough in other respects.

This Neolithic agricultural expansion certainly made possible a strong increase in world population, but it was not, in general, a response to the crisis of predation among preexisting nomadic hunter-gatherer societies. On the other hand, in the centers of origin of Neolithic agriculture, it is probable that the sedentary populations grouped into villages of rapidly increasing size, each exploiting a defined territory, one day or another came up against the limits of this territory's exploitability by simple predation. From that moment, the necessary time to gather and hunt overexploited wild species became greater than the necessary time to cultivate and breed them. As the technical (tools) and ethological (sedentary living) conditions were already brought together, proto-cultivation and proto-breeding became from that moment more advantageous in these places than simple predation. However, it remained for these societies to realize the last and most difficult of the conditions necessary for the development of agriculture, that is, a genuine social and cultural revolution. As necessary as this revolution appears to us after the fact, nevertheless it cannot be explained by nor is it reducible to this necessity.

It is doubtful that the first farmers were able to take special notice of and protect the descendents of plants and animals that presented obvious advantages for them. Nevertheless, domestication appears essentially as the unpremeditated final result, inconceivable a priori, of proto-cultivation and proto-breeding practices applied to populations of specially exploited wild species, of whom some were progressively revealed as "able to be domesticated."

We have attempted to produce a comprehensive representation of the Neolithic agricultural revolution, with its centers of origin, its areas of extension, its secondary areas of domestication, and its mechanisms for domesticating plants and animals, based on current knowledge that is rich in quantity and variety, but also incomplete, confused, and contradictory. Thus conceived, this immense human adventure appears, from the beginning, to be the result of a technical and cultural history that had reached a certain stage and then continued with new means in given geographic and ecological conditions, rather than as the result of a revelation, a fortunate accident, or human free will acting outside of these historically constituted and geographically defined conditions and possibilities.

This reconstruction of the Neolithic agricultural revolution relies on the traces of human activities patiently collected, organized, and interpreted by archaeologists. These traces mainly show the changes that occurred in the material life of humans in each time period. The absence of written sources makes it practically impossible to know the thoughts of these people. However, there is hardly any doubt that these changes were accompanied, for the people who experienced them, by a sort of reversal in their relation to the world and to themselves. Even if it is impossible to understand and recount the Neolithic agricultural revolution, it certainly required, on the part of those who made it, an infinite number of inventions, choices, initiatives, and reflections in every domain of material and social life, as well as in the domains of thought, beliefs, morals, language, and other means of expression.

# 3

# Systems of Slash-and-Burn Agriculture in Forest Environments:

# Deforestation and the Formation of Post-Forest Agrarian Systems

Humanity, disdainful of what was created without it, believes ... that it can develop [the planet] by destroying the slow accumulation of plant wealth that collaboration between the atmosphere and the earth had produced over thousands of centuries. Will the large ... tropical ... forest, this huge laboratory of climates, this humid and warm velvet belt of plants from which rhythmic spirals of atmospheric waves harmoniously soar, be transformed wisely, exploited with respect for humanity and nature, by taking into account its relationship with the soil and the atmosphere, or will humanity give in to the temptation to assault the earth, attack the tropical forest quickly and without thought? In the latter case, if one thinks about it, it is humanity itself which would be endangered, ... because the atmosphere would be unbalanced and instability would be introduced into climates around the whole world.

—F. SHRADER, *Atlas de géographie historique*, 1896

Slash-and-burn agriculture is practiced in diverse wooded environments: amid mature trees, in a copse, shrubby or bushy thickets, wooded savanna, etc. It is established on terrains previously cleared by grubbing, that is, cutting followed by burning but without removing the stumps. The parcels thus cleared are only cultivated for one, two, or three years, rarely more, after which they are abandoned to return to their wooded wild state for one or more decades, before being cleared and cultivated again. Systems of slash-and-burn agriculture, also called *forest agrarian systems*, are thus characterized by the practice of temporary cultivation alternating with long-term wooded idling, forming a rotation with a period varying from about ten to fifty years.

The origin of these systems goes back to the Neolithic epoch. From that time on, they spread to most of the forests and other cultivable wooded environments of the planet, where they lasted for thousands of years. In each region of the world, this pioneer dynamic accompanied a strong demographic growth and was pursued as long as uncleared, accessible wooded terrain remained. When all these virgin reserves were used and the population density continued to increase, the frequency and intensity of clearings increased, thus beginning a dynamic of deforestation of lands cultivated by slash-and-burn techniques. Ultimately, this resulted in the impossibility of pursuing this mode of cultivation. Deforestation generally resulted in deterioration of fertility, development of a more or less serious erosion problem, depending on the biotope, and worsening of the climate, even up to the point of desertification.

The double ecological and subsistence crisis that resulted was overcome only by the development of new and diversified "post-forest" agrarian systems: hydraulic systems in arid regions, systems with fallowing in temperate regions, savanna systems in the tropical regions, systems of aquatic rice-growing in the monsoon regions, pastoral systems extended to secondary grassy formations created by deforestation, etc.[1]

However, even today, diverse forms of slash-and-burn agriculture continue to exist and expand in the tropical forests of Africa, Asia, and South America, where they have various names: *tavy* in Madagascar, *ladang* in Indonesia, *ray* in the Indochinese peninsula, *kaingin* in the Philippines, *milpa* in Central America, *lougan* in Africa, etc. In all of these regions, deforestation progresses rapidly due to the demographic explosion, but also from the exploitation of tropical woods and the extension of plantations and animal breeding operations. The question of the survival and transformation of systems of slash-and-burn agriculture is thus a pressing question today.

The objective of this chapter is to respond to the following questions:

— Where did systems of slash-and-burn agriculture come from, and how were they formed? Behind the diversity of their forms, what are the essential organizational and functional characteristics upon which their identity is based?
— What are the reasons for and consequences of the pioneer expansion of slash-and-burn agriculture and its extinction due to deforestation? What were the effects of this deforestation in different parts of the world?
— Finally, what are the problems of forest farmers today? How can knowledge of systems of slash-and-burn agriculture save us from committing serious errors in formulating projects and policies concerning them?

## 1. FORMATION OF SYSTEMS OF SLASH-AND-BURN AGRICULTURE

Though it is universally accepted that systems of slash-and-burn agriculture appeared and expanded across the world beginning in the Neolithic epoch, this type of agriculture was not, however, the first to be practiced. The results of archaeological work point to the fact that all of the early Neolithic agricultures were established either in types of gardens close to dwellings, already cleared and fertilized by domestic waste, or on land that was freshly covered with alluvial deposits from river floods. But since these privileged zones were by nature so limited, they were necessarily extended to forests and neighboring grasslands when cultivation and animal breeding activities took on a larger importance.

Armed with relatively effective axes of polished stone to cut shrubs and trees, but deprived of a tool to work the soil other than the digging stick, Neolithic farmers were much better equipped to clear and cultivate a forest than clear and cultivate a dense, grassy carpet. This is why the Neolithic populations that occupied wooded regions mainly developed cultivation, while those that expanded into the prairies, savannas, and steppes mainly developed animal breeding.

Few things are known about how slash-and-burn agriculture was carried out in this distant epoch. There exists no written account on this subject, because the first civilizations to use writing developed several thousand years after the beginnings of agriculture and in regions where systems of slash-and-burn agriculture were in the process of disappearing. On the other hand, at the time of colonization, some farming peoples of America, Southeast Asia, and Polynesia still used axes of polished stone and their agricultural practices were often recorded, even though they had modified their practices by the utilization of metals before being made the object of systematic study. In the forests of Papua-New Guinea, there still exist farmers using these types of tools and it would certainly be interesting to study their agriculture before it is too late.[1] Finally, the experiments with slash-and-burn agriculture practiced by some archaeologists using polished stone tools are undoubtedly interesting, but they are too limited to be able to formulate solid hypotheses about the way in which Neolithic farmers really behaved.

Thus, only the study of slash-and-burn agriculture as it is widely practiced today with metal tools in the forests and intertropical wooded savannas represents a sufficient basis for comparative analysis to understand how systems of slash-and-burn agriculture were able to arise, expand widely, and last so long.

## Cultivation in a Forested Environment

### *Slashing, Burning, and Soil Preparation*

In order to create arable land in a slightly dense forested environment, it is first necessary to make room for the soil and the sun by destroying all or part of the natural cover. In order to accomplish this when confronted with a strong primary forest, the forest farmers, equipped with axes and machetes, are generally restricted to carrying out a partial clearing: they cut only the undergrowth and trees that are easy to chop down. In less formidable wooded environments, the clearing is more complete, to the point that almost all of the standing trees can be cut down, save certain useful ones. In every case, whether it be partial or complete, the cutting is not a complete clearing, in the sense that it neither includes stump extraction nor systematic cleaning of the soil.

After the cutting, the surface of the ground is covered with leaves, branches, and dead trunks, which must be cleared before sowing or planting. The standard procedure consists of allowing the plant material to dry, then burning it right before the rains and the sowing so that the cultivated land benefits the most from the nutritive minerals contained in the ashes. In some societies of poorly equipped farmers, the burning is followed directly by sowing or planting, without any particular soil preparation. The seeds, plants, or cuttings are then placed in single holes, dug in the soil with digging sticks or small hoes. These holes are then filled in and packed down to facilitate the germination of the seeds or the revival of the plants. But most often a working of the soil designed to favor the development of cultivated plants follows the burning. This work, done with a hoe, consists of breaking open, turning over, and mixing the soil down to several centimeters in depth, in order to prepare what is called a seed bed or cultivation bed. In order to sow cereals, the loosened soil is left in place to form a flat seed bed of uniform thickness. But to plant tubers or cuttings, the upper levels of the soil are then arranged as either rounded buttes or elongated ridges.

All these activities of clearing and soil preparation are done with largely unspecialized and ineffective rudimentary manual tools. Parcels that are only partially and temporarily cultivable are the ultimate outcome of this long and difficult work. Indeed, the uncut trees, the stumps and the roots that do not have to be pulled up continue to clutter the land, such that the surface area actually seeded and harvested is much smaller than the surface area of the cleared parcel. On the other hand, many of the stumps of the felled trees remain alive, and shoots are rapidly formed. As they grow, these shoots join the standing trees to regenerate into a secondary wooded formation. Finally, even before reforestation has begun,

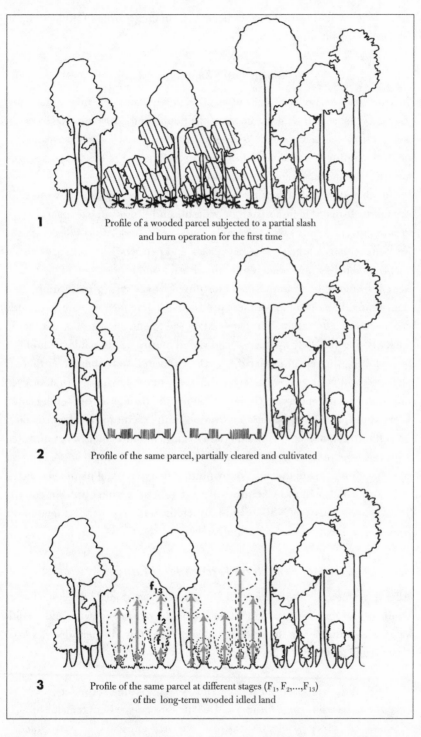

**1**      Profile of a wooded parcel subjected to a partial slash
and burn operation for the first time

**2**      Profile of the same parcel, partially cleared and cultivated

**3**      Profile of the same parcel at different stages ($F_1$, $F_2$,...,$F_{13}$)
of the long-term wooded idled land

*Figure 3.1 Profiles of a Wooded Parcel Cultivated by the Slash and Burn Technique*

an undergrowth of natural grasses benefits from the clearing to proliferate and invade the soil intended for cultivation.

## Temporary and Short-Term Crops

In some systems, only one cultivation is carried out after clearing, which must then cover most of the caloric needs of the population. Most often it is a cereal such as rice, millet, or maize, a tuber such as yams, or a root such as manioc or taro, which provides a staple diet rich in starch. The rest of the diet comes from gardens or animal breeding, hunting, fishing, and gathering.

In other systems, this primary cultivation is followed by one or two secondary cultivations of legumes rich in proteins and lipids such as peas, beans, groundnuts, or soya, as well as various fruits, vegetables, and condiments used for sauces such as tomatoes, okra, squash, and peppers. The primary cultivation is established immediately after the clearing and preparation of the soil, to benefit from the best conditions of fertility. The secondary cultivations, less demanding, less essential, and less productive, come next and are frequently conducted in association. They are juxtaposed to and succeed one another in such a way as to satisfy the regular and varied dietary needs, while methodically exploiting the remaining fertility of the cultivated soil. Sometimes there is also a second cultivation of a cereal, tuber, or root following the first, or even mixed in with secondary cultivations. Among the Baoulés of the south central Ivory Coast, for example, joint cultivation in the second year of maize, taro, groundnut, tomato, eggplant, and okra occupy mounds that were cultivated with yams the preceding year. Finally, it sometimes happens that the last year of cultivation is made the most of by putting in a perennial plant such as the plantain, which will grow amid wild vegetation and will supply, in coming years, a backup harvest.[2] Slash-and-burn cultivation, in general, is temporary, lasting only one, two, or three years, rarely more.

## Long-Term Forested Idled Land

After this short period of cultivation, the parcel is abandoned to become forested idled land for a long period of time, before being cleared again and recultivated. Depending upon the system, the duration of this idle period varies from about ten years to several decades:

— If the idle period lasts for thirty to fifty years, a strong secondary reforestation has time to be formed and, as the forest ecosystem remains predominant, it is possible to speak of a cultivated forest and a forest agrarian system.

— If the idle period lasts for less than 20 years, the new growth can best be characterized as a copse and the parcels must be cleared more completely for cultivation in order to obtain enough ashes and maintain good yields.
— If the idle period lasts no more than 10 years, small-size shrubby thickets make up the new growth and, in this case, it is no longer possible to speak of a forest agrarian system.
— If the idle period lasts from 6 to 7 years, herbaceous vegetation becomes predominant, and cultivation exists in rotation with grassy fallow land that can no longer be cleared by slash-and-burn techniques. Then we have a post-forest agrarian system.

Slash-and-burn agriculture is thus a temporary agriculture of short duration, alternating with a long idle period of full or partial reforestation, forming a rotation that can vary, depending upon the system, from twelve to fifty years. However, even if it is temporary, this type of cultivation must ensure a regular yield from year to year. Every year, each farmer's family must clear a wooded area sufficient to carry out the main cultivation corresponding to its needs. Every year, the cultivation changes locations and, in the same way, the secondary cultivations that follow are moved as well. This is why it is sometimes said of these *temporary* cultivations that they are *itinerant*.

Even so, one should not conclude that slash-and-burn farmers were nomads. On the contrary, they were generally settled peoples who live grouped in villages and whose cultivated fields moved around within a radius of several kilometers from their houses. Each village must then have had at its disposal, at every moment, a reserve of idled lands old enough and extensive enough to establish cultivated lands for every family in the village. In other words, it must have, besides cultivated areas, equivalent idled lands of all ages, which will be cleared one after another over the course of the following years. However, once cleared, some fragile tropical forests are regenerated with difficulty and, after some years of cultivation, grassy savanna is established on a long-term basis. In this case, after having cleared and cultivated the neighboring forests as long as possible and transforming them into savannas, the population must move and found a new village in a sufficiently wooded area. The villages of some peoples in the mountains of northern Laos and Vietnam still move every ten to twenty years.[3] But this type of system, which can, if necessary, be qualified as "nomadic," is rare.

# Fifteen-Year Rotation ($c_1$, $c_2$, $f_1$, $f_2$ ..., $f_{13}$) and Plot Allotment for Fifteen Parcels ($p_1$, $p_2$ ..., $p_{15}$)

## *Parcels*

| Years | $p_1$ | $p_2$ | $p_3$ | $p_4$ | $p_5$ | $p_6$ | $p_7$ | $p_8$ | $p_9$ | $p_{10}$ | $p_{11}$ | $p_{12}$ | $p_{13}$ | $p_{14}$ | $p_{15}$ |
|---|---|---|---|---|---|---|---|---|---|---|---|---|---|---|---|
| 1 | $c_1$ | | | | | | | | | | | | | | |
| 2 | $c_2$ | $c_1$ | | | | | | | | | | | | | |
| 3 | $f_1$ | $c_2$ | $c_1$ | | | | | | | | | | | | |
| 4 | $f_2$ | $f_1$ | $c_2$ | $c_1$ | | | | | | | | | | | |
| 5 | $f_3$ | $f_2$ | $f_1$ | $c_2$ | $c_1$ | | | | | | | | | | |
| 6 | $f_4$ | $f_3$ | $f_2$ | $f_1$ | $c_2$ | $c_1$ | | | | | | | | | |
| 7 | $f_5$ | $f_4$ | $f_3$ | $f_2$ | $f_1$ | $c_2$ | $c_1$ | | | | | | | | |
| 8 | $f_6$ | $f_5$ | $f_4$ | $f_3$ | $f_2$ | $f_1$ | $c_2$ | $c_1$ | | | | | | | |
| 9 | $f_7$ | $f_6$ | $f_5$ | $f_4$ | $f_3$ | $f_2$ | $f_1$ | $c_2$ | $c_1$ | | | | | | |
| 10 | $f_8$ | $f_7$ | $f_6$ | $f_5$ | $f_4$ | $f_3$ | $f_2$ | $f_1$ | $c_2$ | $c_1$ | | | | | |
| 11 | $f_9$ | $f_8$ | $f_7$ | $f_6$ | $f_5$ | $f_4$ | $f_3$ | $f_2$ | $f_1$ | $c_2$ | $c_1$ | | | | |
| 12 | $f_{10}$ | $f_9$ | $f_8$ | $f_7$ | $f_6$ | $f_5$ | $f_4$ | $f_3$ | $f_2$ | $f_1$ | $c_2$ | $c_1$ | | | |
| 13 | $f_{11}$ | $f_{10}$ | $f_9$ | $f_8$ | $f_7$ | $f_6$ | $f_5$ | $f_4$ | $f_3$ | $f_2$ | $f_1$ | $c_2$ | $c_1$ | | |
| 14 | $f_{12}$ | $f_{11}$ | $f_{10}$ | $f_9$ | $f_8$ | $f_7$ | $f_6$ | $f_5$ | $f_4$ | $f_3$ | $f_2$ | $f_1$ | $c_2$ | $c_1$ | |
| 15 | $f_{13}$ | $f_{12}$ | $f_{11}$ | $f_{10}$ | $f_9$ | $f_8$ | $f_7$ | $f_6$ | $f_5$ | $f_4$ | $f_3$ | $f_2$ | $f_1$ | $c_2$ | $c_1$ |
| 16 | $c_1$ | $f_{13}$ | $f_{12}$ | $f_{11}$ | $f_{10}$ | $f_9$ | $f_8$ | $f_7$ | $f_6$ | $f_5$ | $f_4$ | $f_3$ | $f_2$ | $f_1$ | $c_2$ |
| 17 | $c_2$ | $c_1$ | $f_{13}$ | $f_{12}$ | $f_{11}$ | $f_{10}$ | $f_9$ | $f_8$ | $f_7$ | $f_6$ | $f_5$ | $f_4$ | $f_3$ | $f_2$ | $f_1$ |
| 18 | $f_1$ | $c_2$ | $c_1$ | $f_{13}$ | $f_{12}$ | $f_{11}$ | $f_{10}$ | $f_9$ | $f_8$ | $f_7$ | $f_6$ | $f_5$ | $f_4$ | $f_3$ | $f_2$ |
| 19 | $f_2$ | $f_1$ | $c_2$ | $c_1$ | $f_{13}$ | $f_{12}$ | $f_{11}$ | $f_{10}$ | $f_9$ | $f_8$ | $f_7$ | $f_6$ | $f_5$ | $f_4$ | $f_3$ |
| 20 | $f_3$ | $f_2$ | $f_1$ | $c_2$ | $c_1$ | $f_{13}$ | $f_{12}$ | $f_{11}$ | $f_{10}$ | $f_9$ | $f_8$ | $f_7$ | $f_6$ | $f_5$ | $f_4$ |
| 21 | $f_4$ | $f_3$ | $f_2$ | $f_1$ | $c_2$ | $c_1$ | $f_{13}$ | $f_{12}$ | $f_{11}$ | $f_{10}$ | $f_9$ | $f_8$ | $f_7$ | $f_6$ | $f_5$ |
| 22 | $f_5$ | $f_4$ | $f_3$ | $f_2$ | $f_1$ | $c_2$ | $c_1$ | $f_{13}$ | $f_{12}$ | $f_{11}$ | $f_{10}$ | $f_9$ | $f_8$ | $f_7$ | $f_6$ |
| 23 | $f_6$ | $f_5$ | $f_4$ | $f_3$ | $f_2$ | $f_1$ | $c_2$ | $c_1$ | $f_{13}$ | $f_{12}$ | $f_{11}$ | $f_{10}$ | $f_9$ | $f_8$ | $f_7$ |
| 24 | $f_7$ | $f_6$ | $f_5$ | $f_4$ | $f_3$ | $f_2$ | $f_1$ | $c_2$ | $c_1$ | $f_{13}$ | $f_{12}$ | $f_{11}$ | $f_{10}$ | $f_9$ | $f_8$ |
| 25 | $f_8$ | $f_7$ | $f_6$ | $f_5$ | $f_4$ | $f_3$ | $f_2$ | $f_1$ | $c_2$ | $c_1$ | $f_{13}$ | $f_{12}$ | $f_{11}$ | $f_{10}$ | $f_9$ |
| 26 | $f_9$ | $f_8$ | $f_7$ | $f_6$ | $f_5$ | $f_4$ | $f_3$ | $f_2$ | $f_1$ | $c_2$ | $c_1$ | $f_{13}$ | $f_{12}$ | $f_{11}$ | $f_{10}$ |
| 27 | $f_{10}$ | $f_9$ | $f_8$ | $f_7$ | $f_6$ | $f_5$ | $f_4$ | $f_3$ | $f_2$ | $f_1$ | $c_2$ | $c_1$ | $f_{13}$ | $f_{12}$ | $f_{11}$ |
| 28 | $f_{11}$ | $f_{10}$ | $f_9$ | $f_8$ | $f_7$ | $f_6$ | $f_5$ | $f_4$ | $f_3$ | $f_2$ | $f_1$ | $c_2$ | $c_1$ | $f_{13}$ | $f_{12}$ |
| 29 | $f_{12}$ | $f_{11}$ | $f_{10}$ | $f_9$ | $f_8$ | $f_7$ | $f_6$ | $f_5$ | $f_4$ | $f_3$ | $f_2$ | $f_1$ | $c_2$ | $c_1$ | $f_{13}$ |
| 30 | $f_{13}$ | $f_{12}$ | $f_{11}$ | $f_{10}$ | $f_9$ | $f_8$ | $f_7$ | $f_6$ | $f_5$ | $f_4$ | $f_3$ | $f_2$ | $f_1$ | $c_2$ | $c_1$ |

## 2. THE ORGANIZATION AND FUNCTIONING OF SLASH-AND-BURN AGRICULTURAL SYSTEMS

Beyond the periodically cleared forested lands, the cultivated ecosystem includes vegetable gardens close to houses and relatively unimportant breeding of small and large livestock. Often, the territory of each village also includes some "virgin" wooded lands, cultivable or not, which are not yet cleared.

### Periodically Cultivated Forested Lands

In order to understand better how the cultivated and idled lands of various ages are distributed in time and space, let's consider the case of a family established in a village of forest farmers, where they practice temporary cultivations of two years that alternate with forested idled land lasting thirteen years, thereby forming a rotation period of fifteen years.

*Crop Rotations and Plot Allotments.* The "Rotation and Plot Allotment" table above shows how, over the course of many years, this cycle unfolds on each new cleared parcel. In the first year, the newly settled family clears the first parcel p1 for cultivation during the first year c1. In the second year, the family clears a second parcel p2 to undertake cultivation c1, while on parcel p1, cleared the preceding year, it conducts the second year cultivation c2. During the third year, the family clears a third parcel p to undertake cultivation c1, while it conducts cultivation c2 on parcel p2 and abandons parcel p1 to first year idling f1. Thus the cycle continues until the fifteenth year, during which it clears the fifteenth parcel p15 to undertake cultivation c1. It then conducts cultivation c2 on parcel p14 and it abandons parcel p3 to its first year idling f1.

At the end of fifteen years, the first cleared parcel p1 will thus have been through, in order, two years of cultivation c1 and c2 and thirteen years of lying idled f1, f2 ... f13. The sixteenth year, parcel p1 will be cleared again and will then experience the same sequence of cultivations and idlings. The periodic repetition (here every fifteen years) on the same parcel of the same sequence of cultivations and idling forms what is called a *crop rotation.*

The table also shows how, at the end of fifteen years, a comprehensive plot allotment of cultivated land and idled land of various ages is formed. During the fifteenth year, the fifteen parcels cleared earlier (p1, p2 ... p15) are respectively occupied by increasingly younger idled land (f13, f12 ... f1) and by the two cultivated areas c2 and c1. This spatial distribution among different parcels of all types of idled and cultivated land, which together form the crop rotation sequence, is called a *plot allotment.* The following year, the cropping plan will still be composed of the same types of idled and cultivated lands, but each of these types will be displaced in order to occupy the parcel cleared one year later.

However, each family can also clear more than one parcel each year and, in a village composed of several families practicing the same rotation, a large number of parcels are cleared each year, which then go through the same succession of cultivations and idlings. The totality of parcels that at a given moment are in the same state of cultivation or idling is called a *plot*. Thus there is the plot of the main cultivation $c_1$, the plot $c_2$, or the plot of idled land $f_1$ or $f_2$, etc.

In systems where the cultivable lands are abundant, the parcels to clear are allocated to families without restricting the size and without taking into account which family had cultivated them earlier. These parcels are thus scattered and of variable form. Also, when the rotation is long, the choice of parcels to clear depends more on the state of development of the wooded idled land than on its exact age. In this case, the duration of the idling is not absolutely constant. It varies several years more or less from the average duration.

### Regulated Plot Allotment

When there is no unused land in reserve, a strict organization of rotation and plot allotment occurs. This is the case, for example, in some villages of manioc farmers southwest of Brazzaville in the Congo, which we were able to study. The rotation practiced is for twelve years. The entire space around the village is subdivided into twelve equal plots: ten plots of idled land ($f_1$, $f_2$ ... $f_{10}$) and two plots of manioc, one for the first and one for the second year ($m_1$ and $m_2$). These plots all lie together, arranged side by side in the order in which they are cleared and cultivated, so that the plot allotment plan is directly visible on the ground. Each year, the oldest plot of idled land ($f_{10}$) is subdivided into adjoining quadrangular parcels, which are distributed among the families to be cleared and planted with manioc. In such a system, the farmers of the village are obligated to follow a rotation and a plot allotment common to all. This is referred to as *obligatory rotation* and *regulated plot allotment*. An astonishing case of this type of regulated plot allotment was found in the mountains of South Vietnam at the turn of the century.[4] A ma village had at its disposal a territory of 2,250 hectares, of which 700 hectares were granite boulder fields and swamps and 1,550 hectares were cultivable forest divided into thirty-eight plots of 40 hectares each. Each year, all the parcels cleared by the villagers were grouped on one of these plots and, year after year, the cultivated lands were moved in such a way that they returned to their starting point at the end of thirty-eight years.

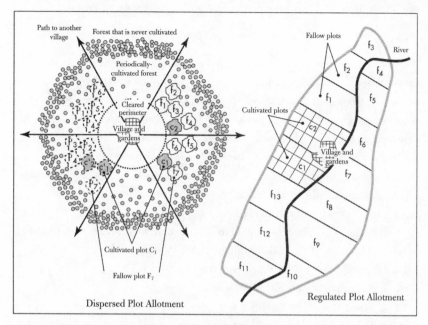

*Figure 3.2 Diagrams of the Organization of a Communal Territory of Forest Farmers*

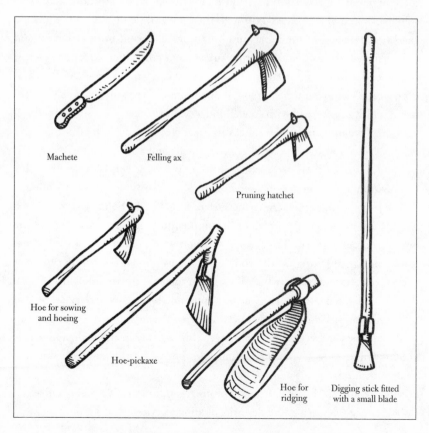

*Figure 3.3 Contemporary Forest Farmers' Tools*

## The Renewal of Fertility

The cultivable forests and other wooded environments of the planet are more or less fertile depending upon the climate, geomorphology, soil, and the nature and strength of the afforestation. However, whatever the differences may be, when a virgin forest cleared by slash-and-burn techniques is cultivated for the first time, the fertility of the soil is then at its height. Indeed, this soil, cultivated for the first time, continues to benefit from the ordinary deposits of minerals coming from the solubilization of the parent rock, the fixation of atmospheric nitrogen, and the mineralization of a fraction of the forest soil organic matter. It benefits further from extremely high mineral deposits coming mostly from the ashes from burning but also from accelerated mineralization of an additional fraction of the soil organic matter. The reheating and aeration of the surface layers of the soil resulting from the clearing and preparation of the cultivation bed cause this accelerated mineralization.

The first cultivation in this particularly fertile soil, carried out in the months that follow a slash-and-burn operation, provides an abundant harvest, thereby removing a portion of the available minerals. In addition, cultivated soil loses some of its minerals through leaching and denitrification. Thus the exceptional deposits of minerals resulting from slash-and-burn activities tend to be exhausted and the yields from subsequent cultivations decline quickly. In the least fertile environments, the option of carrying out a second cultivation, whose yield would be too low, is abandoned. In the most fertile environments, cultivations can be prolonged longer, but they are then in competition with the development of natural cover. After clearing, undestroyed trees and bushes, as well as invasive wild grasses, push their roots into the cultivated soil and absorb a growing portion of the nutritive minerals. To reduce this competition, the farmers uproot the weeds and destroy their roots by hoeing. This hoeing loosens and aerates the surface layers of the cultivated soil, which slightly accelerates the process of decomposition and mineralization of the organic matter and enriches the soil solution with minerals for a short time. Moreover, by breaking up the soil, hoeing slows down the capillary ascent and evaporation of water, which remains in the soil for use by plants. For all these reasons, repeated hoeing makes it possible for the secondary crops that follow to more effectively extract part of the cultivated soil's remaining fertility. But after several years, the declining yields become insufficient, and weeds abound and hoeing becomes impracticable. The land is then left idle until the next cultivation.

A system of slash-and-burn agriculture, then, can only perpetuate itself if, from clearing to clearing, the deposits of minerals, derived from the ashes and the accelerated mineralization of the humus, are maintained at a level sufficient to ensure good harvests. To this end, the wooded idled portion of the rotational cycle must last long enough to produce a substantial volume of biomass that can, in turn, be cut and burned to provide the necessary quantity of ashes. It must also last long enough to provide a large enough litter of dead wood and leaves to rebuild the soil's reserve of humus, which is worn down by the accelerated mineralization that follows each clearing.

As a general rule, when the idle period lasts about twenty years, the regenerated forest is quite strong, and after a slash-and-burn operation the soil's fertility is quite high. It then suffices to partially clear a reduced area to support the needs of a family. When the idling lasts more than ten years, the reforestation between two clearings is thin, the litter and ashes less abundant, the fertility of the soil not as high, and weeds abound. Thus in order to obtain the same volume of production a more extensive area must be completely cleared.

## Residual Virgin Forest

This periodically cultivated forested environment generally forms a ring of several kilometers radius around each village of forest farmers. Next to that are usually more or less significant remainders of virgin forest that has never been cleared. Basically, these portions of the forest are noncultivable or difficult to cultivate, situated in wet, low-lying areas or on land that is too hilly or soil that is too thin or not very fertile. Also, when the population of a village is still not very numerous and the reserves of cultivable forests are not yet used, then the latter are arranged more or less regularly in a ring at the periphery of the village territory.

## Enclosed Gardens and Orchards

Outside of the forested environment, whether cultivated or not, the ecosystem also includes small parcels close to the houses that are enclosed and cultivated continually, i.e., there is no fallow period. These are gardens, or garden-orchards, generally completely cleared, with the stumps removed, where fertility is renewed by domestic waste, by planting fruit trees in the garden, and sometimes by animal manure transported and spread manually. These garden-orchards are planted with annual crops such as maize, sweet potato, groundnuts, tomatoes, or biannuals such as manioc. There are also perennial crops such as banana or sugar cane, and various food-producing trees such as avocado, breadfruit, mango, various citrus fruits, etc.

## Animal Breeding

A dense forest environment such as the humid tropical forest is not hospitable to domestic animals. This relatively impenetrable and sometimes dangerous environment offers limited foraging resources to herbivores (cows, sheep, goats, donkeys, and horses) and to other animals (pigs and poultry). As a result, these animals are dependent on meager agricultural surpluses and by-products of crops destined for human consumption. On the other hand, so long as the cultivated environment remains heavily forested, animals can hardly be of service to agriculture. Cleared parcels encumbered with stumps and roots do not lend themselves to animal-drawn implements for cultivation. As for reproducing the soil's fertility, there is no need to use animals, since fertility is ensured by long-term forested fallowing. Animals even represent a threat to crops, because they are attracted by the cultivated parcels where they can wreak havoc. Since the periods of cultivation are so short, the scattered parcels cannot be effectively enclosed.

Many of the forests cultivable by slash-and-burn techniques are, however, more hospitable to livestock than the dense forests of humid tropical regions. Tropical forests with only a single rainy season act as secondary pasturage during the dry season. In the past, forests in the temperate regions were also used for livestock at times when there was little grass in the natural pastures, in the middle of the summer and during the winter, and to fatten pigs that ate acorns and beechnuts in the autumn.

Nevertheless, livestock, above all herbivores, develop better when a portion of the ecosystem has been deforested. Thus, many villages of forest farmers create a ring of completely deforested, grassy savanna land between their dwellings and the surrounding enclosed gardens, on one side, and the cultivated forest, on the other.

## The Performance of Slash-and-Burn Agricultural Systems

Performance varies greatly as a function of the length of the rotation and the size of the cultivated ecosystem's biomass in slash-and-burn agricultural systems. By performance, we mean the volume of production per unit of surface area (the yield per hectare or square kilometer) and the volume of production per worker (labor productivity).

In order to illustrate this, let us consider, for example, a strong tropical forest whose original supra-soil biomass totaled 500 tons per hectare before any clearing and which is initially cleared and cultivated every fifty years. Let us suppose that after each clearing the biomass is reduced to 50 percent of the original

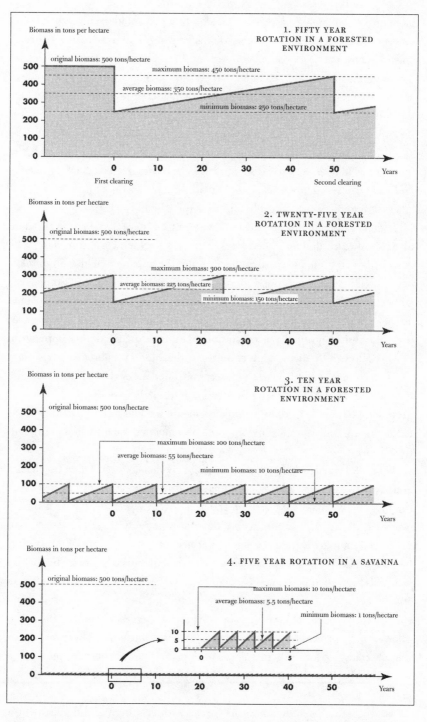

*Figure 3.4 Variations in the Biomass of a Cultivated Environment as a Function of the Length of the Rotation*

biomass (or 250 tons/hectare), and that after fifty years of idling it regenerates 90 percent of the original biomass (or 450 tons/hectare). Thus, each slash-and-burn episode reduces to ashes some 200 tons of aerial biomass per cleared hectare. The biomass of cultivated or idled parcels then oscillates around an average of 350 tons per hectare (or 70 percent of the original biomass). In these conditions, the cultivated soil is well supplied with organic matter and minerals, and it is possible to obtain high yields. But as the area actually sown, between the stumps and the remaining trees, does not exceed half of the cleared surface area, the *apparent* yield does not exceed 10 quintals of grain per cleared hectare, while the *real* yield can reach 20 quintals per hectare actually sown. This yield, without any contribution from external fertilizers, is quite high.

In such a system, however, as we have seen, it is necessary to have about 50 hectares of cultivated and idled land of all ages for each cleared hectare. The real yield of 20 quintals per sown hectare and the apparent yield of 10 quintals per cleared hectare correspond therefore to a *territorial* yield of 10 quintals for 50 hectares of periodically cultivated forest, or 0.2 quintal per hectare, or only 20 quintals per square kilometer. Assuming that the basic needs of the population add up to 2 quintals per person per year, then the maximum population density allowed by this system is on the order of ten inhabitants per square kilometer. On its 1,550 hectares of forest cultivated every thirty-eight years, the village Ma, referred to above, produced 300 quintals of hulled rice each year making it possible to feed the 150 inhabitants of the village.[5] This corresponds exactly to a population density of ten inhabitants per square kilometer of cultivable forest.

Now let us consider that this same originally dense tropical forest is cleared every twenty-five years. The biomass then oscillates between 30 and 60 percent of the original biomass (or between 150 and 300 tons/hectare). The biomass destroyed at each slash-and-burn episode is 150 tons per hectare. The average biomass is no more than around 225 tons per hectare (or 45 percent of the original biomass). The burning produces fewer ashes than the preceding case and the real yield falls from 20 to 14 quintals per actually sown hectare. In order to maintain an apparent yield of 10 quintals (per cleared hectare), the area actually deforested and sown must be augmented, which leads to cutting 70 percent of the forested area as opposed to 50 percent. With a territorial yield of 10 quintals for 25 hectares of periodically cultivated forest, or 40 quintals per square kilometer, the maximum population density allowed by the system is twenty inhabitants per square kilometer of cultivable forest.

Lastly, let us consider that this forest, formerly high and dense, is now reduced to a thicket and cleared and cultivated every ten years. In order to obtain an apparent yield as high as possible, a clear-cutting, which almost entirely destroys the bushy biomass and nearly the whole of the terrain, right up to the stumps, is sown. The total biomass of this periodically cultivated thicket then oscillates between about ten and about a hundred tons per hectare, and the biomass reduced to ashes at each clearing is on the order of 90 tons. The real yield falls to 8 quintals, which corresponds to a territorial yield of 7 quintals for 10 hectares of allotted plots, and makes it possible to support the basic needs of a population on the order of 35 inhabitants per square kilometer of cultivable forest.

If the population density goes beyond this level, the frequency of the clearings increases as well, the length of the idle period falls to less than five or six years, and even the wooded thickets no longer have time to regenerate. The idled land remains at the grassy stage and its biomass oscillates from less than one ton per hectare in the off-season to about ten tons at the most in the high season. Slash-and-burn cultivation becomes impracticable, but it can be replaced by temporary cultivation alternating with grassy idled land of medium duration, on condition, however, of having both the necessary tools to clear a grass cover and a new method of renewing the fertility.

According to this analysis, as long as the population density does not rise above a certain level, a level that is variable depending on the environment, slash-and-burn systems of cultivation do not, in general, entail the destruction of the wooded biomass or a considerable reduction in fertility. They do not by nature involve deforestation or declining fertility. On the other hand, when the population density clearly rises above this level, the result is necessarily deforestation and the impossibility of continuing to practice this type of cultivation. Thus, as long as a growing population of slash-and-burn farmers has access to virgin forest reserves, they subjugate them step by step in such a way as to maintain the population density within the limits that permit a reliable regeneration of the biomass and fertility. And it is thanks to this pioneer dynamic, which is non-deforesting, that these systems of cultivation have been able to last for so long in most regions of the world. But as soon as the virgin forest reserves are exhausted, the continuation of population growth is necessarily characterized by an increase in population density, which rapidly leads to deforestation.

## Social Organization

Villages of forest farmers are composed of "families," related among themselves or not, each of which forms a unit of production-consumption. The village territory is open to the rights of use by every family, except for the developed lands, the enclosed garden-orchards adjoining the dwellings, and any possible perennial plantings, which are subject to regular rights of use, similar to a type of private property. As long as the village territory has a small population and the lands remaining to be cleared are superabundant, the right of use is even easily granted to possible newcomers. The relevant village authority (village head, a council) grants to each family every year cultivable forested parcels corresponding to its needs. The (private) rights of use of a family on the parcels that are allotted to it—i.e., the right to clear, cultivate, and harvest the fruits of its labor—disappears with the last harvest. When the land is abandoned to long-term forested idling, it returns to the common domain.

These temporary rights of use tend to become permanent rights of use when the perennial plantings (coffee, cacao, hevea) are put in, or also when, because of population growth or deterioration of a portion of the land, the duration of the fallowing is reduced to the point that the exploitation of an area by the same family tends to become continual. But then, it is no longer a question of temporary cultivation alternating with a long-term idling. When a good portion of the land is subjected to permanent rights of use, and the temporarily cultivable idled lands are scarce, the right to cultivate each parcel of land is increasingly rationed and strictly allotted to a specific family, such that the transfer of this right to a third party assumes the form of a loss in earnings that requires a compensation, i.e., in fact, the payment of a land rent: a "tenant farm" if the transfer of the rights of use is temporary, a "sale" if the transfer is permanent. By becoming a commodity, this land also becomes a publicly recognized object of appropriation.

But guaranteed access to lands for clearing is not the only arrangement that ensures the food security of each unit of production-consumption. In a system of manual cultivation with low productivity, it is also important that the ratio between the number of persons of working age and the number of mouths to feed does not fall below a certain level in each of these units, on the order of 1 to 3 or 1 to 4. Since this requirement is much easier to realize in a large family than in a small one, societies of forest farmers were, even recently, organized into units made up of several households. Moreover, the regulation of marriage exchanges and the adoption of young outsiders also contribute to maintaining this equilibrium in each unit. Finally, sharing the labor of large projects (clearings, hoeing), the cultivation of common fields, and the formation of village food reserves are likely to compensate for possible disequilibria.

Furthermore, the low productivity of labor in systems of slash-and-burn agriculture reduces the possibilities of social differentiation. Artisans, merchants, and warriors continue to participate in agricultural tasks. Political and religious functions are filled by a very small number of persons, whose consumption level is barely above that of other villagers.

## 3. THE DYNAMICS OF SLASH-AND-BURN
## AGRICULTURAL SYSTEMS

### The Pioneer Dynamic

Beginning in the centers of origin of the Neolithic agricultural revolution between 10,000 and 5,000 years before the present, systems of slash-and-burn agriculture gradually spread to most of the cultivable forested environments of the planet. The population densities that these systems could support were several times higher than those possible with the systems of predation that were practicable on the same land. Over millennia, the geographic expansion of these systems thus acted as a medium for the strong population growth that occurred between the beginning of the Neolithic age and the appearance of the first post-forest agrarian societies of early antiquity. Recall that between 10,000 and 5,000 years ago, the world population grew from around 5 to around 50 million inhabitants. This pioneer movement then continued everywhere where forest reserves that had never been cleared remained. Even in our day, this movement continues on the frontier of the last "virgin" forests of Amazonia, Africa, and Asia.

If it is difficult to know with precision how this pioneer dynamic was organized in the past, it is known how these things happen today. In the vicinity of a pioneer front and abundant virgin forest reserves, one notes first of all that the villages of the slash-and-burn farmers are generally established a good distance from one another (5 to 6 kilometers, or a one-hour walk), which enables them to make use of a cultivable forested area on the order of 30 square kilometers, assuming that all the forested territory is cultivable. Moreover, these populations today experience high rates of population growth, on the order of 3 percent per year. It nearly doubles in size with each generation, or every twenty to thirty years. Now, despite that, it should be noted that village populations rarely surpass a thousand inhabitants as long as virgin forest reserves exist. This is explained by the fact that beyond this number, the population density exceeds thirty inhabitants per square kilometer. In order to support such a density, the length of the rotation becomes shorter and then, consequently, idled land old enough to provide good harvests becomes scarce. In order to avoid such a situation, a portion of the village population begins to clear and cultivate new, more fertile lands in the nearby virgin forest, located beyond the pioneer front. They build new shelters there and, after some time, set up and found a new village of several dozen then of several hundred inhabitants coming from the old village, whose population is reduced in an amount equal to the population of the new village. The population of each of the two villages can be increased again over several decades, up to the point where it reaches the maximum size of around a thousand inhabitants, after which it is subdivided again.

Thus the population of villages of forest farmers generally oscillates between a minimum of several hundred inhabitants and a maximum of around a thousand inhabitants, such that the population density varies between ten and thirty inhabitants per square kilometer of cultivable forest. This mechanism of subdivision-migration acts, then, as a regulator for the villages. It maintains the population density and the length of the fallowing within the most appropriate limits for the reliable functioning of a slash-and-burn agricultural system and its continuation. When a portion of the village territory is not cultivable, because it is too rocky, too wet, etc., the villages are generally farther apart from one another, or smaller, and the population density is thus smaller.

The division and migration activity of villages beyond the pioneer front is as rapid as the population growth is strong. Thus, since the middle of the twentieth century, the population has increased so rapidly in the pioneer fronts of Africa, Asia, and South America that most of the villages must subdivide and migrate at least once each generation. Since such disruption each time requires several years of preparation in the original village and several years of setting up the new village, the population hardly ever experiences stability. In these conditions, the length of the idle period varies all the time, so no rotation of definite length can ever be established, which makes these systems difficult to understand. On the other hand, in the past, when the rate of population growth was decidedly less than 1 percent per year and took one or several hundred years to double, this subdivision-migration activity occurred less than once per century. The pioneer front moved around one kilometer per year and systems of slash-and-burn agriculture were able to survive for hundreds of years with few modifications. On the scale of a single generation, such a system appeared relatively stable in the eyes of those who practiced it.

## Deforestation

But whether rapid or slow, the pioneer movement of slash-and-burn agricultural systems necessarily encounters an insurmountable frontier. This could be a natural frontier, such as an ocean, an uncultivable forest such as the taiga, a grassy formation, or a mountainous barrier. It could also be a political frontier, such as the territorial limit of another population, of a state, or of a natural reserve. In any case, from the moment there is no longer any accessible virgin forest, the surplus population can no longer be absorbed by the process of subdivision-migration if the population continues to grow at the same rate as during the pioneer phase. The population density increases and consequently the area cleared each year must be extended, which necessarily leads to cutting on increasingly younger idled land. In order to compensate for the lower real yield that results, the clearings must be increasingly comprehensive. Trees that had been spared up until then are cut, thereby extending the area actually sown. Quite quickly, clear-cutting becomes a common practice and, in order to compensate for the lower yields, the only thing left to do is to extend the area subjected to clearing each year. From this time on, the length of the idle period diminishes quickly and deforestation is accelerated.

This acceleration of deforestation, which occurs from the moment the population density exceeds a certain point, is the reason why tropical forests that were still untouched in the middle of the century, and subsequently colonized by a population that doubles every generation, are today practically destroyed. In such circumstances, the deforestation phase is linked closely with the pioneer phase, to the point of almost being confused with it. This is why many observers have concluded that slash-and-burn agricultural systems are by nature "deforesting." Except for unstable forested environments that are too fragile to regenerate after clearing, this is not true. In general, it is rather the increase in the population density and the resulting reduction in idling time that are the causes of the deforestation.

But this double process of an increase in both population density and deforestation does not occur only when the geographical limits of slash-and-burn agricultural systems are reached. It also occurs in regions colonized and cultivated long ago, which one day end up too far from the pioneer front for the subdivision-migration process to take place. In order to escape deforestation and its consequences, to reach some new "promised land" located hundreds of kilometers away, surplus populations must organize increasingly risky, distant expeditions, which end up becoming impossible. In these regions, which were conquered and cultivated long ago, the population increase leads eventually to a more or less total deforestation.

For several millennia, systems of slash-and-burn agriculture continued to expand thousands of kilometers from the centers of origin of Neolithic agriculture. At the same time, deforestation had been under way for a long time in these same centers and in neighboring regions brought under cultivation in the past. Then, deforestation spread gradually in all directions, following the advance of the pioneer fronts from a great distance and with several centuries' delay.

However, the proximity of the center of origin is not the only variable determining the age of deforestation in the different regions of the world. The nature of the original plant population also plays a large role. The more penetrable and exploitable a particular region's forested formation, the more quickly the pioneer front advances into it. As for the deforestation that will subsequently occur, the less resistant the ecosystem is to ax and fire, the earlier that deforestation will happen. Thus in the expansion area of the agriculture originating in the Near Eastern center, the first deforested environments were the most penetrable and fragile sparse forests and wooded savannas, which extended into the hot and relatively dry subtropical zone of Saharan Africa and of the Arabo-Persian Near East. In these regions, the deforestation began in the eighth millennium before the present, and it undoubtedly contributed to the drying up of the climate, which led, in the fifth millennium, to the desertification of a large portion of these regions.

Less fragile than the preceding regions, the broadleaf forests of the hot temperate regions of the Mediterranean area resisted longer. Nevertheless, the destruction of these forests began early enough, before 2000 c.e. on the eastern rivers of the Mediterranean, and spread gradually toward the west, into southern Europe and North Africa up to the last few centuries B.C.E. The deterioration and destruction of entire forested areas of middle Europe, denser and more resistant than the Mediterranean forests, began in this epoch. The deforestation of this area continued until the first centuries of the common era. During this time, the deforestation also spread south of the Sahara. Forests of deciduous trees in the tropical zone characterized by a single rainy season began to be transformed into savannas, and this process continued up until the recent past. The evergreen forests of the humid equatorial zone began to recede much more recently and remain partly standing even today.

## The Consequences of Deforestation

In general, deforestation entails not only a reduction in the fertility of the soil, but also the beginning or worsening of erosion and, in some cases, a drying up of the climate. These phenomena are quite variable. Depending on the environment, they can be more or less profound and serious.

### The Reduction of Fertility

We have seen that the evolution from the long-term forested idle period to the medium-term or short-term grassy idle period leads, first of all, to the disappearance or reduction in the quantity of ashes obtained after burning, as well as a reduction in the litter, which decreases the organic matter content of the soil. This reduction lowers the soil's storage capacity for water and mineral salts and diminishes the quantity of minerals resulting from mineralization of the humus. The consequent lowering of fertility varies greatly depending on the climate. After deforestation, the hotter the climate, the lower the level of residual organic matter in the soil. In the cold temperate regions, it can be maintained at 1 or 2 percent, while in the hot regions it falls to less than 1 percent.

What is more, under some hot climates with a pronounced dry season, the clay-like colloids are dehydrated in the denuded and overheated soils, which again reduces the soil's storage capacity for fertilizing minerals. This dehydration also leads to a hardening of the soil that is unfavorable for the rooting of cultivated plants. Finally, on some deforested lands, as the water is no longer absorbed by the roots of trees, groundwater collects deep in the soil. During the dry season, this groundwater, along with the iron oxides it carries, rises to the surface by capillary action. These oxides are crystallized on contact with the air at the moment of evaporation, which then cement all the hardened materials of the soil into a sort of carapace. This carapace, or lateritic duricust, is completely sterile.[6]

## Erosion

In a deforested environment, rainwater directly strikes the soil without the softening effect of vegetation. Moreover, its flow over the surface of the soil generally encounters few obstacles. In these conditions, the rapid runoff of water increases, while its seepage diminishes. In hilly regions that receive strong rains, the runoff is so great that it causes catastrophic floods, which tear the soil loose and transport enormous masses of earth down to lower valleys and into deltas where it accumulates. However, erosion does not only have destructive and negative effects. The deposits of sediments formed at the bottom of slopes and in valleys can also contribute to enlarging and enriching cultivable lands.

The first manifestation of such a change in water flow characteristics appeared in the Tigris and Euphrates valleys, in the vicinity of the Near Eastern center, in the sixth millennium before the present, following the deforestation of the sides of the basins. While the rainfall had hardly increased, a regime of violent floods took place in this time period, a veritable "deluge" that lasted for more than a thousand years. The magnitude of these phenomena is well-known in the hot temperate regions of the Mediterranean. The plant cover is fragile and the rains, while insubstantial, are violent because they are concentrated in a few months. The stripping of top soil from the slopes, the formation of skeletal soils and gullies, the overdeepening of the high valleys, the silting up of the low valleys and deltas, and the filling of gulfs in the sea have been going on since antiquity. Hence, many ancient ports in the Mediterranean region today are found far inland. In these regions, erosion and degradation of fertility combine to make the most exposed or most fragile of the deforested lands uncultivable. Those lands are thus used as pasture. Only the zones retaining a soil that is deep enough, rich enough and wet enough continue to be cultivated (see chapter 6).

The ancients, moreover, had an acute consciousness of this ecological disaster. Thus, in the *Critias*, Plato makes a comparison between the Athenian countryside of his epoch (fifth century B.C.E.) and the same more or less mythical countryside 9,000 years earlier. This comparison gives a luminous account of the phenomena we have just mentioned:

> And the soil was more fertile than that of any other country and so could maintain a large army exempt from the calls of agricultural labor. ... So the result of the many great floods that have taken place in the last nine thousand years (the time that has elapsed since then) is that the soil washed away from the high land in these periodical catastrophes forms no alluvial deposit of consequence as in other places, but is carried out and lost in the deeps. You are left ... with something rather like the skeleton of a body wasted by disease; the rich, soft soil has all run away leaving the land nothing but skin and bone. But in those days the damage had not taken place, the hills had high crests, the rocky plain of Phelleus was covered with rich soil, and the mountains were covered by thick woods, of which there are some traces today. For some mountains which today will only support bees produced not so long ago trees which when cut provided roof beams for huge buildings whose roofs are still standing. And there were a lot of tall cultivated trees which bore unlimited quantities of fodder for beasts. The soil benefited from an annual rainfall which did not run to waste off the bare earth as it does today, but was absorbed in large quantities and stored in retentive layers of clay, so that what was drunk down by the higher regions flowed downwards into the valleys and appeared everywhere in a multitude of rivers and springs.[7]

In the cold temperate regions, where the rains are better distributed and where the plant cover and the soil remain more substantial, these erosion phenomena are generally much less pronounced. On the other hand, in tropical regions with a single rainy season and hilly terrain, it is possible to observe erosion phenomena analogous to those of the Mediterranean region. In the humid tropical regions, just as in the monsoon regions that receive several meters of water per year, the deforestation of the slopes only reinforces erosion phenomena that are already gigantic, which explains the particular geomorphology of these countries. The strong-flowing rivers of these regions carry enormous quantities of sediment that accumulate up to dozens of meters in depth in broad, flat valleys and in vast, weakly sloping deltas. These valleys and deltas are submerged a good part of the year and are perfect lands for aquatic rice growing.

## A Drier Climate

Finally, when deforestation expands over a wide area, there is a tendency for the climate to dry out. By destroying several hundred tons of forested biomass per hectare, enormous reserves of water found in the vegetation and in the upper levels of the ground are made to disappear at the same time. This mass of water, which represents several times the dry biomass itself, can reach thousands of tons per hectare, such that deforestation results in drying up ground water to a depth of dozens of centimeters.

After deforestation, the water reserves of the soil and vegetation, which were built up again during each rainy season, are greatly reduced. They are exhausted more quickly and evapo-transpiration stops earlier, at the beginning of the dry season. The soil is dried out and the lower layers of the atmosphere are no longer humidified nor cooled down. As a result, the cloud systems that pass above these regions during the dry season no longer encounter the cold and humid atmospheric front that previously triggered late rains. Consequently, the lower rainfall is intensified and the dry season is lengthened.

The fall in the pluviometry and the lengthening of the dry season have quite diverse consequences depending on the nature of the initial climate. In the wet equatorial zone, which receives more than 2,000 millimeters of rain per year, the decreased rainfall has few effects. But in the drier tropical zone (less than 800 millimeters of rain per year) with a pronounced dry season, the reduction in rainfall and the shortening of the cultivation season quite noticeably affect the yields. The relative dryness of the ecosystem also has negative consequences in the hot temperate zones (Mediterranean climate) and significant consequences, although less pronounced, in the cold temperate zones.

Yet it is in the dry and hot subtropical zones, receiving less than 500 millimeters of rain per year and with a long dry season, that the consequences of the drying out of the soil and climate are the most serious. After deforestation, the rainfall can diminish by at least 250 millimeters per year in these areas. The "dead season" is extended to the entire year and results in desertification, an ecological catastrophe like those that occurred for similar reasons in the Sahara, Arabia, Persia, and in other regions of the world over thousands of years.

In addition, the effects of deforestation are characterized not only by a drop in local rainfall but also by a reduction in precipitation in regions far from the deforested areas. Thus the cloud systems coming from the Atlantic that pass above hot and humid forests, such as the Guinean one, are primarily supplied by the evaporation of the ocean, but also by evapotranspiration of the water reserves in the soil and forest vegetation. After the destruction of the forest, the rainwater is no longer reserved, flows rapidly toward the sea, and does not supply the cloud formations passing through. Consequently, the rainfall also drops in regions usually watered by these clouds. Thus it is possible to conclude that the diminishing rainfall registered in the Sudanian and Sahelian regions in the course of the last few decades results not so much from their own deforestation, which is generally much older, but from the recent decline of the equatorial forest of West Africa.

## 4. EMERGENCE AND DIFFERENTIATION OF POST-FOREST AGRARIAN SYSTEMS

Systems of slash-and-burn agriculture have been among the most extensive and most long-lasting that ever existed. After having penetrated the forests and cultivable wooded environments, they persisted for centuries, until population increases and the too frequent repetitions of cultivation led to the destruction of the forests. This process of deforestation, which affected most of the formerly forested and cultivated environments of the planet, was undoubtedly the greatest ecological destruction in history.

By destroying megatons of biomass, as well as reserves of water and humus on the continental level, deforestation created new ecological conditions of great diversity. It opened the way to a whole range of post-forest agrarian systems, as different from one another as are the hydraulic systems of the arid regions and those of the monsoon regions, the systems characterized by fallowing and associated animal breeding in the temperate regions or the varied agrarian systems of the tropical savannas. It also made possible the enlargement of pastoral systems. But the progress of the post-forest agrarian systems was not immediate. Effective and sustainable exploitation of the diverse ecosystems originating from deforestation demanded, in each region of the world, the development of new tools, new modes of clearing and renewing fertility, and of course new modes of cultivation and animal breeding, all of which had to be appropriate to the new ecological conditions and other characteristics of each of the agrarian systems in gestation.

It is the emergence and evolution of these great post-forest agrarian systems that we are going to outline now, by focusing a little more on those that will not be studied in depth later in this book, i.e., systems of tropical savanna agriculture and systems of aquatic rice growing in the monsoon regions.

### Desertification and Formation of Hydraulic Agrarian Systems in Arid Regions

In arid regions, or regions that became arid after their deforestation or the deforestation of peripheral areas, vegetation becomes rare, the soil deprived of organic matter becomes skeletal, and cultivation based on rainfall becomes

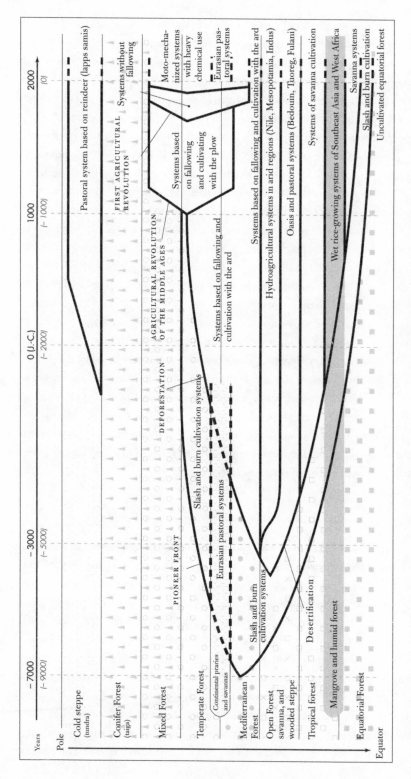

Figure 3.5 Agrarian Geneology: Historical Sequence and Geographical Differentiation of Eurasian and African Systems Originating in the Near Eastern Center

impossible. The only zones that remain cultivable are those that benefit from an external source of water. These privileged zones form more or less extensive verdant oases, provided with water either by large rivers that are supplied with water in distant rainy regions, by nearby streams coming from mountains, resurgent subterranean groundwater supplied from the outside, or by fossil groundwater. The cultivation of these oases and valleys is not always easy. It often requires preliminary hydraulic equipment and installations, sometimes small, but also sometimes quite large. It was in this type of environment, beginning in antiquity, that new forms of agriculture not based on rainfall (instead, based on floodwaters and irrigation) gave birth to the first large hydro-agricultural civilizations in history. Thus, in the sixth millennium before the present, the farming and animal breeding peoples of the Sahara, Arabia, and Persia, driven by the dryness that began to predominate in these large areas, moved toward the low alluvial valleys of the Indus, Tigris, Euphrates, and Nile. Coming from all directions, these diverse peoples, who had long pastured their herds in these valleys, began to cultivate their fringes. Then they had to build the necessary installations to protect their crops from untimely floods, ensure a sufficient supply of water, and, if need be, drain any detrimental excess water.

All kinds of hydraulic works contributed to this essential control of water: dikes, supply canals, outlet canals, basins, locks, dams for raising the water level, storage dams, not to mention wells, culverts, and all kinds of machines to elevate the water level. Depending on the morphology and the hydrologic regime in each valley, as well as the time period in question, these vast installations were constructed and arranged quite differently. They gave rise to a very particular hydraulic architecture in each case.

## Floodwater Farming and Irrigation Farming

The essential problem in a valley submerged several months a year by a massive flood, as in the Nile Valley, was, as we will see in the next chapter, to save the floodwaters in basins designed for this purpose, then to drain these waters at the requisite time and practice floodwater farming, and finally to protect the crops against a possible return of a late flood. But in valleys that were not regularly inundated, floodwater farming was hardly possible. The essential problem in that case was to use water from a river or from other bodies of water to practice irrigation farming, while protecting the crops if necessary from occasional floods. However, most of the valleys included some parts set up for floodwater farming and other parts arranged for irrigation farming. The system of basins and floodwater farming, initially predominant, coexisted for more than 5,000 years with an irrigation farming system in the Nile Valley. The latter was gradually extended and ended up coming into general use in the twentieth century, after the construction of the reservoir dams at Aswan (see chapter 4).

As we will see in chapter 5, the development of the important hydraulic agrarian civilizations in America, in particular that of the pre-Incan and Incan civilizations in the oases of the Peruvian coastal deserts and the arid valleys of the Andes, presents analogies with the development of the Near Eastern hydraulic civilizations. In the Andes as in the Near East, it took centuries to develop a social and political organization capable of building the hydraulic installations, maintaining them and also ensuring the coordinated management of the water. Some convergences between these civilizations, which were formed independently of one another thousands of kilometers and many centuries apart, are bound to be unsettling.

## Deforestation and the Development of Systems of Cultivation with Fallowing and Associated Animal Breeding in the Temperate Regions

In the hot temperate regions of the Mediterranean area, deforestation led to the formation of grasslands, steppes, and moors with more or less clumped vegetation (bush and scrub land). Hilly lands subjected to heightened erosion became uncultivable and were then reserved for pasturage, while cereal crops were concentrated in low-lying zones, which benefited from more substantial deposits of sediments. From that point on, these crops alternated with grassy fallowing of short duration, forming a generally biannual rotation.[3]

But in order to carry out this type of rotation, farmers had to use tools that made it possible to clear the grass cover of the fallow land and develop a new method of renewing fertility. The question of clearing was resolved by the adoption of new manual tools, the spade and hoe, and of a light, animal-drawn implement, the ard, that was borrowed from the ancient hydraulic civilizations of the Near East (see chapter 6). The reproduction of fertility was undertaken by the development of pastoral animal breeding that exploited the peripheral pasture lands. Part of the animal manur went onto the fallow land. Thus agrarian systems based on cultivation using an ard, fallowing, and associated animal breeding were formed in the hot temperate regions (chapter 6). The performance of these systems, limited by the dryness of the climate and by erosion, was subsequently ameliorated in various ways: by terracing the slopes, which makes it possible to enlarge the area of cultivable lands, greatly reduced by erosion; by arboriculture, because trees with deep roots suffer less from dryness than annual plants; and by the development of irrigation. Systems based on fallowing and cultivation with an ard were also extended to cold temperate regions after their deforestation. In these regions, which suffer less from dryness and erosion, the performance of these systems is strongly limited by the cold, the interruption of plant growth, and the lack of forage in winter. The latter condition limited the size of the herds and thus resulted in an insufficient supply of animal manure. This is a problem that was resolved only in the Middle Ages, by the development of cultivation with a heavy animal-drawn plow (chapter 7).

## The Change into Savanna and the Formation of Systems of Cultivation Using the Hoe in the Tropical Regions

In the intertropical regions, deforestation led to the development of predominantly grassy plant formations, sometimes coexisting with relics of the forested environments. These grassy formations, which range from the savanna with tall grasses to the broken steppe, may include scattered bushes, shrubs and some trees. The pursuit of agriculture in these new environments supposes that the double problem of clearing the grass cover and renewing the fertility of cultivated lands had been resolved. In these "savanna" regions, the word being taken in the largest sense, these problems were surmounted by the development of a great variety of systems of cultivation using a hoe. These systems can be classified into five main types:

1. Systems of cultivation with a hoe, without animal breeding, using mounding, ridging, and possibly burning
2. Systems of cultivation with a hoe, without fallowing, and with associated animal breeding, in high-elevation tropical regions
3. Systems of cultivation with fallowing and associated animal breeding of the Sudanese and Sahelian regions
4. Systems associating cultivation, animal breeding, and arboriculture for producing fodder
5. Mixed systems, of savanna and forest

### Systems of Cultivation Using a Hoe: Mounding, Ridging, and Burning

There are equatorial savannas practically devoid of livestock, as for example the vast savannas of Central Africa which are isolated from the important animal herding regions of West and East Africa by a relatively impenetrable encircling forest. The large metal hoe makes it possible to clear the ground cover's dense roots after burning the tall grasses of these savannas in the dry season. Then it is possible to practice temporary cultivations alternating with a grassy idling of medium duration (four to seven years). But the soils of these well-watered savannas are often not very fertile because of the rapid mineralization of organic matter, brush fires, and intense leaching. The small amount of not–yet mineralized raw organic matter and non-leached fertilizing minerals that these soils contain are concentrated in a superficial layer of a few centimeters on top of thick layers of sterile sands.

In order to cultivate savannas of such low fertility successfully, the farmers of the Congolese and Zairean plateaus break up the top layer of the soil into small clods of earth that they gather and pile into mounds or ridges around one-half meter high and one meter long or more. The cultivation bed thus formed concentrates into a small area all the fertile soil scraped from the top layers of a vast area of savanna. This cultivation bed is well suited to a crop that requires few inputs over a long period of time, such as manioc, which can benefit over two or three years from the decomposition of the organic matter. But to cultivate more demanding plants over a shorter period of time, such as maize or potatoes, a portion of the minerals contained in the organic matter is exploited more rapidly by burning these mounds slowly with a covered fire. This technique, which was formerly practiced in Europe,[8] is today practiced in various savanna regions of Africa and Asia. Moreover, some populations of the Congolese plateaus, such as the Koukouyas, who practice mounding, ridging, and burning, refine the process up to the point of reconstituting small islands of forest in order to cultivate with slash-and-burn techniques special crops such as coffee, cacao, palm oil, etc. In order to do that, they move their villages every thirty or forty years, abandoning the circle of aging garden-orchards with enriched soil, on which a secondary forest can then develop.[9] These clever expedients allow the Koukouyas to prolong the exploitation of a savanna no matter how infertile.

Even if there are some savannas without livestock, most of the intertropical savannas are, however, exploited as pasturage. In this case, animal manure is generally used to ensure the renewal of fertility on cultivated lands. This combination of agriculture and animal breeding can be organized in very different ways depending upon the region.

### Systems of Cultivation Using a Hoe, without Fallowing and with Associated Pastoral Animal Breeding, in High-Altitude Tropical Savannas

One system of this nature exists in the high altitude savannas of the Great Lakes region of Africa (Rwanda, Burundi, etc.). In these hilly regions, the settlements are dispersed over the hills. Each settlement is surrounded by an enclosure, where the cattle that pasture during the day in the surrounding savanna are penned at night and deposit a portion of their manure. The fields cultivated with a hoe are grouped below each enclosure. They are almost constantly cultivated during the long rainy season. Moreover, manioc, sweet potatoes, and bananas continue to occupy the land during the short dry season. There is no fallowing, which would make it possible to collect the animal excrement directly onto the cultivated land, as in the systems of fallowing in the temperate regions. The crops, however, benefit directly from animal manure because the latter is transported by the water streaming from the enclosures higher up or even collected each day and carried to the crops by hand or in baskets. This rapid recycling of animal excrement is well adapted to the climate, topography, and the lack of means of transport.

It is doubtful, on the other hand, that saving the animal manure and mixing it with crop residues into compost heaps, as is sometimes imposed by extension services inspired by European models, is effective. In reality, in such a climate and in such a system the crop residues are rapidly decomposed and recycled by the current crops, so putting these residues into compost heaps is only occasion for losses of fertility through leaching and denitrification, and a considerable excess of work. The tools, the topography, and the climate of Rwanda and Burundi are after all not those of northwest Europe. In the same way, it is easy to understand that where crops overlap or follow one another quite closely on every small parcel, attempts to introduce the plow generally fail. The cart, which these artisans and peasants do not have the means to make or buy, would provide a great service by transporting wood and forage and delivering the harvests. In order to be convinced of this, it is sufficient to see the pathetic implements (wheelbarrows and bicycles with entirely wooden wheels, made with the means at hand) with which agricultural produce is transported locally, not to mention the still quite widespread use of human transport.

Lastly, in these highly populated regions, where people and settlements are still multiplying, it is notable that cultivated land is expanded to the detriment of pasturage and cattle, while the development of perennial plants such as the banana and other nutritive and fodder trees makes it possible to maintain production and ensure the renewal of the fertility of cultivated land. New systems combining annual crops with arboriculture have been introduced. These types of garden-orchards form, in fact, a forested ecosystem, completely domestic and productive, and endowed with the same capability of reproducing its own fertility as a forest. Equivalent garden-orchards are found in other heavily populated tropical regions (Haiti, Yucatan, Southeast Asia) and formerly in the Mediterranean region.

## Systems of Cultivation with Fallowing and Associated Animal Breeding in the Sudanese and Sahelian Regions

A second system of cultivation with associated animal breeding is found in the Sudanese and Sahelian regions, which have distinct rainy and dry seasons and where deforestation led to the formation of uncultivable savannas and steppes on the most eroded and degraded soils, which can only be used as pastures. The cultivated lands are thus concentrated on the deepest soils, in the midst of which the settlement is located. In these regions, the relatively lengthy dry season forces an interruption in cultivation for several months. During this period, the village's animals, possibly joined by transhumant herds belonging to shepherds from the north, are led each day to nearby pasturage. In the evening, they are led to the fallow lands where they leave their manure. On the other hand, in the cultivating season, most of the animals are taken away to pasturage situated some distance away. Those animals that remain near the village are penned at night close to the dwellings. The earth from these animal pens, mixed with manure from the animals, is then transported to the closest cultivated lands.

We note that these systems of cultivation with fallowing and associated animal breeding strongly resemble their homologues in the temperate regions. But here, the rotation is annual and the dry season fallowing lasts only a little more than six months, while in the temperate zones the rotations (biannual or triannual) include a long fallowing of more than one year. This difference can be explained by the fact that in the Sudanese and Sahelian zone, a long fallowing of eighteen months would not be effective. Animal manure gathered at the beginning of a fallowing that long would be quickly mineralized due to the heat, and the soluble minerals thus produced would be leached during the rainy season, which occurs between two seasons of cultivation.

## Systems that Combine Cultivation, Animal Breeding, and Fodder Arboriculture

In the Sahelian zone, however, the dry season pastures are often insufficient to feed the livestock and manure the cultivated lands. Thus a third system consists of preserving or planting scattered trees on these lands, which, by completely drawing out the minerals, produce a supplemental biomass contributing to the renewal of fertility. Several species of trees can be used, but the most interesting is undoubtedly *Acacia albida*, a legume that enriches the soil with nitrogen and produces quite opportunely an abundant fodder in the dry season. This providential tree is fond of deep and well-drained alluvial soils of ancient valleys and fossil dunes. Sometimes spontaneous and protected, sometimes introduced deliberately as well, *Acacia albida* forms an integral part of a whole series of systems that combine cultivation, pastoral animal breeding, and fodder arboriculture, extending along the Sahelian rivers of the desert between Senegal and Sudan.

## Mixed Systems of Savanna and Forest

In the wooded savannas, which incorporate remnants of cultivable residual forests, like those, for example, in the region of the Pool in the Congo, mixed systems are generally found, or more exactly, composite systems. These consist of cultivation using a hoe, with or without animal breeding, on the grassy parts and slash-and-burn cultivation in the wooded parts.

## Development of Systems of Wet Rice Growing

In the humid tropical regions, which receive several meters of rain per year, flooding rivers, water from runoff or even rainwater, periodically submerge valleys and low-lying areas. It is in this type of environment that wet rice—rice that sprouts in flooded land—began to be cultivated more than 6,000 years ago in several monsoon regions of Asia, from India to southern China. The cultivation of Asiatic rice (*Oryza sativa*) next spread to all of the tropical and subtropical regions of Asia, and then to the hot temperate regions of Asia, Europe, and America. On the other hand, around 3,500 years ago, another species of rice of African origin (*Oryza glaberina*) was domesticated in the central delta of the Niger. Numerous varieties of this species were then cultivated in the valleys of the Niger, Senegal, Gambia, Casamance, and on the Guinean coast.[10] Furthermore, "dry" rice growing also developed in wet tropical regions, but it hardly went beyond the stage of slash-and-burn cultivation and its importance remains secondary. Wet rice, on the contrary, due to the progress of hydraulic installations, agricultural practices, and different varieties, experienced an immense development making it, next to wheat and maize, one of the three most consumed cereals in the world. Nearly one half of humanity eats it every day.

*Natural Lakes.* At the beginning, wet rice was cultivated in areas that were naturally submerged several months per year. Varieties of floating rice are particularly well adapted to these lakes, which have an uncontrolled and thus variable level. In fact, their stalk can grow by several centimeters per day as the water rises, reaching four or five meters in length and curling over when the water level falls.

*Rice Paddies.* Apart from natural lakes, wet rice growing expanded by the establishment of artificial lakes, due to the construction of small basins, or rice paddies, formed from a piece of relatively flat land surrounded by an earthen dike of a few dozen centimeters in height. Beyond simple isolated paddies, the complete development of more widespread land took the form of a grid pattern of dikes separating contiguous paddies, with naturally flat or leveled bottoms, spread out along the contour lines. These constructions were at first built on easily drained high ground (such as piedmont and interfluve areas) in regions where rainwater was sufficient to fill them for the time required for rice cultivation. In these high rainfall zones, wet rice growing can be a rainfed crop (non-irrigated). In order to control the water level in the paddies, it was sufficient to drain the excess water from paddy to paddy, from the higher to the lower, by breaching the dikes at the desired height. Possibly, the excess water was captured by collecting canals, which carried it toward a natural outlet. From that time on, it was possible to cultivate non-floating rice that was less tolerant of water-level variations.

*Terracing of Slopes.* The spread of paddy rice growing to hilly piedmont areas and the slopes of high valleys took place through the construction of step-like terraces, which could be stretched out along long contour lines. This type of monumental and quite spectacular construction spread gradually to the mountainous regions of the Philippines, Indonesia, China, Vietnam, and elsewhere.

*Developing Flooded Valleys and Deltas.* The conquest of low valleys and deltas required another type of hydraulic architecture. On these relatively flat and frequently flooded lands, even before laying out a grid of rice paddies, it was necessary to guard the cultivation area against floods by erecting high dikes around the riverbed and its branches. It was also necessary to dig a network of slightly sloping, long and wide canals to drain excess water at the right time. Lastly, it was sometimes necessary to build dikes to protect against tides in the lower part of deltas facing the sea and along the branches of the river. The complete development of these vast geomorphological areas necessarily took much time. It began in zones that were the easiest to protect and drain, situated above and on the periphery of these valleys. Then it was extended gradually to the lowest parts.

In these low valleys, the rice paddies could be fed by rainwater and by the more or less well-controlled spreading of floodwaters. The insufficiently drained lowest parts were only cultivated during the low water season, while the insufficiently irrigated highest parts were only cultivated during the rainy and high water season. Only the intermediary zones, well supplied with water in the low season and well drained in the high season, could support two harvests of rice per year. In other words, the development of systems with two or three harvests per year in the low valleys was at first conditioned by the advances in irrigation and drainage works, making it possible to control the water level in all seasons.

*Irrigation, Extension of Rice Growing, and Multiplication of Harvests.* However, irrigation not only made it possible to multiply the number of harvests in the wet tropical regions, it also allowed rice growing to be expanded into subtropical and hot temperate (Mediterranean) regions where rain and floods are insufficient to practice wet rice growing.

In valleys and deltas, irrigation is not always easy because, in the low season, the water in the rivers and canals is generally at a lower level than the water in the paddies. Thus it is necessary either to raise the water using elevators operated by humans, animals or motors, which is costly, or construct a vast irrigation network beginning far upstream and lead the water to the paddies through canals that are elevated above the rice-growing plain. Today pipes under pressure are often used.

In different ways, these stages of development of irrigation made it possible to expand wet rice growing to more varied lands and climates, lengthen the cultivation season and multiply the harvests.

*Transplanting, Animal Traction, Selection and Multiplication of Harvests.* Many other improvements contributed to this tremendous development of wet rice growing. Transplanting rice, previously sown and raised in small nurseries, shortened the time needed in the rice fields and made it easier to increase the number of annual harvests. The use of animal and later motorized traction to plow, mix, and level the soil before transplanting made it possible to save precious time. Lastly, selection of non-photoperiodic varieties (not as sensitive to the relative length of day and night and thus cultivable in all seasons in various parts of the world) and of varieties with a short reproductive cycle made it possible to achieve more than three rice harvests per year.

From natural lakes to huge installations in valleys and deltas, a whole range of hydraulic systems were created, combining in diverse ways rice paddies, terraces, dikes, locks, diversion dams, reservoir dams, irrigation canals, and draining canals. The architecture of the large hydraulic works of the rice-growing regions was different from that found in the large valleys of the arid regions. But the works were of comparable scale and gave rise to comparable forms of social and political organization (see chapters 4 and 5).

Notice, however, that the great aquatic rice-growing civilizations of monsoon Asia began to develop more than two millennia after the hydro-agricultural civilizations of the Indus, Tigris, Euphrates, and Nile Valleys. In China, the very first hydraulic city-states appeared in the second millennium B.C.E., in the middle Yellow River region situated close to the Chinese center of origin. These cities were united into the first embryonic empire under the Shang Dynasty (seventeenth to eleventh centuries B.C.E.). However, historians speak of a true wet rice-growing civilization beginning only in the following period (eleventh to third centuries B.C.E.), during which ten hydraulic and wall-building kingdoms were formed and fought one another until the most powerful among them, the Qing (from 249 to 206 B.C.E.) imposed its supremacy and administration to all of China, from the Great Wall to Canton.

In India, while rice was cultivated in the east since at least 2000 B.C.E., the first hydro-agricultural civilization of the middle Ganges Valley did not appear until around 800 B.C.E. The emergence of this civilization followed the Aryan penetration which had begun several centuries earlier, around 1,500 B.C.E. Coming from the north of Iran where their herds exploited the steppes, which were relatively unproductive and not very propitious for cultivation, tribes of Aryan herders had previously invaded the Indus Valley where they had, it is thought, precipitated the collapse of the preexisting large hydraulic cities (Mohenjo-Daro and Harappa). Then, traversing the Punjab, they colonized in successive waves the quasi-intact great forests of the Ganges Valley and north-west India, still occupied by communities of hunters and fishers who occasionally practiced a temporary form of slash-and-burn agriculture. As a result, the immigrants were compelled to abandon pastoral nomadism, become settled, adopt the cultural complex of the humid tropical forest (including *Oryza sativa*) and, over several centuries, extend their clearings before reaching the end of the arboreal ecosystem. After that, the first post-forest agrarian systems and first hydraulic city-kingdoms of the Ganges Valley were formed. In the sixth century B.C.E., one of these kingdoms (the Moghada) began to subjugate and unify its neighbors forming, in the fourth century B.C.E., an empire that occupied at first the whole Ganges Valley and which, two centuries later, extended from the Indus to the Gulf of Bengal, and from the Himalayas three-quarters of the way down the Deccan peninsula.

During the first millennium C.E., a whole series of other hydraulic and rice-growing city-states were formed, either autonomously or by expansion, in the Indochinese peninsula, Japan, Indonesia, and even Madagascar. If China exerted a technological and commercial influence over most of these civilizations, India often provided them with certain cultural elements (writing, religion, art, government, and administration).

## 5. PROBLEMS OF DEVELOPMENT IN TODAY'S FOREST AGRARIAN SYSTEMS

In most of the formerly forested regions of the world, systems of slash-and-burn agriculture were superseded a long time ago by all sorts of other systems: hydro-agricultural systems in arid regions, systems with fallowing and associated animal breeding in temperate regions, systems of cultivation using a hoe with or without animal breeding in tropical regions, wet rice-growing systems in humid tropical regions, and so on.

Nevertheless, systems of slash-and-burn agriculture continue to exist in still-standing intertropical forests today. But because of poor tools and low productivity, these systems are today threatened by economic competition from stronger agricultural systems. Moreover, their very existence is called into question by the rapid advances of deforestation. The question of their survival and improvement and of the development of post-forest systems capable of replacing them is thus posed in a pressing manner. The knowledge gained in this chapter should make it possible for us to analyze the current problems of forest systems and to devise development strategies that suit them.

## Current Problems

### *Lack of Equipment*

The first handicap of slash-and-burn agriculture systems arises from the rudimentary character of their poorly diversified and weak tools. For each farmer, the equipment consists of no more than an ax, a machete, and some hoes, generally less than US$40 in value. Such tools hardly make it possible to cultivate more than one hectare per worker, and, despite the fertility of forest soils, the productivity of labor rarely rises above 10 quintals of cereal-equivalent per worker per year, or hardly more than the needs of a family of four or five persons. The result is a law marketable surplus and monetary income. Increasing the manual implements to include axes, hatchets, brush hooks, saws, hoes, weeding hoes, dibbles, dryers, shellers, or hullers, etc., would make it possible for the forest farmers to double their labor productivity. Such a tool set is worth five to ten times more, but an investment of this size is beyond the reach of most farmers.[11] However, these systems suffer additionally from many other handicaps.

### *Dispersion*

The frequent dispersion of the cultivated parcels, which obliges the producers to make long daily trips, is another factor limiting the productivity of labor. The temporary character of the cultivated lands makes the building of stable country roads suitable for motor vehicles prohibitive in practical terms. Moreover, because of the dispersal of the population and produce, the cost of the infrastructure and transportation serving these regions is high. Thus it is not possible to promote heavy, bulky, or perishable marketable products nor to set up large processing plants with a large collection radius. And it is contraindicated to develop, as is often done, technical services and concentrated, specialized supply and collection services, because the personnel and vehicles clearly lose time in traveling. However, it is even worse to consolidate villages along the service roads forcibly, which leads to deforestation in the surrounding area, simply to avoid the disadvantages of dispersion.

## Difficulties with Mechanization and Synthetic Fertilizers

Moreover, forest agriculture lends itself poorly to the adoption of the most conventional—and for "developers" most tempting—means of agricultural development: mechanization and use of fertilizers. The use of animal-drawn or motorized equipment is nearly impossible, unless it is for removing the tree stumps and thus destroying the forest and forest agriculture itself. The experiments with moto-mechanization of agriculture after complete clearing by bulldozer failed. This method of clearing partly scrapes away the most fertile top layers of the soil, gathering them into huge piles that remain uncultivated, while, after several years of costly cultivation, the meager residual fertility of mechanically cleared lands collapses, and a quasi-sterile savanna is established in the long term. As for mineral fertilizers, these are generally not profitable. Their supply cost is high, and they are often quickly leached after spreading. Their effect is less noticeable in soils constantly supplied with minerals through decomposition of forest humus.

## Deforestation

Sooner or later, the threat of deforestation is added to these handicaps and barriers to improvement. Mechanical clearings, the expansion of large plantations and livestock ranches, and the overexploitation of tropical woods consistently reduce living space for forest farmers, while, as we have shown, the demographic explosion entails, in less than one generation, the accelerated conversion of the cultivated forests into savannas.

## Development Strategies

The protection and continuation of systems of slash-and-burn agriculture implies, first of all, that the forest regions of peasant agriculture be protected from the growing ascendancy of agricultural, livestock, and wood extraction industries that do not belong there. The development of these peasant agricultures also implies that their own pressure on forest resources be reduced and that the productivity and income of forest farmers be significantly improved.

## Improvement of Equipment, Perennial Plantations, and Garden-Orchards

In these conditions, the imperative direction of development in the short term is to help forest farmers acquire more effective manual equipment, through loans, subsidies, or, even better, by donations in kind. In fact, the development of compact and well-maintained commercial plantations (coffee, cacao, palm oil, etc.), preferable to wide-open, so-called nonintensive plantations, which occupy too much land, is dependent on the improvement in equipment. These plantations can provide peasants with the monetary income they need, when well served by small family or village processing facilities and by marketing and comprehensive training services.

However, in the long term, deforestation is undoubtedly unavoidable and it is advisable as of now to orient research and development toward implementing new, appropriate forms of cultivation and animal raising, capable of replacing slash-and-burn agriculture when it finally fails. Forms of agriculture and animal raising that have already succeeded in comparable regions of the world, and on which this research and development should be based, are organized around three main axes:

1   The first axis consists, as we have seen, of gradually replacing the soon-to-be destroyed forest with entirely human-made productive plantations and garden-orchards, such as those in Southeast Asia, Central America, and the Caribbean.

2   The second axis rests on the development of savanna systems that closely combine cultivated crops with the breeding of small and large livestock. The animals consume the crop by-products, graze on the pastures in the deforested and noncultivated parts of the ecosystem, contribute to the renewal of the fertility of the cultivated lands, and participate in agricultural labor.

3   The third axis consists of developing the resources of relatively impermeable and wet valleys, until now little or unexploited, by employing different forms of hydro-agriculture and possibly aquaculture. Initially, these new systems will complement the food-producing output of declining cultivated forest lands and, if suitable shallow water areas are adequately expanded, could even replace the latter. A suitable shallow water area for wet rice growing and fish farming could feed several hundred inhabitants per square kilometer, or at the very least ten times more than slash-and-burn agricultures.

## Short-Term Preservation and Improvement of Forest Systems

Since slash-and-burn agriculture is, as we have seen, not amenable to improvement, it is only in the short term that these strengthened forest systems would have to be implemented, thereby giving animal-drawn cultivation, mechanization, and motorization time to develop. In the meantime, it is preferable to avoid harmful technological transfers. But it should be remembered that the development of post-forest systems will take time. It will require labor, equipment, and substantial investments. This agricultural transformation will only be possible if systems of slash-and-burn agriculture are protected right now from being destroyed too rapidly and if they are adequately strengthened. The short-term improvement in equipment, productivity, and income of forest farmers determines, in the long term, the development of post-forest systems that are sustainable and capable of development.

# 4

# The Evolution of Hydraulic Agrarian Systems in the Nile Valley

At the same time that it discovered and improved its agricultural technology, humanity had to control water, to struggle against an excess that was as harmful as a scarcity, to force back swamps as well as the desert by digging and maintaining a network of drainage or irrigation canals, in short, to conquer the land in order to subject it to a disciplined fertility. ... In Egypt and Mesopotamia, the joint influence of three factors was felt then: natural conditions, of course, but used by a collective organization in close connection with religion. How do these last two factors, both human, appear and how are they generalized to the point of acquiring such force? That is the great mystery, probably always unfathomable, because the birth of a religion remains irreducible to the conviction of material utility. And this conviction is not enough to account for the lasting acceptance, by the masses, of a sometimes very heavy burden.

—ANDRÉ AYMARD, *L'Orient et la Grèce antique*

One of Neolithic agriculture's most ancient centers of origin, the Near Eastern center of Syria–Palestine, was formed 8,500 years ago. At that time, most of the cultivated plants (spelt, wheat, barley, peas, lentils, flax) and livestock (goats, pigs, sheep, cattle) originating in this center were already domesticated. Starting from there, populations of Neolithic farmers and herders gradually spread these domestic species in all directions and into the most favorable environments over several millennia. New species were later added to those originally domesticated, in the Near Eastern center itself or in its area of expansion (the donkey in the Near East, oats and rye in Europe, sorghum, millet, earth peas, yams, rice, in tropical Africa, etc.).

Thus, 5,000 years ago, when Neolithic agriculture of Near Eastern origin had scarcely reached the Atlantic, the North and Baltic Seas, Siberia, the Ganges Valley, and the great equatorial forest of Africa, the regions closest to the center, in western Asia, eastern Europe, and northern Africa, were long cultivated and traversed by herds, to the point that the Saharan and Arabo-Persian regions with the least rainfall, originally occupied by deciduous forests, savannas, or wooded steppes, were already deforested. As a consequence, no doubt, they were also in the process of drying up.

In these dry regions, rainfed agriculture gradually became impossible and pastoral activities decisively regressed. Farmers and herders then slowly flowed either toward peripheral regions that remained more humid or toward privileged areas that were well supplied with water from subterranean groundwater or from rivers of distant origin. In these green oases lost in the middle of the desert, they developed diverse forms of hydro-agriculture: cultivation based on receding floodwaters, on irrigation, or on surface groundwater. The valleys of the Tigris, Euphrates, Nile, and Indus, in which the extension of cultivated lands demanded vast hydraulic installations, formed the largest of these oases. The first great hydro-agricultural civilizations of early antiquity were born in this context.

The object of this chapter is to retrace the emergence and development of hydro-agricultures in the Nile Valley, from the sixth millennium B.C.E. to the present. Broadly speaking, it can be said that two main types of agrarian systems developed, existed side by side, and succeeded one another in this valley: systems of winter cultivation based on receding floodwaters and systems of irrigated cultivation in different seasons.

## Systems of Winter Cultivation Based on Receding Floodwaters

The Nile, a river of equatorial origin but largely supplied by tropical rains in the northern hemisphere, formerly overflowed its banks each year between July and October. The flood covered most of the valley and delta for several weeks, with the exception of levees and natural promontories. The height of the water was variable depending on the place and the size of the flood, but it could reach several meters. The crops were sown after the retreat of the floodwaters, when the soil was saturated with water and enriched with alluvia, and harvested in spring. Cereal crops (wheat, barley, millet in the south) and flax, which require lots of mineral elements, alternated with food legume crops (peas, lentils) or fodder legumes (Egyptian clover), which enrich the soil.

Beginning in the sixth millennium before the present, systems of floodwater cultivation were expanded in several steps, in conjunction with the development of basins to hold the receding floodwaters. At the time of the first villages, it can be assumed that, without the development of basins, only the fringes of the flooded zone were cultivated after the retreat of the water. In the next time period, rudimentary basins to hold the receding waters were built by erecting simple dikes that enclosed noncontiguous natural depressions along the border of the flooded zone. These dikes first made it possible to retain the water as long as necessary to moisten and deposit alluvium in the soil and then to protect the contents of these basins against the possible return of the floodwaters. In the following time period, the construction of transversal series of basins, terraced from the riverbanks to the edge of the desert, then the construction of longitudinal series of basins, terraced from upstream to downstream, made it possible to develop increasingly wider sections (in fact, half-sections) of the valley.

Finally, the gradual construction of large protective dikes along the river and large feeder canals or sluices, gradually linking series of basins from the upper valley, middle valley, and delta, made it possible to distribute insufficient floodwaters more equitably and also to absorb excessive water by spreading it out. Large feeder canals moreover made it possible to carry floodwaters onto "new lands," rarely or even never reached by natural flooding. These large hydraulic works led not to an integrated set of installations for the valley and delta and a unified management of the flood but to a set of linked local and regional installations with coordinated management of the flood, thanks to rules of water use and to a centralized and hierarchical system of control.

Although this is only a hypothesis, neither proven nor shared by all Egyptologists, it is still tempting to think that the important stages in the development of these hydraulic installations and the coordinated management of the floodwaters on increasingly more extensive parts of the valley coincided with stages in the development of increasingly more powerful forms of social and political organization, capable of extending their hydraulic power to corresponding territories: villages strung out along the valley and on the fringes of the delta; rudimentary city-states dominating a small section of the valley, then, toward the middle of the sixth before the present, more powerful city-states dominating an entire alluvial plain ranging between two narrow passages in the valley; large kingdoms that unified several cities and dominated several alluvial plains, then, in the second half of the sixth millennium, two kingdoms (Upper Egypt corresponding to the valley properly speaking and Lower Egypt corresponding to the delta); finally, a little more than 5,000 years ago, formation of the pharoanic state that unified these two kingdoms. After that, over the next 3,000 years, some 200 pharaohs belonging to thirty dynasties reigned more or less completely over these two kingdoms. The prosperous periods (Old Kingdom, Middle Kingdom, New Kingdom) coincided with a strong concentration of power, and the decadent periods (intermediate periods, Late Period) coincided with a weakening and breaking up of central power.

Whatever the case may be, these very ancient systems of winter cultivation based on receding floodwaters were built around *collective* hydraulic installations, made up of sets of basins, dikes, and canals built, maintained, and used under the aegis of hydraulic authorities who operated, depending on the time

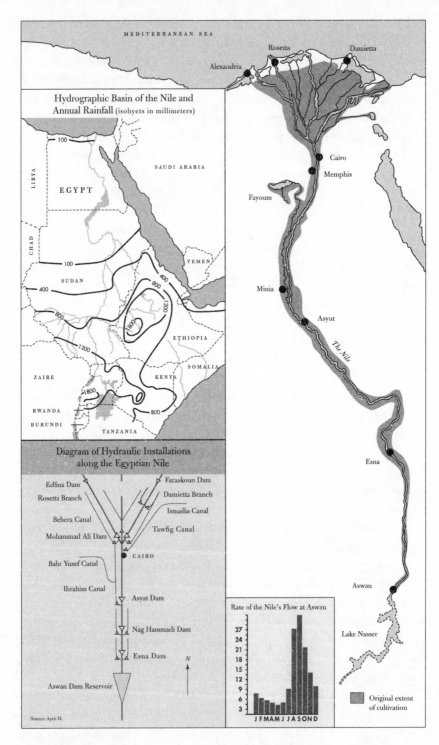

Figure 4.1 Egypt and the Nile

period, at the level of village, local city, or kingdom. A close-knit peasantry, grouped into villages situated on promontories, levees, and dikes, carried out these systems of cultivation, for which they were granted plots of land. Further, they were subjected to heavy forced labor on estates belonging to the state, temple, and high dignitaries. Products from these estates and a tax in kind were used to support the needs of the pharaoh, his palace, administration, clergy, soldiers, state workers, and artisans and to construct palaces, temples, tombs, and pyramids. But, in large part, they were also set aside as security reserves in order to be able to deal with any irregularities in the flood or harvests, as well as support the expansion and maintenance of hydraulic installations and other works of public utility.

## Systems of Cultivation Based on Irrigation in Different Seasons

Irrigated cultivation, whether by water drawn from the river, lakes, and other temporary or permanent bodies of surface water or even drawn from shallow subterranean groundwater, is as old as cultivation based on floodwaters. At the time of the first villages, city-states, and pharaohs, manual irrigation with pottery jars could not be extended beyond the immediate vicinity of sources of water. Beginning in the fourteenth century B.C.E., however, irrigated cultivation took over more land thanks to the adoption of the *shadouf*, of Mesopotamian origin.[1] But irrigated cultivation developed above all after the Greek conquest (333 B.C.E.), thanks to the use of new, much more effective machines for drawing up water: the Archimedian screw and the bucket wheel (*saqiya*) which, in antiquity, were generally operated by slave labor. In the Middle Ages, notably during the Arab epoch, irrigated cultivation developed due to the growing use of animal traction, windmills, and water mills to operate these machines, bucket wheels in particular.

In contrast to cultivation based on floodwaters, which was always carried out in winter, irrigated cultivation could be practiced in different seasons depending on the situation: at the end of winter and in spring (between two floods) in low-lying, floodable zones; in summer and autumn (during the flood) on raised ground; in all seasons (so-called perennial irrigation) in zones protected from the flood by natural or human-made levees. Until the end of the eighteenth century, systems of irrigated cultivation were essentially based on private investments (wells, machines to draw and raise water). In these conditions, it remained confined to the delta, depressions close to sheets of surface water or groundwater, and low riverbanks, which were no more than 20 percent of the total cultivated area.

Beginning in the nineteenth century, irrigated lands began to benefit from public hydraulic installations. The development of year-round irrigable areas made it indeed possible to plant sugarcane (a perennial crop) and above all cotton (a crop that occupies the land from the end of the winter to autumn). These two tropical export crops were sources of profit, foreign currency, and raw material for industry, which, in turn, inspired the policies of redeveloping the valley in order to extend irrigation. In the first half of the nineteenth century, the reconstruction for irrigation of old canals that had been used for distributing floodwaters was unsuccessful. But, at the end of that century, the construction on the arms of the delta of dams for raising the water level allowed irrigation to be extended to nearly the whole of the delta.

In the first half of the twentieth century, with the construction of the reservoir dams at Aswan (the first dam was built in 1905), situated very far upriver, and the dams for elevating the water level in Middle and Upper Egypt, irrigation could be extended to the whole valley. Lastly, in the second half of the twentieth century, the construction of the Aswan Dam made it possible to enlarge the cultivable domain, generalize irrigation to all seasons and develop double and even triple annual crops, as well as perennial plants. Numerous and talented, the Egyptian peasantry has been able to take advantage of these new possibilities, while adopting fertilizers, treatments, and, to a lesser extent, motomechanization in order to develop complex systems of production, combining cereal, fodder, and animal products as well as vegetable and fruit products.

Why and how, in a land not very distant from the civilizations north of the Mediterranean, and more than two thousand years before the latter's emergence, did several million peasants and government officials, confined to a microcosm exploitable only with the support of a large number of hydraulic installations, build a succession of hydro-agricultures and social and political organizations of such inexhaustible originality and richness? These are the basic questions to which we try to respond in this chapter.

## 1. THE ORIGINAL ECOSYSTEM AND THE FIRST OCCUPANTS OF THE VALLEY

### The Formation of the Egyptian Desert

Outside of the Nile Valley, Egypt is today a desert sprinkled with a few oases. But 10,000 years ago, the two plateaus and the landscape framing the valley were still occupied by a shrubby savanna, composed of various grasses and thorny shrubs, while the valley and the delta, outside of swamps populated by papyrus, reeds, and other aquatic plants, was occupied by a type of gallery forest of tamarisks, acacias, date palms, doum palms, sycamores (fig trees with soft and rotproof wood), and terebinths (pistachio trees providing resin). Beginning in the sixth millennium before the present, the climate dried out and it became desert-like toward the middle of the fifth millennium. Today, in Upper Egypt (from Sudan to the point where the water of the Nile disperses into the delta), the climate is very hot and dry. Aswan, for example, receives only 3 millimeters of rain per year and the average monthly temperatures vary from 15 to 33°C. In the delta, it rains more (24 millimeters of rain in Cairo, 190 millimeters in Alexandria), and the temperatures are much lower. The climate becomes more Mediterranean closer to the littoral.

## An Oasis in Winter Produced by a Flood in Summer

The Nile Valley appears as a long, threadlike oasis of more than 1,200 kilometers in the middle of the Egyptian desert. The Nile, whose source is in Burundi, south of the equator, 6,700 kilometers from its mouth, receives tributaries from equatorial and tropical regions. It was the waters from the tropical regions, coming from Ethiopia in particular (Blue Nile, Atbara, and Sobat), which supply most of the flow of the river and which, until the construction of the Aswan Dam in the twentieth century, were the origin of the summer flood. The flood began in mid-July and peaked in September. The whole valley was then submerged in red water, rich in silt taken from the catchment area of the river. In autumn, the waters of the While Nile coming from equatorial regions took over and sustained the flow of the river. From mid-November, however, the river returned to its original channel and its flow decreased until May, the month with the lowest water level.

The Nile carries some 80 billion cubic meters of water per year on average, but this volume is quite variable from one year to another (from 44 to 129 billion cubic meters). During an average flood, the river level at Aswan attained (before the construction of the high dam) a height of 9 meters above the low-water mark. The flood spread more broadly in the delta, where the variation in the water level did not go over 3 of 4 meters. The flood maintained the groundwater, added to the water in the soil, and deposited a thin layer of silt each year. The sedimentation of this silt over millennia, at the rate of 1 millimeter per year on average, formed a slightly convex alluvial plain at the bottom of the valley, whose width varies from a few hundred meters to ten kilometers depending on the place. The soil texture is much finer the farther one is from the riverbed or the closer one approaches the delta. Thus in the delta, which is 250 kilometers wide, the alluvia supplied by the different arms of the river are often clayish, too. The accumulation of larger materials on the edge of the river formed two bulging riverbanks longitudinally. Although gentle, the transversal incline of the valley bottom on both sides of the river, starting from the bulging banks, is generally greater than the longitudinal incline.

After the floodwaters receded, the vegetation became abundant for several months. Then, in spring, the lowering of the groundwater and the drying out of the soil extended to almost all of the valley, except the low parts of the delta, a few swampy depressions, some permanent sheets of surface water as well as the permanent riverbed and the arms of the river that had not dried up.

## The Return of Neolithic Farmers and Herders to the Valley

At the end of the Paleolithic epoch, populations of gatherers (of rhizomatic plants such as the earth nut and the bulrush from marshes), fishers (catfish, tilapias), and sometimes also hunters (aurochs, antelopes, gazelles, hippopotami, onagers, wild boars, aquatic birds, etc.) already frequented the valley.

In the Neolithic epoch, between 10,000 and 5,500 years ago, Egypt went through a relatively wet period, though already interspersed with dry phases. In the course of this period the first herders and farmers appeared. The oldest traces of cereal crops in Egypt go back to some 8,000 years ago. These crops, barley, emmer wheat, and spelt, had been domesticated in the Near Eastern center. Traces are located on the plateaus and are associated with the remains of domestic cattle, which could be descended from cattle that had been domesticated in the Near East around 8,400 years ago.[2] But a few cattle bones were also found on the plateaus, in sites dated from 9,800 to 9,000 years before the present. The small size of these bones leads one to think that they could be from domestic cattle. Some people have concluded that there could exist an African center for domestication of cattle, a center that would then be older than the Near Eastern one.[3] But the existence of this center is not proven, and it is indeed possible that these bones belonged to wild cattle of small size, adapted to a semiarid environment and possibly moving to summer pastures along the edges of the Nile in the dry season. Whatever the case may be, starting from 8000 before the present, the presence of small groups of herders, who were incidentally farmers, migrating between the valley and the Egyptian plateaus is well attested.[4] Much later, from 5500 before the present, the return of Saharan and Near Eastern populations to the valley, driven from all over by desertification, began. Herders (with dark skin) of cattle with short horns came from the southwest, herders (with light skin) of cattle with long horns came from the east, and herders of sheep and goats came from the north. They were the origin of the establishment of more and more numerous villages on the sides of the valley and on the edges of the plateau.[5] These were the villagers who began to clear, develop, and cultivate the valley, thus gradually destroying the original gallery-forest. Lastly, around 4,500 years ago, the plateaus became utterly barren and were no longer occupied by any permanent population.

## 2. BASIN AND WINTER FLOODWATER CULTIVATION SYSTEMS

Neolithic farmers, having taken refuge on the perimeter of the valley and no longer able to cultivate the desert plateaus, were confronted with a distinctive environment. The life of the inhabitants was governed by the rhythm of the three main seasons: that of the flood (*akhet*) which submerged, filled with water, and deposited silt on all or part of the alluvial lands favorable to cultivation (the black lands or *khemet*) for a few weeks between July and October; that of the after-flood, of renewal or resurgence of the lands (*peret*), season for cultivation based on the receding floodwaters of "winter," which occupied the land from November to "spring"; lastly, the dry season (*shemou*), which came to an end with the arrival of the next flood, in mid-July.

The old systems of winter cultivation were an integral part of the seasonal hydrological cycle of the Nile Valley. At the end of October, after the retreat of the flood, the groundwater underlying the valley was fully recharged and rose to the surface of the soil up to the edge of the desert. The soil was then muddy and it was necessary to wait for a few days before sowing. From the end of autumn to the next flood, this groundwater was supplied only by the low water that flowed in the regular bed of the river and its arms. Its level gradually fell to the point that it could no longer supply water to the cultivated fields. As a result, the growth of vegetation ceased at the end of winter or beginning of spring, which happened earlier in Upper Egypt than in Lower Egypt. Then, from spring to the next flood, the soil dried out and cracked to the point of accomplishing a sort of natural plowing.

### Step-by-Step Development of Basins for Holding Floodwaters

#### *The First Villages and Floodwater Basins*

The first villages, made up of small huts of dried mud, were situated on the edges of the desert on both sides of the valley, or in the valley itself but on natural levees to escape the flood. During the flood, the latter appeared as islands emerging from the inundated valley. Hastily protected by human-made levees, these were sometimes completely carried away by large floods.

On the fringes of the flooded zone, bordering the desert and high ground in the valley, the farmers began to plant crops of Near Eastern origin (winter barley, emmer wheat, spelt, lentils, peas, flax, two kinds of vetch), using the corresponding Neolithic tool set (axes of polished stone, blade sickles, sickles fitted with microliths, grinding stones and rollers, pottery). Sowing was done just after the floodwaters receded, the plants grew over the course of the following winter months, drawing on the reserves of water in the soil. Barring exceptions, these crops were not watered and required hardly any other work than watching over the fields in order to protect them, mainly from birds. The harvest took place between March and May.

This cultivation was practiced initially without any preliminary preparation, except for some clearing to the boundaries of the gallery-forest.[6] Yet this way of working presented many disadvantages. On the elevated areas situated along the edges of the flood zone, silting was not very great and the groundwater level fell rapidly. In the lower areas, however, the waters receded slowly and the risk that the flood might return after sowing was very real. It is precisely in order to guard against returning floodwaters and ensure a water supply and sufficient silt in the higher areas that, probably from the beginning of the sixth millennium before the present, the villagers developed the first floodwater basins.

These basins were formed from small natural depressions situated along the edges of the flood zone: localized enlargements of the valley, small valleys formed at the outlet of dried-up, ancient tributaries of the Nile, hollows at the base of natural levees. These basins were easily made: a simple earthen dike made it possible to enclose and isolate them from the flood zone, thereby controlling the rise and fall of the floodwaters in the small basin. In summer, the dikes were pierced with one or several openings to let the floodwaters enter the basins. Then these breaches were sealed so as to retain the floodwaters long enough to allow the silt to be deposited, the soil to be saturated with water and the groundwater to be recharged. In the autumn, new openings were made to drain the floodwaters at the desired time. Then the dikes were sealed again in order to protect the newly sown ground from an untimely return of the floodwaters.

## City-States and the Construction of Basins Made Over Small Sections of the Valley

It can be assumed that the construction of basins from larger parts of the valley began toward the middle of the sixth millennium, under the aegis of the first city-states. Small sections of the valley, or more precisely, semisections, situated on one side or the other of the river, were arranged into a sequence of imperfectly quadrangular basins, separated by dikes and positioned according to the slope of the land. These purposefully laid-out sections of the valley could be protected from strong floods by a longitudinal dike raising the lip of the bank and, if necessary, by a reinforced transverse dike situated upstream.

*Series of Transverse Basins.* These sequences, or series, of basins could be transverse. In this case, the basins were lined up, with a downward gradient, from the ridge on the riverbank and the thick silt in the center of the valley to the edges of the desert. Openings made in the ridge of the bank and emptying directly into the first basin, the highest one, could supply the basins spaced out below with floodwaters. But the order in which these basins were filled was not random: by always beginning with the first basin, the most central, the latter functioned as a settling basin and the silt accumulated there to the point that it placed the basin out of the water during years of small floods. In order to avoid this deterioration of the hydraulic system, the remote basins had to be filled first by beginning with the last one.[7] But the depositing of silt in and the heightening of the remote basins certainly had its limits. Ultimately, it was the entire string of basins that could end up being heightened, to the point of being usable only on the occasion of exceptionally high floods, thereby provoking a crisis for the hydro-agriculture.

The arrangement of the basins and the method of distributing the floodwaters, then, not only organized the annual distribution of water, they also governed the spatial distribution of the silt and, in the end, its differential accumulation and, consequently, the alluvial architecture of the valley.

*Series of Longitudinal Basins.* In order to supply water to those basins heightened by alluviation, including during times of weak floods, it was necessary to divert the water from the river much higher upstream and lead it into the basins through long feeder canals. These canals started from the top ridge of the banks, and the intake from the river was just below the level of the lowest floods. Each of these canals thus supplied a longitudinal series of basins spread out from upstream to downstream. In places where the valley was wide and rounded, these basins were sometimes subdivided transversally also. The weak incline of the canals made it possible to extend the flooded area as far as possible. Yet, this incline had to be great enough to ensure that the water flowed rapidly to prevent the canals from silting up with alluvia.[8]

During the rising flood, the canals transported the water into the basins where it reached, in good years, a height of 1.2 to 1.5 meters. It was retained in the basins for forty to sixty days. Then, around mid-October, the dikes were opened and the water flowed in the direction of a drainage canal or a natural drainage ditch before rejoining the riverbed. As a general rule, the filling of the basins started downstream. However, when the flood looked weak, the basins situated upstream were filled first, and then were emptied into the basins downstream. Thus the floodwaters were of use several times.

The gradual development of these different basin systems made it possible to extend the cultivated surface areas to most of the land lying between the top ridges of the riverbank and the edges of the desert, except for non-dried-up

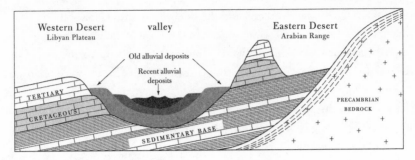

Western Desert
Libyan Plateau

valley

Eastern Desert
Arabian Range

Old alluvial deposits

Recent alluvial
deposits

TERTIARY

CRETACEOUS

SEDIMENTARY BASE

PRECAMBRIAN
BEDROCK

*Figure 4.2 Geomorphological Section of the Nile Valley*

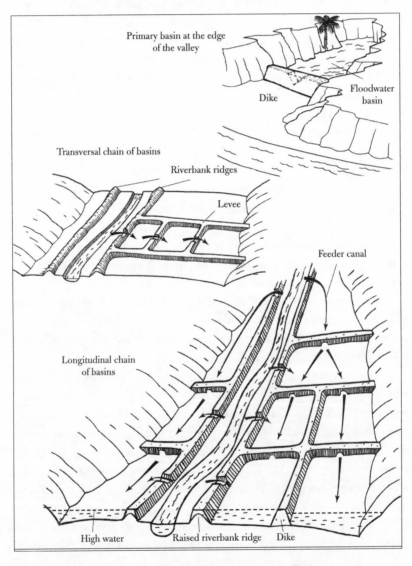

Primary basin at the edge
of the valley

Dike

Floodwater
basin

Transversal chain of basins

Riverbank ridges

Levee

Feeder canal

Longitudinal chain
of basins

High water          Raised riverbank ridge     Dike

*Figure 4.3 Diagrams of Water Basin Installations*

marshes and permanent bodies of water. Thanks to these installations, the population increased and the villages, constructed on the natural levees and on human-made dikes, multiplied even in the heart of the valley. But this progress was certainly not continual. Expansion phases were interspersed with large-scale hydraulic and demographic crises.

If the construction of the first basins, isolated from one another, was undoubtedly within the capacity of village peasant communities, it was certainly otherwise with the systematic organization of a segment of the valley. This kind of organization demanded a mobilization of numerous workers from villages distant from one another, the supplying of construction sites with provisions and tools, and thus a certain planning of work, which presupposes the existence of a central authority for decision making and coordination. Is it this necessity that led, in the course of the second half of the sixth millennium, to the formation of city-states, each of which dominated peasant communities and administered the hydro-agriculture of a section of the valley? It can be assumed that that is true. But the whole question is to know how such city-states could have been formed, having not only political, military, and religious power but also true competency in hydraulics.

It can certainly be imagined that neighboring villages formed a group and consulted each other in order to build complex hydro-agricultures on a small scale, and that the social groups charged with exercising organizational functions for this work gradually arrogated political power to themselves as they monopolized hydraulic knowledge. But it can also be imagined that such villages or federations of villages were conquered militarily and subjected to an external political and military power, which thus appropriated preexisting hydraulic skills and knowledge. In any case, it is highly improbable that an already-constituted political power could itself invent hydraulic techniques and impose them on peasant communities without experience in the matter. From what innate or revealed science could a technocracy really derive more knowledge than from practice itself? It is certain, however, that, once constituted, an hydraulic power can progressively acquire a growing experience at designing installations, organizing work and managing the water and that its capacity to govern the development of increasingly larger hydraulic installations is going to increase.

Thus, starting in the 6th millennium before the present, peasant communities, federations of communities, and pre-dynastic principalities undoubtedly contributed to developing the economic, political, military, and religious organization, the elementary hydro-agricultural techniques, the administrative methods, and maybe even the writing that was subsequently widely used in the pharaonic epoch.

## The Unified Pharaonic State

Gradually, the best-organized and most powerful of the city-states that were spread out along the valley conquered and subjugated the weakest ones. This concentration of authority led, in the last centuries of the sixth millennium before the present, to the formation of the Two Kingdoms: Upper Egypt or the Kingdom of the South, corresponding to the valley, and Lower Egypt, or the Kingdom of the North, corresponding to the delta. This regrouping of city-states into vast kingdoms gave to the now concentrated hydraulic institutions potential for increased investments and new possibilities for more widespread development. For example, there was the possibility of linking together over a great distance the dikes protecting each section of the valley, elevating dikes in local areas when necessary, so as to ensure for the whole valley a better defense against powerful floods. There was equally a possibility of linking together natural outlets and flood drainage canals in order to drain every section of the valley more completely and rapidly. Also, there was the possibility of increasing the security of the water supply as well as the capacity of intake canals, by raising their intake levels upstream and sustaining them downstream. Nevertheless, it remains the case that when the flood was weak, the unrestricted use of water in Upper Egypt could cause serious damage in Lower Egypt, and when the flood was strong, insufficient retention of water in Upper Egypt could cause dramatic flooding in Lower Egypt. That could help explain the strategic superiority of the kingdom of the South which, around 5,200 before the present, imposed its domination on the kingdom of the North.

Three or four generations later, Menes, legendary king of unified Egypt, founded the first of thirty Pharaonic dynasties that were going to rule over the Two Kingdoms for three millennia. He installed his capital at Memphis, at the point of articulation of Upper and Lower Egypt. It can be assumed that this political unification made it possible, if necessary, to balance the distribution of water between the north and the south better.

## Systems of Winter Cultivation Based on the Receding Floodwaters

There is no methodical description of the systems of cultivation based on the receding floodwaters of winter during Pharaonic times. But fragmentary information relative to this distant past and the perpetuation, until the beginning of the twentieth century, of hydro-agricultures of this type, well studied by nineteenth-century engineers, makes it possible to show, by way of assumption, the principles of the organization and functioning of these ancient systems.

Sowing after ard tillage

Sowing after hoeing

Uprooting flax

Harvesting wheat

Carrying ears of wheat

Milking a cow

Force-feeding a crane

*Figure 4.4 Tools and Scenes of Agricultural Work in Ancient Egypt*

The basins were filled, one after another, from mid-July, and each remained submerged under more than a meter of water for nearly two months. They were then emptied at the end of October. From November to spring, it was possible to cultivate in these basins a whole series of crops, thanks to the water stored in the soil and the mild winter: cereals (six-rowed winter barley, emmer wheat, spelt), food legumes (lentils, peas), fodder legumes (vetch, vetchling), and a textile plant (flax). Subsequently, other fodder crops (Egyptian clover, alfalfa) came to be added.

Agricultural work began at the end of October or at the beginning of November, a little after the floodwaters receded. As long as the soil was still wet, the sowing was carried out without preparing the ground. However, because the top layer of the soil dries out and hardens very quickly under this climate, the last land sown had to be broken up with a hoe first, the largest clods being broken with a mallet. The seeds, sown broadcast, were covered over with earth by traversing the field using a bough with many branches, pulled by hand, or by making one or two passes with an ard. In order to facilitate the germination of the seeds, the sown earth was packed down by herds trampling on it, or possibly by rolling a palm trunk. The ard, undoubtedly borrowed from Mesopotamia, is a scarifying tool that, unlike the plow, does not turn over the earth (see chapter 6). Up until around 4000 before the present, the ard was a simple tool of wood pulled by hand. Then it was harnessed to draft animals (oxen, cows, and donkeys), the point possibly outfitted with a flint. Once the sowing was over, the crops were generally left to themselves until the harvest. But it also happened that they received additional irrigation because of their unique requirements (e.g., crops planted late, crops with a long cycle such as Egyptian clover) or because of insufficient flooding.

Depending upon the crop and the latitude, the harvesting was done between March and May, with a sickle for cereals and by simple uprooting for flax. The harvesters cut the cereal stalks quite high, followed closely by the gleaners, who collected the ears and gathered them into large straw baskets. The latter were then transported by donkey to the threshing area close to the village. The threshing was done either by means of a long stick (palm ribs, for example) or by the tramping of animals (oxen and donkeys) or, later, by repeatedly passing back and forth a threshing tool formed from a wooden frame equipped with stone teeth or wheels. Next, the grain was separated from the straw using pitchforks, winnowed, and then stored in the village's high cylindrical silos made from beaten earth or mats. After the harvest, the land was turned into common grazing ground for goats and sheep until the next flood.

Though the natural pastures situated in depressions, on the riverbanks, and in the vicinity of marshes were originally relatively abundant, they were gradually reduced in size by the spread of both hydraulic installations and cultivated land, to the point that they became insufficient. The rotations of cultivated land then had to provide food, on a long-term basis and in a relatively well-balanced manner, not only for the people but for the animals as well. Thus in addition to cereals, legumes, and flax, the principal products of which were intended for human consumption and the by-products (straw and tops) consumed by the animals, the crop rotations included  fodder legumes. In these conditions, a convenient practice, still largely attested in recent epochs, consisted of establishing different biennial rotations of the following kind:

| YEAR 1 | | | YEAR 2 | | |
|---|---|---|---|---|---|
| November | May | July → October ⟶ | November | May | July → October → |
| •winter cereals | short dry | flood | •food legumes | short dry | flood |
| —barley | fallow | | —lentils | fallow | |
| —wheat | period | | —peas | period | |
| •textile plant | | | •fodder legumes | | |
| —flax | | | —clover | | |
| | | | —vetchling | | |
| | | | —vetch | | |

The presence of legumes in these rotations made it possible to overcome the principal factor limiting the yields of cereals, which was the shortage of nitrogen. In fact, the silt and floodwaters supplied barely more than 20 kilograms of nitrogen per hectare per year, while an Egyptian clover crop supplied between 40 and 80 kilograms of atmospheric nitrogen to the soil.[9] Later, fodder legumes and the remains of food legumes were distributed to fettered animals, whose excrement was mixed with the earth and then transported by donkey to the cultivated lands. Moreover, the other crop residues, straw from cereals in particular, were grazed on right in the field after the harvest and the animal excrement was then directly returned to the soil. In any event, in a system without an effective tool for burying the legumes and crop residues, only the animals were able to transform these plant materials into immediately usable fertilizers.

These biennial rotations, which alternated "demanding" cereals with "enriching" legumes, are at the heart of the Egyptian agronomical tradition that was brought to Europe by Greek, Latin, and Arab agronomists and finally by the partisans of the "new agriculture" in the sixteenth, seventeenth, and eighteenth centuries. These rotational systems prefigure, in fact, the intensive rotations that were developed in western Europe during the "first" agricultural revolution of the sixteenth to nineteenth centuries (see chapter 8).

The diet of the Egyptian people was essentially based on cereals, specifically wheat and barley, consumed in the form of bread, biscuits, or beer, and on dry legumes, specifically lentils, peas, and later broad beans, originally from India. It also included fish, fruits (grapes, figs and dates), legumes, and various types of vegetable oil (castor, olive and later sesame and safflower oils). Wine and meat were reserved for the privileged classes of the population.

## The Performance of the System

By assuming that around two-thirds of the cultivated area was devoted to cereals and food legumes (subtracting fodder, textile, and other crops) and that the average yield of grain from these crops was on the order of 10 quintals per hectare, it is possible to estimate that food production was from 6 to 7 quintals per hectare within rotation, or 600 to 700 quintals of grain-equivalent per square kilometer. At the rate of 2 quintals per person per year, that would make it possible to support the basic needs of a population of more than 300 inhabitants per cultivated square kilometer. This density was at the very least ten times higher than that of systems of slash-and-burn cultivation and systems of cultivation using fallowing in Mediterranean and European antiquity. Thus with 1.5 million hectares (or 15,000 square kilometers) organized in basins, the valley could feed at best some 4 to 5 million inhabitants. This estimate corresponds to the maximum population ancient Egypt attained in its most prosperous periods.[10] What is more, with fodder legumes and crop residues, it would have been possible on 2 hectares in rotation to feed one bovine, or two donkeys, or even five to six small ruminants, and sometimes a pig. The valley could then feed more than a million animals of all types, which contributed to the renewal of fertility.

But, as well conceived and carried out as it was, the system of cultivation using basins and floodwaters was confined within the relatively inelastic limits of the floodable space. Furthermore, it was planned and organized with the techniques and administrative methods of the moment, and finally remained at the mercy of the irregularity of the floods. A weak flood led to a reduction of the cultivated surface area and harvest, a flood that was too strong could damage the hydraulic installations, and a late return of the flood could ravage established crops. The powerful Egyptian civilization could neither have developed nor survived without adequate stored provisions, a well-conceived, implemented, and maintained hydraulic system, and good management of water. The Pharaonic state was effectively in charge of the food security of the country by administering the hydraulic system and collecting and redistributing food stocks.

## Social Organization and Role of the Pharaonic State

### *Pharaoh, Scribes, Priests, and Peasants*

Ancient documents (papyrus, tomb frescoes, engravings, etc.) of Pharaonic Egypt essentially show the material and spiritual preoccupations of a small fringe of this society, including the pharaoh, his court, administration, and clergy. Rare are indications of the everyday condition of the immense majority of the population, principally the peasants, but also artisans, soldiers, and some slaves.

Eminent owner of all the land of Egypt, of the water in the Nile, of all living beings and of all the goods that existed there, the pharaoh was the absolute master by divine right of the entire country. Proclaimed son of Ra (the Sun God) since the Fifth Dynasty, he was the earthly executant of the divine will, the organizer and guarantor of the flood, of production, and of life. Surrounded by his numerous relatives and his favorites, he lived luxuriously in his palace in the capital. Supreme head of the army and the clergy, in the last analysis disposing of everything and everyone, the pharaoh ruled, assisted by a "vizier" who relied on a large specialized and hierarchical administration and an army of scribes.

This vizier, a sort of prime minister by delegation, was master of the granaries (reserves of provisions), treasures (reserves of metals, cloth, and other goods), and royal corvées. He was organizer of large projects, responsible for the workshops, for distant expeditions for obtaining supplies of various kinds (stone, wood, minerals, tropical products, etc.), distributor of the reserves, and grand master of justice. He exercised these high responsibilities through the medium of the specialized services of the central administration—made up of his project leaders and provincial administrations. Scribes, trained in schools where they learned to read, calculate, write, inventory, and draw up documents, were the mainspring of this omnipotent administration. Charged with transcribing orders from above and with keeping the central power informed of activities in the whole empire, they were omnipresent during all transactions involved in surveying, recording harvests, inventorying goods, population, and herds, calculating and paying taxes, recording contracts and legal proceedings, etc.

Remunerated in kind from the tax yield or by produce from the land over which they had usufructuary rights (money was not used in ancient Egypt), the functionaries and scribes were an envied part of the population. Without any control other than that of their own hierarchy, they were in a position to commit numerous and serious abuses (fiscal exaction, repression) as attested by written royal reprimands, endlessly repeated and thus relatively ineffective, addressed to them.

The clergy formed another privileged category freed from manual labor. Endowed with numerous personnel in a hierarchical organization, it had its own administration, schools, and artisanal workshops. It exploited the lands the pharoah provided for its use by means of peasant corvées. The clergy thus formed a strong power group that was sometimes a rival of the administration.

In fact, priests and functionaries belonged to the same social category. The same people successively occupied high positions in one or the other of these established bodies, and each contributed, in its own way, to the functioning of the social system. How could the administration have imposed such heavy labor on the population without the valuable cooperation of religion, maintained by the clergy? And inversely, how could the clergy have imposed heavy corvées on its own estates and be protected during periods of revolt without the help of the repressive apparatus of the state?

The immense majority of the population was made up of peasant families, grouped into large villages, with little or no social differentiation. Each family had a house of crude earthen walls built by hand, a small plot of basin ground, a barely improved Neolithic tool set (sickles, hoes, straw baskets, pottery, sometimes an ard), poultry, and, in the best of cases, some small livestock (cows, donkeys, goats, sheep). A portion of the land was reserved for the pharaoh and the administration and another part was given in usufruct to the clergy and high functionaries. But, over time, since the administrative functions tended to become hereditary, usufruct of the lands granted was transformed, in fact if not in law, into a sort of inheritable private possession. Let us equally note that beginning in the Thirteenth Dynasty, hired soldiers, whether Egyptian or foreign, were granted cultivable plots of land in exchange for military service alone. These plots were transmissible to the next generation, as long as one of the sons was also hired as a soldier. Thus a warrior caste was gradually formed, which exploited part of the land of Egypt in a hereditary manner.

Artisans (carpenters, potters, basket makers, weavers, masons, water carriers, barbers, embalmers, etc.) shared the same miserable condition as the peasants. The specialized workers who worked in construction sites for temples, palaces, tombs, and pyramids, or in the royal or ecclesiastical workshops (architects, stonecutters, plasterers, designers, sculptors, painters, ceramists, cabinetmakers, goldsmiths, etc.) experienced slightly better conditions.

## Tribute in Kind and Tribute in Work

The peasantry was subjected to a heavy tribute in work, in the form of corvées, for the purpose of cultivating the royal estates, as well as those of the clergy and high dignitaries, and for constructing large public works. The nonagricultural corvées systematically occupied all of the peasantry's time not scheduled for work in the fields. The flood season was put to good use by organizing the transport of heavy materials by boat (large pieces of wood, cut stone, etc.) from one part of the valley to another and by carrying out distant expeditions (principally in Nubia). As for the dry season that preceded the flood, it was above all dedicated to the maintenance and extension of hydraulic works.

Taxes in kind (head tax, tax per head of livestock, tax proportional to harvest, various other taxes) were collected under the strict control of scribes and stocked in the numerous granaries of the state. These reserves constituted the main part of the Royal Treasury, an important administrative department that kept a precise accounting of available stocks and their origins and destinations. Naturally, it was a question of feeding the pharaoh, his relatives, his favorites, and his household gathered at court, but also feeding the army, administration, and artisans, who were paid in kind. The food reserves were also used and replenished in order to deal with shortages in bad years and to feed the large number of peasants sent to building sites or assigned to maintain the hydraulic installations, defensive works, or extravagant buildings.

In summary, taxes and corvées were so heavy that they did not leave any surplus for the peasantry nor any possibility of its growing richer and investing on a purely private basis in order to improve its means of production.

## A State-Controlled Tributary Society

Foreign slaves (Libyans, Nubians, Syrians), prisoners of war, or captives delivered as tribute by kingdoms subjugated by Egypt carried out the most thankless tasks of quarrying and mining or were bound as mercenaries. For all that, one cannot say that the ancient Egyptian economy was based on slavery: the slaves represented only a small fraction of the population and, in principle, none of these slaves was Egyptian.

On the other hand, although it is true that various sorts of servitude developed in certain periods, imposed on local peasants by some high personage, ancient Egyptian society was not one of serfs belonging to lords in a *personal* capacity, as in Medieval Europe. Rather, ancient Egypt's was a despotic, bureaucratic, and clerical society, based on a relatively undifferentiated peasantry that was subjected to a very heavy tribute in work: a sort of state-controlled tributary society.

## The Role of the State

Beyond the religious power that it more or less well controlled, the Pharaonic state concentrated judicial, administrative, and military power. It carried out extensive technical and economic functions. It devised, organized, and supervised the expansion and maintenance of the hydraulic and transport infrastructures,

with concern for extending the cultivable area and increasing the population subject to unpaid labor. What is more, it managed the very important food reserves,
which came from its estates and taxes in kind. This made it possible both to control the distribution of agricultural produce among the different social classes and
to ensure the food security of everyone, in case of need.

With a vast number of people subject to corvées and an experienced administration controlling the logistics and organization of building sites, the Pharaonic
state built grandiose works. These included hydraulic works such as the large
protective dikes along the Nile, the famous canal of the pharaohs that linked the
delta to the Red Sea, or the canal allowing the diversion of a part of the flood
toward the Fayoum depression, thereby extending cultivation to the latter. Also
there were defensive works such as the "wall of the regent," protecting the delta
from invaders coming from the East, or extravagant works, such as the pyramids,
temples, or palaces. In order to illustrate the properly "pharaonic" character of
these great works, we cite S. Sauneron in *Histoire générale du travail,* who
reports a commentary of Herodotus on the construction of a pyramid:

> Some had the task of dragging to the Nile the stones extracted from the quarries
> which are found in the Arabian Range; other teams were appointed to receive
> these stones, transported by boat to the other bank of the river, and to drag them
> to the Libyan plateau. There were consistently 100,000 workers at the building
> site who were replaced every three months. Ten years of work were necessary to
> build the causeway on which the stones were dragged.... The pyramid itself
> required twenty years of exertion.[11]

Further, the state retained a monopoly over external commerce. As a result of
the deforestation of the valley, Egypt imported structural timber and wood for
naval construction from Phoenicia (pines and cedars from Lebanon). It imported from the Aegean world and from Sinai iron, silver, and copper minerals,
principally for the purpose of making luxury objects, and a small amount for
improving tools. From Africa came ivory, obsidian, gold, livestock, and exotic
animals, while precious stones, essential oils, and perfumes came from Arabia.
Beginning in the eighth century B.C.E., Egypt regularly imported wine, oil,
ceramics, and metallurgical products from Greece. In exchange, it provided
wheat, which Greece severely lacked, and craft and art objects. The delta was
the crossroads of these exchanges, as the point of contact between the
Mediterranean and the Nile, that great waterway into the interior. Some "Syr-
ian" merchants lived there, serving as intermediaries between the Egyptian
state and foreign powers. For this reason, metallic money was introduced into
Egypt, but its use remained limited until the Greek conquest (333 B.C.E.).

## A "Despotic Oriental" State

Montesquieu was the first to identify this type of social and political organization, which he called "Asiatic despotism," because it suited the descriptions that travelers and merchants made about the states of the Near East, India, and China. The classical economists (Smith, James Mill and John Stuart Mill, Jones, and Marx) brought these essential traits to light and Wittfogel dealt with them in a masterly fashion in his work, *Oriental Despotism*. It seems indeed that the Sumerian, Pharaonic, Indo-Gangetic, Chinese, Vietnamese, Angkorian, Sukothai, Inca, and Malagasy Merne hydraulic societies shared a sociopolitical structure and functioning close to that we just outlined. One should note that there is nothing particularly *Asiatic* or *Oriental* about this type of organization since it is found in Africa and America as well as in Asia. On the other hand, if it is true that this organization is well adapted to the necessities of managing large hydraulic systems, with which it is often found associated, the link is not obligatory. A hydro-agriculture can function without a tributary despotic state. The remarkable hydraulic democracy of the Valencian *huertas* in Spain and the lineage organization of the rice-growing Diolas of Casamance demonstrate this quite well.

Conversely, this type of state indeed appears to have existed in former times outside of the main areas of hydraulic civilization, as in Knossos in Crete and in Mycene in the Peloponnesian peninsula (second millennium B.C.E.), where, perhaps, a small-scale hydraulic agriculture was practiced. It also appears to have existed in Sardinia, in Lydia (at the beginning of the first millennium B.C.E.), where the economic functions of the state extended to control over the extraction, transformation, and circulation of gold. Croesus, king of Lydia, remains celebrated among others by his wealth, which came from the tribute imposed on village communities, from the exploitation of state-owned estates and from the gold extracted from the sands of the Pactole River at Sardis, and from the mines of the Lydian mountains.[12] Perhaps the large hydraulic empires of the Near East influenced these societies, but it is not impossible that they were formed entirely as a result of the conquest and domination of relatively undifferentiated peasant communities by some more evolved tribal leadership.

## Alternating Periods of Apogee and Decline

The most prosperous periods of Pharaonic civilization correspond to the periods where this power was the best organized (Old Kingdom, Middle Kingdom, New Kingdom). At those times, the effective taming of the river, the extension of organized and cultivated areas, and the control of the flood's irregularities made it possible to bring agricultural production and the population subject to the corvée to their maximum, taking into account the hydraulic techniques and administrative methods of each of these epochs. During these favorable periods, invaders were contained outside the limits of the valley, and some pharaohs even succeeded in expanding the empire into Nubia, Libya, and Syria. Egypt was a power whose wealth impressed all the neighboring countries and that exercised influence in the whole eastern Mediterranean.

But these phases of prosperity alternated with periods of crisis and decline. In fact, the expansion of basins, cultivated lands, and population inevitably ran up against the relatively unelastic limits of the space available to organize and exploit with the techniques and methods of the moment. Thus production attained a sort of unsurpassable ceiling and, while the population continued to increase, shortages appeared and the continued imposition of tax levies provoked all types of resistances and revolts.

Another hypothesis, of great significance, can be drawn from the remarkable works of G. Alleaume: each phase of prosperity was the result of the deployment of a new hydraulic system, whose functioning entailed, in the long term, an ecological crisis.[13] The decline that appeared at the end of the expansion phase did not then result only from limits to the expansion of the existing hydraulic system, but a real crisis in the functioning of the latter and its regression. This crisis could only be overcome by a "hydraulic revolution," i.e., by the deployment of a new hydraulic system that corrected the dysfunctions of the preceding system and made it possible to surpass them. Depending on the epoch in question, these dysfunctions in the hydraulic system could originate, as we have seen, from a topographic disequilibrium due to differential silting, which gradually put the basins "outside the water." Or they could result from an expansion of the basins pushed too far in relation to the ability of the existing canals to supply water or even from an uncontrollable silting of the distribution canals from a flood that was too weak, not rising high enough to carry the water as far. Or the dysfunctions might have come from a general elevating of the alluvium deposits in the basin and a deepening of the dammed-in riverbed. The latter type of crisis seems to have affected Upper Egypt more than the delta.

These periods of decline began with the hydraulic crisis, characterized by a weakening of the social system's rules of operation and administrative discipline (First, Second, and Third Intermediate Periods and Late Period). Taxes, increasingly difficult to collect, became insufficient to preserve the state officials' and clergy's way of life and became increasingly subject to misappropriation for private ends. The weakening of the central power led to the formation of veritable local principalities governed by warlords, more concerned about pillaging than hydraulic agriculture. Centrifugal forces prevailed over centralization, with a return to the pre-dynastic fragmentation into multiple city-states. This breaking up, favored by the geography of a valley that stretches for more than 1,200 kilometers, led to the degradation of the main hydraulic installations, retreat from the coordinated management of the floods, the lowering of agricultural production, reduction of security reserves, famine, epidemics, war, and demographic collapse. The Libyan, Bedouin, and Nubian incursions could then be transformed into invasions and lasting occupations. Wars between principalities multiplied. Alliances were formed among the latter leading to a replacement dynasty, capable of reorganizing the unitary state and rebuilding the hydraulic system of the valley.

Thus, periods of prosperity (Old Kingdom, Middle Kingdom, and New Kingdom) were followed by periods of chaos and decline (First, Second, and Third Intermediate Periods and the Late Period from the Twenty-sixth Dynasty to the Macedonian conquest). The decline during the Late Period opened the door to a whole series of "eastern" invasions (Hebrew, Assyrian, Persian). For the first time, invaders from the north, Greeks, joined this group and ended up beating them: Alexander of Macedonia conquered Egypt in 333 B.C.E., beginning a period of Hellenistic domination that lasted until the year 30 B.C.E., when Egypt was integrated into the Roman Empire. After the fall of Rome, Egypt passed under the influence of Byzantium, capital of the Eastern Roman Empire.

By 333 B.C.E., Egypt had been living through a period of decline for several centuries. But that does not suffice to explain the easy conquest of such a great civilization. Perhaps the hierarchical, centralized, totalitarian, personalized, and deified character of Pharaonic power made of this country a sort of colossus. It was sufficient to take the head in order to govern the body. But it is also necessary to say that, for nearly 3,000 years, Egypt, which had only little to fear from the still embryonic civilizations of the northern Mediterranean, had remained a relatively unmilitarized society. As soon as the military-oriented Greek cities had acquired sufficient experience from colonial expeditions, Egypt, which supplied these cities with the grain that they chronically lacked, became easy prey for them. Thus colonizers from the north seized Pharaonic power, occupied the key administrative posts, and reproduced the Egyptian methods of government by improving them. They also brought innovations to many areas and exercised a definite influence, notably on irrigated agriculture.

## 3. SYSTEMS OF IRRIGATED CULTIVATION

### A Marginal System in High Antiquity

At the time of the Greek conquest, the systems of basin and floodwater cultivation still largely dominated the whole of the valley. But irrigation was not completely unknown. Indeed, if crops such as wheat, barley, and lentils did not get any watering during good flood years, other crops that lasted from the beginning of spring, such as vetch, peas, garbanzos, and flax, frequently received an additional irrigation before being harvested. And during poor flood years, it was all the winter crops, including cereals and lentils, that were sustained by irrigation.

Otherwise, outside of the constructed basins, there existed special places in the valley, delta, and the Fayoum depression, situated in proximity to surface water (pools, ponds, marshes) or groundwater at a relatively shallow level that could be irrigated by drawing water and distributing it with pitchers. Depending on their location, these lands were irrigable either in all seasons (so-called perennial irrigation) or only in certain seasons. Perennial irrigation could be carried out on land that received very little floodwater or none at all, because it was protected from the flood by the landscape or natural or human-made levees, and was close to a permanent body of water, such as Lake Moeris in the Fayoum depression, or certain marshes in the delta. Common plantings were grapevines, date palms, figs, and other fruit trees, and two or three seasonal crops in sequence: cereals, onion, garlic, cucumber, lettuce, leeks, fennel, cumin, coriander, etc. Irrigated cultivation at "the end of winter and in spring" could also be carried out between the recession of the flood and the next flood, on floodable land that had not been arranged into basins: either local depressions and lowlands of the delta, close to a body of surface water or to a source of groundwater that would persist after the flood, or low banks of the river and its branches, between the low-water channel and the lateral levees. Lastly, irrigated cultivation "in summer and autumn" could be practiced during the flood, on promontories and natural levees, on human-made dikes and on the main banks of the flooded valley.

In order to renew the fertility of the irrigated lands, in particular those that were protected from the flood and did not receive any silt, it was necessary to carry to them large quantities of animal manures and sediments from cleaning the irrigation ditches. Without any means of heavy transport, this development was limited. Besides, until the Greek conquest, the expansion of irrigated cultivation remained limited because of deficiencies in methods for drawing water. These methods were confined to pitchers, which were sometimes carried in pairs with the help of a shaft of wood curved over the shoulders; baskets with two ropes, handled by two people, which made it possible to bring water up from around 50 centimeters in depth; and wells with balancing poles (the *shadouf*). The latter type of well, originally from Mesopotamia and used in Egypt from the fourteenth century B.C.E., made it possible to bring water up from 1 to 3 meters in depth and to water around a tenth of a hectare.

## The Development of Irrigation in Depressions and on Riverbanks

The Greeks brought to Egypt new machines to draw water. A barrel, or Archimedean screw, moved by a handle that raises the water from around meters, could water a third of a hectare daily.[8] The *saqiya*, a vertical wheel with buckets driven by an horizontal toothed wheel moved by a person or animal, could draw water from four or five meters deep (and even down to ten meters when extended downward by a rope ladder holding the buckets) and irrigated from .4 to 2 hectares daily, depending on the depth of the water.[14] In order to draw water more deeply, it was necessary to install several water-raising machines on steps that raise the water from one level to the next.

Under Greek, Roman, and Byzantine domination, the administrative organization, copied from the Pharaonic model, was sometimes even improved. The administration always aimed to increase the productivity of Egyptian agriculture, thanks to an expanded and well-maintained hydraulic system. But from then on, it also aimed to extract the greatest possible exportable wealth for the greatest profit of the colonizers. Numerous roads with relay stations were built to take products, agricultural or otherwise, to Alexandria, from which they were shipped to the metropolis. Egypt was thus one of the wheat granaries of Rome: "Who held Egypt, held Rome." Cultivable land remained, for the most part, subjected to the old system of tribute in kind and in work. However, private property in land developed. The sovereign sold estates when he had need for money. Thus large private estates were formed, belonging to functionaries, foreign colonizers, etc., which were partly irrigated and used slave labor.

### New Sources of Energy: Animal, Wind, and Hydraulic

After conquering Egypt in 640, the Arabs continued to impose a heavy burden on the Egyptian peasantry and diverted the enormous tribute in wheat previously directed at Byzantium (and earlier toward Rome) toward Medina. However, due to the collapse of the Western and Eastern Roman Empires, the European and Oriental sources of slaves dried up, while African sources, still poorly established, could not maintain the shipment. The uses of animal, wind, and hydraulic sources of energy progressed in the Middle Ages, during the Arab (640–1250) and Mameluke (1250–1517) epochs.

Machines to draw water, notably water wheels (*saqiya*), gained in efficiency due to the improvement in mechanisms for reducing force and transmitting movement (toothed wheels of different diameters placed in perpendicular planes) and due to the development of animal traction, which effectively replaced slave labor. In the Fayum depression, the lower elevation of 70 meters made it

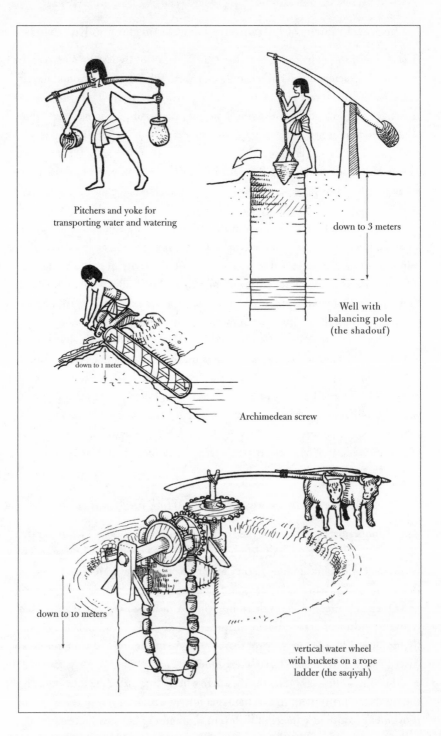

Pitchers and yoke for
transporting water and watering

down to 3 meters

Well with
balancing pole
(the shadouf)

down to 1 meter

Archimedean screw

down to 10 meters

vertical water wheel
with buckets on a rope
ladder (the saqiyah)

*Figure 4.5 Equipment for Watering and Machines to Draw Water for Irrigation
in Ancient and Medieval Egypt*

possible to use the force of the current coming from the Nile diversion canal to operate a whole series of mills set up in tiers on the sides of the depression. This effective addition to the methods of drawing water made it possible to improve and extend irrigation. Animal traction was also used to operate oil and wine presses and different types of mills. The exploitation as fertilizer mines of human, animal, and vegetable organic sediments, rich in mineral fertilizers, which were deposited and mineralized over thousands of years on the sites of ancient villages of the valley (the *koms* or *tells*), experienced limited development, to the benefit of those irrigated lands deprived of silt. However, cultivation with the ard, transport by boat, portage, and manual labor remained largely predominant until the twentieth century.

## New Irrigated Crops: Rice, Sugarcane, Cotton, Maize

The Arabs also introduced new crop species from Asia. There were annual crops, paddy (flooded) rice in particular, and perennial crops above all, such as sugarcane and indigo, which can only be cultivated in the valley if they are irrigated. Sugarcane experienced a large development in the twelfth and thirteenth centuries and, with rice, occupied first place among irrigated crops. After the great discoveries, Arab voyagers introduced plants such as tobacco, cotton, and maize into Egypt, brought back from America by Spaniards and Portuguese.

## The Expansion of Irrigated Cultivation in the Nineteenth Century

More than 2,000 years after the Greeks lost control of Egypt, Europe again burst into Egypt with the Napoleonic expedition (1798–1801). It interrupted a long period of Ottoman domination during which the hydraulic installations had deteriorated and cultivated areas, agricultural production, population, and commerce had declined due to lack of support. The taxes imposed on the peasantry had, however, quadrupled during the same period. In 1800, Egypt, in decline and crushed by taxes, had no more than 2.5 million inhabitants.

Muhammad Ali, new pasha of an Egypt subjected once again to Ottoman protection, led the government from 1806 to 1847. His objective was to modernize the country and provide it with industry and a well-equipped army, capable of resisting European colonial enterprises. To this end, Muhammad Ali restructured the administration and committed the Egyptian economy to the path of interventionism and protectionism, i.e., "state capitalism": monopoly of land, state farms subjected to production plans from the administration, deliveries at fixed prices to government stores, state monopolies in industry, particularly textiles, and reduction in the corresponding private activities.

In this context, agricultural policy aimed first at restoring the cereal base of the country in order to restore its population. But beyond this first objective, it also aimed at extricating an exportable surplus of grain and developing irrigated crops directed at the export market, sugarcane and above all cotton, in order to obtain the necessary currency to finance modernization. The hydraulic policy of Muhammad Ali was composed of two parts. The first consisted of restoring and improving the ancient floodwater basin system, which corresponded to the state's administration of water and hydraulic installations for the previous 5,000 years. The second consisted in undertaking development designed to expand cultivated lands under irrigation. This last part was relatively new for the Egyptian state since, up to that time, investments for irrigation generally came from the private sector.

From the first attempts at extending irrigation in the years 1810–20 up to the construction of the high dam at Aswan (1960–70), successive advances in the new forms of hydraulic construction conditioned the development of irrigated cultivation and the concomitant decline of floodwater cultivation:

— From 1810 to 1843, there was an attempt, without great success, at using the old floodwater distribution canals for irrigation.
— From 1843 to 1891, diversion dams were constructed on the arms of the delta.
— From 1902 to 1970, other diversion dams were built in Middle and Upper Egypt as well as the reservoir dams at Aswan, which dominate the entire valley.

### 1810–1843: The Attempt to Use Old Floodwater Distribution Canals for Irrigation

From 1810 to 1825, there was an attempt to extend irrigation principally by deepening the floodwater distribution canals and by reshaping the old hydraulic installations in the delta. As we have seen, these distribution canals began from the ridges of the riverbanks and took the water to a level just below that of the lowest floods. Each of them controlled and supplied a chain of several basins spread out from upstream to downstream. At periods of low water, the Nile flowed at a lower level than the water inlets and canals. In order to use these canals for irrigation in low-water periods, it was necessary to deepen them by 3 to 6 meters, so that their water intakes were below the low-water level of the Nile and they sloped gently to the boundary of the lands accessible to irrigation. But in this way, water often circulated at a lower level than that of the basins. In order to raise it, it was necessary to install numerous and costly elevating machines and expend much energy only to obtain, in the end, limited outflow. Indeed, as these canals had gentle slopes, the speed of the water flow was reduced, which at the same time led to their silting up. In order for the water to be available for irrigation, it was necessary to reconfigure the basins and level some dikes as well as lands intended for irrigation. These irrigation installations were not very profitable and, in order to compensate for insufficient private investment, the state had to install no less than 38,000 water wheels between 1805 and 1838.

*Small Diversion Dams and the Silting Up of the Canals.* To limit the expense of drawing water, the next developmental stage consisted (beginning in 1825) in raising the water level in these irrigation canals by building a series of small diversion dams. But this plan of action again slowed the water flow and accentuated the silting up of the canals, making them increasingly difficult to maintain. It was necessary each year to mobilize hundreds of thousands of corvée-laborers for two to four months in order to clean them out and conduct other excavation work. Moreover, Egypt did not have a large population at the time and thus confronted a growing need for laborers because of the new policy. One hundred thousand conscripts were mobilized each year and the fast-expanding factories absorbed more and more workers of rural origin. Beyond that, the work schedule for irrigated cultivation, heavier than that for floodwater cultivation, left little time for corvées in the off season. The country lacked workers to such an extent that part of the land in Upper Egypt was no longer cultivated. It became necessary to change the irrigation method.

## 1843–1891: The Era of the Diversion Dams in Lower Egypt

The disadvantages of the irrigation hydraulic system in the years 1810–40 and the limits of its expansion were due to the fact that the water intake levels on the Nile, which were situated at the low-water mark, were too low. In order to avoid having to deepen, close off, and endlessly clean out canals that had been designed for another purpose as well as draw water for irrigation at great expense, it was necessary to raise the water inlets and intake canals. Diversion dams had to be constructed on the riverbed itself, which is certainly much more difficult than blocking off a simple irrigation canal. The construction was, however, much easier in the delta where the gap between the low-water level and the level of the flood was only 3 to 4 meters, whereas this gap was up to 9 meters in Upper Egypt.

Construction of an initial dam on the Damietta branch, immediately downstream from the bifurcation of the river, had limited success. Established on insecure foundations, this small dam, 1.5 meters in height, was insufficient to expand irrigation noticeably. Subsequently, in 1843, a diversion dam 4 meters in height, the Muhammad Ali dam situated in Saida, 20 kilometers north of Cairo, was built at the head of the delta, upstream from the bifurcation.[15] Its water surface dominated a vast irrigable perimeter covering almost all of the delta lands. Its basin was filled in 1861, but it had to be remodeled and entirely rebuilt and only began operation beginning in 1891. In order to correct inadequacies in the Saida dam, two complementary dams (Benha in 1901 and Ziftah in 1902) had to be built on the Damietta branch.

The irrigable surface area was increased to more than 1,400,000 hectares due to the construction of these dams. Sheltered from the flood, lands set up for irrigation could be cultivated year-round. Aside from sugarcane and garden-orchards, the latter occupying a relatively small amount of land, this surface area was first devoted to irrigated summer crops, to cotton principally, but also to the cereals (rice in the center of the delta, maize, and sorghum) which began to replace barley and wheat in the diet. There were also autumn crops (maize, vegetables) and the ancient winter crops (wheat, barley, lentils, broad beans, Egyptian clover). However, double annual cultivation and *a fortiori* triple cultivation still remained limited. As for the old system of basins and winter cultivation using floodwaters, it still occupied 800,000 hectares. At the beginning of the twentieth century, as a result of these developments, Egypt counted 2.2 million hectares of cultivable lands, of which more than 1.4 million were irrigable.

However, the average yield of cereals barely surpassed 10 quintals per hectare because, livestock being less numerous (the size of cattle, buffalo, donkey, sheep, and goat herds was between 0.5 and 1 million head per species, or a total "stocking rate" under 0.5 "units of large livestock" per hectare), organic fertilizer was less abundant, while mineral fertilizers were still not used very much.[16] Egypt could thus feed 10 million inhabitants, which corresponds to a population density of 450 inhabitants per square kilometer of cultivable land, or nearly double that allowed formerly by winter floodwater cultivation. But Egypt, which had for 2,000 years served as the wheat granary for its successive occupiers, to the detriment of its own population and development, had also become the nearest base to supply Europe with tropical products: rice, sugarcane, and above all cotton.

### The Failure of State Capitalism and the Development of Large Cotton-Growing Estates

In the nineteenth century, the benefits of modernization were undoubtedly compromised by the difficulties and failures of the new hydraulic system and by inadequacies in management. Above all, pressure from Europeans, hostile to protectionism and statism in the economy, was increased. With the Treaty of London in 1838, Egypt agreed to reduce the size of its army, submit to the Anglo-Ottoman free-trade accord and dismantle its state monopolies. Nevertheless, the policy of modernizing the country was continued. Other large projects were undertaken (e.g., the Suez Canal, dug between 1859 and 1869 and of which Egypt retained 44 percent of the shares, as well as bridges, railroads, the telegraph, scholarly institutions, and universities). But this ambitious policy exceeded the financial capacity of the country.

The financial difficulties precipitated changes in the property ownership system. In the middle of the nineteenth century, almost all the land was still granted by the state and subjected to tribute in kind (fixed at one-quarter of the harvest), without counting the corvée imposed on the villagers. After 1850, the land was distributed as quasi-private property among peasant families who could prove that they paid the tribute for five consecutive years. But this tribute increased gradually until it reached half of the harvest, and families who could not pay it had to give up their land. These lands returned to the public domain and were granted to the sovereign, his family, or high functionaries. In 1874, the state began to sell peasant tenures, over which it still held eminent domain for a cash payment equal to six years of tribute. This measure, combined with selling state lands by auction (1878), favored the rapid development of a new class of large property owners, who oriented their estates toward the cultivation of cotton. The Civil War in the United States (1861–65) reduced the supply of American cotton, which created a large outlet for Egyptian production and led to a significant price increase.

Egypt was deep in debt and close to bankruptcy, so much so that it was forced to sell its share in the Suez Canal to England. In 1876, the European creditor powers imposed on it the creation of an "office of the public debt," charged with controlling the state's receipts and their priority allocation to debt service. The following year a "council of ministers" was established, a triumvirate composed of an Englishman, a Frenchman, and an Egyptian, but this council had only a brief existence. Not long after, in 1882, England, having succeeded in excluding France, militarily occupied Egypt and imposed on it a growing specialization in cotton production destined for the English textile industry, while its local factories collapsed. Large agro-exporting estates were expanded, while the dispossessed and ruined peasantry swelled the ranks of the workers on the large plantations, in light industry, and in other sectors of urban business, but also added to the ranks of the unemployed.

# The Reservoir Dams and the Spread of Irrigation in the Twentieth Century

At the end of the nineteenth century, all the available water at the time of the low-water level was already used for irrigation. Nevertheless, the cotton interests continued to put on pressure to enlarge the irrigated area. Thus it was necessary to mobilize other resources. A new hydraulic era, the era of reservoir dams, was inaugurated at the beginning of the twentieth century. These dams, capable of storing floodwater for release during periods of low water, were built far upstream on the principal bed of the river, close to Aswan, at a height that towered above the whole floodable valley. As a result, it became possible to extend irrigation upstream from the delta to the middle and upper parts of the valley.

## 1902: First Reservoir Dam at Aswan, Diversion Dams, and Expansion of Irrigation in Upper Egypt

The first Aswan dam, built in 1902, had a capacity of 1 billion cubic meters. Heightened in 1912 and in 1934, its capacity was successively increased to 2.4 then to 5.6 billion cubic meters. Its function was not to store all of the summer floodwater, which reached several tens of billions cubic meters, but to form a more modest water reserve that was released into the riverbed to meet the irrigation needs of the developed perimeters downstream. In order to increase the supply to these perimeters, several diversion dams were built across the middle and upper valley, on the same principle as those of the delta. The first, built at Assiut in 1902, fed an old diversion canal (the Ibrahimiah Canal, built in 1873 to supply the irrigated perimeter of the left bank and the Fayum depression). The second was built far upstream at Esna in 1906 and the third was built midway between the first two at Nag Hamadi in 1930.

Due to these installations, in 1950 the irrigated land area reached 2.2 million hectares, while the rebuilt floodwater basins occupied no more than 0.34 million hectares. However, rotations that included two irrigated crops per year took time to expand because they demanded much labor, livestock, and preliminary expenditures. In fact, annual double cultivation was truly dominant only in the second half of the twentieth century.

## Agrarian Reform and Nasser's State Capitalism

In 1950, on the eve of Nasser's agrarian reform, a third of the cultivable land was in the hands of 0.4 percent of the property owners (large property owners each possessed more than 21 hectares), while at the other extreme, 94 percent of the property owners (each possessing less than 2.1 hectares) held only a third of the land. Also, it is necessary to note that half of the peasant families had no access to land either as owners or renters. The agrarian social structure of Egypt was thus characterized by the preponderance of "landless peasants" and peasant holdings too small to inhabit and support a family (*minifundia*) and by the existence of a minority of large holdings. Nevertheless, these large holdings were not, as in the Latin American countries, large latifundia of several million hectares, and they did not occupy the majority of the land. Poverty was immense and the average food intake was less than 2,000 calories per person per day.

The agrarian reform occurred in three time periods: in 1952 property ownership was limited to 84 hectares per person; in 1961 it was limited to 42 hectares; and in 1969 to 21 hectares. But the application of the reform remained incomplete. In the end, only 400,000 hectares were redistributed to 340,000 families, or less than 10 percent of the families in need. It should be emphasized that the reform instituted a status for tenant farming and sharecropping that was clearly more advantageous for the peasants.

Nasser's policies, like Muhammad Ali's, were nationalist and without any doubt were aimed at accelerating the modernization and industrialization of Egypt. Taking note of the insufficiency of private investment, these policies committed the Egyptian economy to the path of state capitalism and interventionism, under the cover of "scientific socialism" this time. Beginning in 1961, nationalizations multiplied and the state monopolies were reconstituted. The administration set up "cooperatives" in the country, charged with ensuring the supply of fertilizer, improved seeds, and pesticides at low prices (because they were subsidized) to the farms. Subsequently, they were gradually endowed with modern agricultural equipment (motor-pumps, tractors, instruments for working the soil, crop dusters, threshing machines, etc.) and they functioned as centers for renting machines and providing assistance for agricultural work, carried out cheaply for the farmers. Moreover, the cooperatives were the "transmission belt" for the production plans of the government. The farmers, in particular those who benefited from the agrarian reform, were compelled to dedicate a specific part of their land to industrial crops (cotton in the delta, sugarcane in the middle and upper valley) or to basic food crops (rice, wheat, maize, broad beans, lentils, and even onion and garlic). They were obliged to deliver their produce, at low prices, to the wholesale trade and to state industries. All of these measures aimed at ensuring the supply of agricultural raw materials at low cost and at lowering the food cost for the reproduction of the labor force. This was really a policy of transferring value from agriculture toward other sectors of the economy, and in particular toward national industry.

In certain cases, the cooperatives even went so far as to organize a system of regular plot allotment and compulsory rotation for all the farmers of the same village: cotton-based triennial rotation in the delta, sugarcane-based quinquennial or sextennial rotation in the valley, for example. For all that, these were not production cooperatives (of the kolkhoz type). Each peasant family farmed its parcels and raised its animals for its own account, freely disposed of quantities produced beyond the quotas for compulsory delivery, while vegetable, fruit, and animal production was not controlled and was sold on the free market. Even if the effectiveness of the relatively undemocratic administrative management of these cooperatives is strongly debated, it remains fact that they were entrusted with supplying fertilizers, seeds, and pesticides, they granted seasonal credit to the peasants at low interest rates, and agricultural production increased. If it is indeed possible to speak of a new phase of state capitalism in Egypt during the Nasser era, it was far from the semitotal state control of the economy, as in the Soviet Union. Even subjected to state interventionism, agriculture, artisanal activities, small and medium-sized commerce, and most services remained largely private.

Nevertheless, with the policy of opening and liberalization begun in 1973, state monopolies, the role of cooperatives, obligatory crops, and even agrarian reform were gradually called into question. Private Egyptian or foreign capital played a growing role in the economy, including agriculture. Large estates growing fruits and vegetables and even livestock estates were established on newly irrigated lands conquered from the desert. However, in the mid-1990s, a significant number of state enterprises and cooperatives were still in operation, and even if tenant farming and sharecropping again tended to become precarious and too costly, small farms in the valley and delta were still in place.

In addition, the Nasser era left to Egyptian agriculture a work of truly Pharaonic dimensions, the Aswan high dam, which crowned the already well-advanced replacement of the ancient system of winter cultivation using floodwater and basins by the system of year-round irrigated cultivation.

## The Aswan High Dam and the Generalization of Year-Round Irrigated Cultivation

The Aswan high dam, built between 1957 and 1970 several kilometers upstream from the first one, has a capacity of 168 billion cubic meters, of which 30 billion are set aside to store alluvium and 48 billion designated to meet the challenge of exceptionally large floods. There remain 90 billion cubic meters, which almost corresponds to the average annual flow carried by the river, of which around 15 billion are lost by evaporation in the reservoir upstream from the dam. Of the

74.5 billion cubic meters actually available at Aswan, Egypt receives 55.5 (Sudan receiving the rest), to which are added 3 billion cubic meters pumped from the groundwater and 2.3 billion cubic meters of reused drainage water, or a total availability of 60.8 billion cubic meters, of which 3.7 billion go toward human consumption, 2.9 billion to industry, and 39 billion cubic meters for irrigation. The remainder is lost to evaporation or drainage or flows into the sea without having been used.

Theoretically, then, very little water henceforth reaches the sea without being used. There is no longer a Nile flood and the river is no more than the dorsal spine of a generalized system of irrigation by canals. Only some very small and discontinuous zones of the upper valley are still subjected to the flood system. The gradual filling of the reservoir in the 1960s and at the beginning of the 1970s made it possible to complete the expansion of year-round irrigation to 2.4 million hectares of formerly cultivated lands and obtain from the desert some 400,000 hectares of supplementary irrigated lands. But the cultivable area gained by the expansion of "new irrigated lands" is in great part offset by the losses to the ancient fertile lands of the valley, and above all the delta, due to the growing encroachment of cities, factories, stockpiles, quarries providing clay for bricks, and infrastructure.

*Perennial Cultivation, Double and Triple Annual Cultivation.* Today, irrigated lands use on average 14,500 cubic meters of water per hectare per year. These lands occupy 2.7 million hectares, of which more than 300,000 hectares are devoted to perennial crops (sugarcane in Upper Egypt, vines, and orchards) and of which 2.4 million hectares can support two, and sometimes three, crops per year.

The most widespread sequence of two crops per year consists of the ancient winter crops (clover, wheat, barley, broad beans, lentils, flax) that occupy, in varying proportions depending on the regions and the farm, the largest part of the cultivated area at this season. It includes summer crops as well (fodder maize and grain-maize almost everywhere, sorghum in the upper valley, rice or cotton in the heart of the delta).

Cotton is a "demanding" crop that lends itself to annual double cultivation only with difficulty, since it occupies the land for eight months, from March to October, leaving very little time to carry out a winter crop. At most, it is possible to cultivate clover for three to four months before cotton, which provides one to two cuttings of fodder, while a clover of six months provides four to six cuttings.

The sequence of two irrigated crops per year appears under two forms, depending upon whether or not it includes cotton:

| Annual Sequence of Type 1: Without Cotton | | |
|---|---|---|
| November | May | October |
| winter cereal: *wheat, barley* | summer cereal: *rice (center of the delta)* | |
| or food legume: *broad beans, lentils* | or maize *(fringes of the delta and lower middle valley)* | |
| or fodder legume: *clover (4 to 6 cuttings)* | or sorghum *(upper middle valley)* | |
| or textile: *flax* | | |

| Annual Sequence of Type 2: With Cotton | | |
|---|---|---|
| November | March | October |
| *clover (1 to 2 cuttings)* | *cotton* | |

On the other hand, cotton is a risky crop, both for commercial reasons (the market price of cotton fluctuates greatly) and agronomical reasons (leaf worm and bollworm destroy the harvest). It is also a crop that exhausts the soil and is labor intensive. For all these reasons, one year with cotton (sequence of Type 2 in the table) must alternate with a minimum of one year without cotton that includes a food or fodder legume (sequence of Type 1). Thus, cotton is generally included either in a biennial rotation or in a triennial rotation in the following manner:

| Biennial Rotation | | | |
|---|---|---|---|
| November | October | November | October |
| annual sequence of type 1 *without cotton* | | annual sequence of type 2 *with cotton* | |

| Triennial Rotation | | | | | |
|---|---|---|---|---|---|
| November | October | November | October | November | October |
| annual sequence of type 1 *without cotton* | | annual sequence of type 2 *with cotton* | | annual sequence of type 1 *without cotton* | |

The sequences with a double annual crop, with or without cotton, are still the most widespread: they occupy around 70 percent of the irrigated area. But more and more there are sequences with a triple annual crop in which varied short-cycle crops (autumn maize, potatoes, beans, tomatoes, squash, eggplants, cucumbers, watermelons, melons, onions, garlic, lettuce) come to be interspersed between basic crops (Egyptian clover, wheat, broad beans in winter, maize, rice, cotton in summer), or even sometimes come to replace them.

*Increase in Production, Population Growth, and Food Dependency.* In total, Egypt today counts nearly 2.3 million hectares cultivated in cereals, of which the average yield is greater than 55 quintals per hectare. This production, which is close to 130 million quintals, does not suffice to support the needs of a population that today is greater than 55 million inhabitants. In fact, in forty years, from 1950 to 1990, the average food intake increased by more than one half, going from less that 2,000 to more than 3,300 calories per person per day. This average level, moreover, masks enormous disparities. Except for rice, Egypt is a net importer of cereals. It imports more than half of its wheat consumption and 10 percent of its maize consumption. It also imports vegetable oils, sugar, and, to a lesser degree, animal products. In total, agricultural exports (cotton, citrus fruit, potatoes) are far from offsetting imports, of which they represent less than 10 percent in value.

*Relatively Unmechanized, But Intensive Vegetable and Animal Mixed Farming.* With some 5 million agricultural workers, the area per worker has fallen to less than a semi-hectare, which has not only favored the development of the double or triple annual crop, but also the development of vegetable, fruit, and animal productions, which are labor intensive and have a high added value per hectare. Thus today fodder and animal products form the Egyptian peasantry's first branch of activity. Irrigated fodder crops (clover and maize-fodder) represent a quarter of the area harvested. Moreover, the by-products of other crops (straw, tops, canes, leaves, stalks, bagasses, cane ends, and other crop residues) are almost entirely consumed by animals. There remains, then, practically no non-transformed plant litter. In that way, Egyptian agriculture feeds four times more livestock than at the end of the nineteenth century (6 million cows and buffalos, 2.5 million donkeys, 10 million sheep and goats, 0.13 million dromedaries, without counting a few thousand horses and mules, which corresponds to a stocking rate clearly greater than 2 units of livestock per hectare).

Beyond the products that it furnishes (meat, milk, wool, skins), this numerous livestock produces a large quantity of manure that is collected on a bed of silt, with which it is mixed before being returned to the fields. Even though perennial irrigation and the multiplication of crops and methods of farming strongly accelerate the speed of humus decomposition and tend to impoverish the soil, the abundance of animal manure makes it possible to maintain an adequate level of organic matter in the valley's sandy and silt-laden soils as well as improve their structure. What is more, a good part of the fertilizing minerals incorporated in the cultivated biomass are recycled from season to season through fodder and animal manure. But that is not sufficient to explain the high level of yields attained today by Egyptian agriculture. In fact, the supply of fertilizing minerals from solubilization of the existing silts and fixation of atmospheric nitrogen would not even make it possible to attain the yield of 10 quintals of grain achieved with cultivation based on floodwaters of long ago. It would still lack the annual contributions coming from the silt that now accumulates behind the Aswan high dam. The elevated yields of today result above all from the use of strong doses of mineral fertilizers (urea, ammo-nitrates, superphosphates). In addition, a non-negligible portion of these yields results from imported animal feed whose minerals are largely recovered in animal manure.

Moreover, the advance of motomechanization (water pumps, soil work, threshing), thanks to the cooperatives and public works, is indeed real. But because of the limited size of the vast majority of agricultural holdings, most of the work (sowing, hoeing, spreading manure, fertilizers and pesticides, harvests, feeding the animals, milking) is carried out manually. As for internal transport on the farms (of fodder, manure, harvests), it is most often effected by donkey or with the help of carts. Thus, although the yields per hectare of irrigated Egyptian agriculture are today the same size as those of the developed temperate countries, the productivity and the earned income remain incomparably much lower.

### Ecological Consequences of the Construction of the Aswan High Dam

*Salinization.* In the old floodwater and basin systems, a large leaching of the soil took place each year at the moment of the flood, which explains how these systems could be maintained for nearly 5,000 years without leading to salinization of the soil. But with the generalization of irrigation, salinization spread to valley lands. Salinization is the scourge of irrigated cultivation in arid regions. It has destroyed numerous irrigated areas and maybe even led to the decline of entire hydraulic civilizations (Mesopotamia, Indus).

Indeed, under the dry and hot climate of Egypt, evaporation is intense, above all in spring and summer. Consequently, a non-negligible fraction of the water retained at Aswan and in other reservoirs as well as a portion of the water of the Nile and in the large irrigation canals evaporate each year. This water, collected by runoff in the catchment basin of the Nile, already contains a large quantity of salts in solution that, because of evaporation, are concentrated in the remaining water. The level of salts in the water that is used to irrigate cultivated land is consequently quite high. These lands are generally irrigated by runoff from one end of the year to another (except during a few weeks in winter set aside for cleaning the canals), the overhead irrigation towers being put to work every four to twenty days depending on the crop and the season. The crops absorb a fraction of the irrigation water and the salts that it contains, but another important fraction of this water evaporates, so that the level of salts in the water that penetrates into the soil (the soil solution) increases even more. However, in most regions of Egypt today, the level of these salts in the water does not reach the toxic level that would prevent the planting of certain crops, because irrigation by flooding is most often overabundant and in well-drained sandy or silt-laden permeable soils, excess water (not evaporated and not absorbed by plants) seeps in deeply, thus diluting and carrying away a portion of the salts concentrated in the soil solution down to the groundwater and beyond, toward its outlets.

But the permanent presence of large quantities of water throughout the whole valley causes, in certain areas, large seepage and a general rise in the groundwater level, a rise particularly clear in the vicinity of reservoirs and large irrigation canals. In these areas, as well as in poorly drained depressions, the capillary rising of water from the groundwater close to the surface sustains a large evaporation during a good part of the year. As it rises, this water brings with it the salts and the concentration of the soil solution increases in proportion to the evaporation. If this upward movement is not completely offset by an inverse movement of draining salts through seepage of irrigation water, these salts are increasingly concentrated and sometimes even end up crystallizing on the surface. This phenomenon is aggravated in water-deficient areas, where irrigation water is used several times. Non-drained residual water is reused, and is particularly full of salt because of the evaporation sustained during the earlier use. Soil salinization is also more frequent and more serious in the north delta, where the groundwater becomes brackish because it takes in water from the Mediterranean and coastal lagoons.

Even if, remember, the salinity of Egyptian soil rarely reaches toxic levels that would prohibit the growing of certain crops, it remains the case that this salinity is elevated enough to form the principal factor limiting the fertility of much of the soil. The level of salt in mediocre soil is generally around 0.8 percent, in average soil around 0.5 percent, while in good soil the average level of salt is around 0.3 percent. However, the sensitivity of crops to salinity is quite variable. Wet rice, though relatively sensitive to salinity, can be cultivated in soil that is too saline at the beginning because it is submerged for several months by great quantities of irrigation water that dilute the soil solution. As a result, in the saline areas of the delta, rice cultivation carried out every two or three years is considered a desalinating crop, which opens the way to planting other crops that are relatively sensitive to salinity.

Beyond that, the rising of groundwater too close to the surface of the soil, even when that water is not saline, is also harmful to crop growth, particularly for plants with deep roots, like cotton. From the end of the nineteenth century, several areas of the delta were affected by the rising of the groundwater and sometimes by salinity. Such difficulties also occurred in middle Egypt, on lands in proximity to the large irrigation canals of Ismailia and Ibrahimya. In order to resolve these problems, it was necessary to install a drainage network making it possible to push the groundwater level down more than a meter deep. Moreover, in the low areas of the delta without natural outlets, it was necessary to drain the water by pumping it toward the sea. In some situations, in order to lower the groundwater level, it even became necessary to remove or lower some irrigation canals that were too high, thus making it necessary to pump the water in order to irrigate.

With the generalization of year-round irrigation, the problems of water saturation and soil salinity have multiplied, to the point that in 1973 the fertility of two-thirds of the cultivated land was significantly affected. Even if, since then, three-fourths of this land has been drained, there remains much to do in terms of irrigation and drainage in order to reduce the wasting of water, lower the groundwater level, reduce evaporation, and avoid salinity, such as expansion, intensification and if possible burial of drainage networks; covering, concreting, and waterproofing the large canals; using pipes to distribute water.

*Other Consequences.* The silt that accumulates behind the dam no longer contributes to the reproduction of the fertility of cultivated lands, whose need for manuring has increased because of the development of double and triple annual cultivation. But the fertilizing contribution of silt was less important than generally thought and it has been more than compensated for by the growing use of mineral fertilizers and the development of fodder legumes and animal raising. Furthermore, the constantly hot and humid atmosphere in the Nile Valley has favored the proliferation of insects and other plant parasites. But, here again, pesticides have made it possible to control this phenomenon. Beyond salinity, a serious problem resides henceforth in the excessive concentration of nitrates and pesticide residues in certain areas and in certain products of consumption resulting from the multiplication of irrigated crops and the use of fertilizers and pesticides. Lastly, without the flood and its sediments, the lands of the delta have stopped spreading into the sea and are even worn down by marine erosion.

But in the last analysis, the principal danger for irrigated Egyptian agriculture is, as always, inadequate floods, based as it is on an enormous hydraulic system whose function comes down to reserving the floodwaters and redistributing them in all seasons and in all places as a function of the electric, urban, and agricultural needs of the country. Beginning in 1979 and continuing through part of the 1980s, a succession of weak floods led to a huge reduction of the water reserves of Lake Nasser. In July 1988, the stock of useful water fell to ten billion cubic meters, and had it not been for the providential return of good floods beginning in the summer of 1988, Egyptian agriculture would have been almost destroyed for a time. As great as human works are, there always remain, upstream or downstream from the latter, even greater works to accomplish.

## 4. CONCLUSION

Enclosed in the middle of an immense desert, within the narrow limits of the carefully organized and cultivable lands of a slender valley and delta submerged each summer by the Nile flood, Egyptian civilization has always rested on hydraulic agriculture. During nearly 5,000 years, the system of winter cultivation using basins and the receding floodwaters remained predominant, while coexisting with the system of irrigated cultivation at different seasons.

Ancient Egyptian civilization was the fruit of the perpetual labor of an impoverished and relatively undifferentiated peasantry, equipped with barely sufficient manual, and for a long time essentially Neolithic, tools. A centralized and hierarchical political, administrative, and religious organization forced this peasantry, by means of tributes and corvées, to build and maintain gigantic collective works, both utilitarian and extravagant. Until the nineteenth century, Egypt experienced a succession of prosperous epochs, characterized by centralization, hydro-agricultural progress and demographic growth, alternating with periods of decline, marked by political fragmentation, hydro-agricultural regression, and a fall in the population, which then oscillated between 2 and 5 million inhabitants.

After several external shocks (the Greek, Roman, Arab, Ottoman, and European invasions), irrigated cultivation benefited from new means of using water and progressed. But the ancient system of winter cultivation using basins and floodwaters was only truly destabilized at the end of the nineteenth century, after which the Egyptian state adopted the modern hydraulic system of irrigation in order to expand irrigated cultivation to the whole valley year-round.

Then an agriculture developed based on private property in land, open to trade, consisting of a multitude of small—often too small—family farms and a small number of large holdings using wage labor. While continuing to practice plant and animal mixed farming intended for self-consumption, the majority of farms turned partially to irrigated crops for export, primarily cotton, but also sugarcane and rice. More recently, in order to respond to the growing demand from cities, many farms turned more toward vegetable, fruit, dairy, and meat products. Some fruits and vegetables were also exported.

In the twentieth century, the expansion of year-round irrigation to the entire valley, beyond the area dominated by the ancient floodwater basins, and the move to double, even triple, annual cultivation, made it possible to triple the harvested area each year. The combined use of mineral fertilizers, high-yield varieties, and pesticides made it possible, depending on the crop, to triple, quadruple, and sometimes even quintuple the yields. But the population, which reached a maximum of 5 million inhabitants in 1850, increased more than ten-fold (in 1992, it surpassed 55 million inhabitants, and it continues to grow more than 2 percent per year) and because of the increased standard of living and a firm policy of low agricultural and food prices, consumption per inhabitant grew by more than half since 1950. As a consequence, Egypt is mired in food dependency and, despite its exports of cotton, fruits, and vegetables it has even had a large commercial agricultural deficit since the end of the 1970s.

Egypt is a good example of the great hydraulic civilizations of early antiquity, which were built at the heart of the Near Eastern arid regions more than 2,000 years before the first European civilizations. However, from their beginnings, these hydraulic civilizations made use of only rudimentary Neolithic tools and knew nothing of iron or the wheel.

Although ancient Egypt experienced over the course of its very long history notable hydraulic, political, and cultural progress, it has been anchored for a long time in a hydraulic, social, and political system that, after each crisis and after repeated external shocks, always tends to reconstitute itself on the same basis. But it would be rash to see in this failure to change, more apparent than real, the cause of the weakness of contemporary Egypt's economic heritage. To do so would be to forget that Egypt was colonized for more than two millennia, during which its successive occupiers appropriated a good part of its agricultural production, to the detriment of its own population and development.

Egypt began to leave behind its ancient "hydraulic frontier" only in the middle of the nineteenth century, progressively replacing the formerly predominant basin and floodwater system of cultivation by the system of cultivation based on year-round irrigation, extending private property in land, and developing commercial production and trade with the West. In the twentieth century, the Egyptian peasantry, in the space of a few generations and soon as it was in a position to do so, took advantage of irrigation hydraulics, developed complex and adaptable combinations of crops and animal raising, and adopted the use of improved varieties, fertilizers, and pesticides at a comparable level to that of the developed countries.

However, Egypt has not industrialized to any great extent. The development of nonagricultural employment is inadequate considering the exodus from agricultural areas and the demographic explosion. It is a country hit hard by unemployment and emigration, where the excess agricultural labor force also reduces the cultivated area available per worker and limits the advances of mechanization, productivity, and income from agricultural labor.

# 5

# The Inca Agrarian System:
# A Mountain Agrarian System Composed
# of Complementary Subsystems
# at Different Elevations

The common ways mainly employed by the Spaniards who call themselves Christian and who have gone there to extirpate those pitiful nations and wipe them off the earth is by unjustly waging cruel and bloody wars. Then, when they have slain all those who fought for their lives or to escape the tortures they would have to endure, that is to say, when they have slain all the native rulers and young men (since the Spaniards usually spare only the women and children, who are subjected to the hardest and bitterest servitude ever suffered by man or beast), they enslave any survivors. ...

Their reason for killing and destroying such an infinite number of souls is that the Christians have an ultimate aim, which is to acquire gold, and to swell themselves with riches in a very brief time and thus rise to a high estate disproportionate to their merits. It should be kept in mind that their insatiable greed and ambition, the greatest ever seen in the world, is the cause of their villainies. And also, those lands are so rich and felicitous, the native peoples so meek and patient, so easy to subject, that our Spaniards have no more consideration for them than beasts.

BARTOLOMÉ DE LAS CASAS, 1552,
*The Devastation of the Indies: A Brief Account*

On the eve of Spanish colonization, the Inca Empire occupied vast territories that today belong to Ecuador, Peru, Bolivia, and Chile. This empire extended along the desert Pacific Coast, into the high, semi-arid, and cold Andes Mountains and over to the hot, humid, and forested Amazonian side.[1] It was the heir of the city-states and hydro-agricultural civilizations that had begun to develop 1,000 years before our era, in the oases of the desert coast and in the arid valleys of the Andes range.

The Inca agrarian world formed a heterogeneous, fragmented, and dispersed archipelago of coastal oases, irrigated Andean valleys, high-altitude fields and pastures, and cultivated clearings of Amazonian forest, separated by vast distances, which were almost empty of people, arid, cold, or forested. As in many mountain systems, the Inca agrarian system was composed of complementary subsystems, each exploiting a particular ecological niche.

In order to increase its population, power, and wealth, the Inca state continually expanded irrigated cultivation and tried to take advantage of the diverse resources of the territories it had conquered and unified. In order to do that, it created large storehouses of provisions, constructed large hydraulic and road works, and organized transportation and trade between the different regions. It accomplished all this by relying on a hierarchical administrative and religious organization and on a vast system of corvees imposed on relatively undifferentiated peasant communities. As with the Pharaonic state and other hydraulic states of the ancient world (Mesopotamia, the Indus, China, Vietnam, etc.), the Inca state, from the beginning of the American Bronze Age, created a type of centrally administered economy, commonly called "Oriental despotism."[1] Through massacres, imported diseases, enslavement of the population, but above all by dismantling and corrupting the social and administrative structure of the Inca Empire, colonization brought about the collapse of its economy and caused four-fifths of the population to perish from hunger or disease in a half-century, all for the purpose of pillaging and exploiting gold and silver. This demonstrates, in the negative, that the Inca state filled essential economic functions for this type of society. The organization of a satellite colonial economy by the Spanish based on exporting raw materials from mining, then of an agro-exporting economy based on large specialized estates (the *latifundia*) and on the marginalization of the peasantry, led Peru, like most of the Latin American countries, into a type of economic and political impasse that persists to this day.

The study of the Inca agrarian system is justified because it represents an archetype of mountain system, composed of complementary subsystems at different elevations. Moreover, although this system was practically destroyed by colonization, its study demonstrates the exceptional contribution of the Indians of America to the agrarian heritage of humanity, a contribution that can be assessed by the number and economic importance of the plants they domesticated: maize, potatoes, manioc, beans, cotton, tobacco, tomatoes, etc. Finally, the Inca system is an American example of a post-forest hydro-agriculture in an arid region, very different from the Egyptian system, but which nevertheless presents, though separated by thousands of years and kilometers, astonishing convergences of social and political organization with the latter.

## 1. HISTORICAL SUMMARY

As we saw in chapter 2, the manipulation and domestication of plants began independently in America in three regions: in southern Mexico around 9,000 years ago (expanding Central American center), in the Peruvian Andes around 6000 years ago (South American center), and on the middle Mississippi around 4,000 years ago (North American center). Around 4,500 years before the present, the Neolithic agricultural wave coming from the Mexican center reached South America, and then expanded, incorporating the Peruvian center in the process. Floodwater cultivation developed in some valleys, while temporary slash-and-burn cultivation expanded to the forested formations that were the most accessible and easiest to cultivate, leaving aside the more dense forests, which were more difficult to clear, particularly the large Amazonian forest.

Thus in Central America and in the Andes range, just as in the Near East, the Sahara, and Iran, the process of deforestation began during the Neolithic epoch, and also led to erosion, deterioration in soil fertility, and drying out of the climate. Were some of the desert regions of America formed or enlarged during this epoch? This is a question with no certain answer. We note, however, that a part of the Andes, today covered with scraggly grassy formations, was originally partially occupied by formations of shrubs or trees and that, as in the Near East, the first hydro-agricultural civilizations of America appeared after several centuries of Neolithic agriculture.

In Central America, the Olmec civilization developed beginning 1500 B.C.E. They developed the first irrigation systems, the first religious cities, the first pyramids, and the first forms of writing in the New World. Starting from the coastal plains around the lower part of the Gulf of Mexico, this civilization extended its influence toward the west (the central plateau), the south (Pacific Coast), and the east (Guatemala, Honduras, Nicaragua). After the collapse of the Olmec Empire around 300 B.C.E. and following the increase of regional cultures over a long period of time, two great civilizations arose and asserted themselves: to the west, Teotihuácan, metropolis of the central plateau whose influence lasted for a few centuries (from 300 to 600 C.E.); to the east, the Maya cities whose longer lasting influence (from 300 to 900 C.E.) developed across all of southern Mexico, as far as Yucatán and Guatemala. From the nineth century of our era, groups of nomadic hunter-gatherers, originating in the semidesert plains of northern Mexico, began to penetrate into the central plateau and the territories of the ancient Maya cities. The last of these immigrants to arrive, the small Aztec tribe, was at the origin of a brilliant civilization. Its economy, based on agriculture, artisanal activities, and commerce, was prosperous. Diet rested primarily on maize and beans, supplemented by squash and peppers. Animal breeding, not very important, was limited to turkeys and dogs, raised for food. The Aztec state, whose capital Mexico (Tenochtitlán) was considered by the Spanish upon their arrival to be "the most beautiful city in the world," imposed a heavy tribute in kind on the peoples it subjugated, in the form of food products, gold, cacao, cotton, and cotton goods.

## The First Hydro-Agricultural City-States of South America

In South America, the first hydro-agricultural civilizations were formed a little later, beginning in the year 1000 B.C.E., in the Andes and on the desert coastal plain along the Pacific Ocean. The population was thus concentrated in the valley bottoms and at the mouths of streams descending from the high Andes.

The first of these civilizations to see the light of day was in the region of Chavín, at an elevation of between 2,000 and 3,500 meters in the Andes. This was a civilization of peasant maize farmers, remarkable for its large stone edifices, sculptures in bas-relief or full relief, and pottery. A whole string of agrarian societies practicing cultivation on irrigated lands, centered on populous and magnificent city-states, was then established (from 300 B.C.E. to 700 years C.E.) along the Pacific Coast: Salinar, Vicus, and Mochica in the north, Lima in the center, Nazca in the south. The hydro-agricultural city-states of the less arid, high Andean valleys developed between 700 and 1000 C.E. and they rapidly became expansionist. Tiahuanaco, on the edge of Lake Titicaca, momentarily extended its influence toward the Ayacucho region and, from there, toward the southwest side of the range. Beginning in the year 1200, Chimú kingdom, a powerful military state, dominated a large area of the north coast. In all of these civilizations, cities and arts (textile, pottery, architecture, metallurgy and goldsmith and silversmith trade, etc.) underwent remarkable development.

## The Formation of the Inca Empire

The rise of the Inca tribe, which began around the year 1200, was an integral part of the whole complex of emerging hydraulic agricultural civilizations and their grouping into empires in South America. Over the course of two centuries, this tribe occupied only a modest territory around Cuzco, and it is only from the beginning of the fifteenth century that the Incas conquered and unified under their rule the largest, the most fertile and the best situated of the high Andean valleys: the Valley of the Urubamba, high tributary of the Amazon, which became the sacred valley of the Incas.

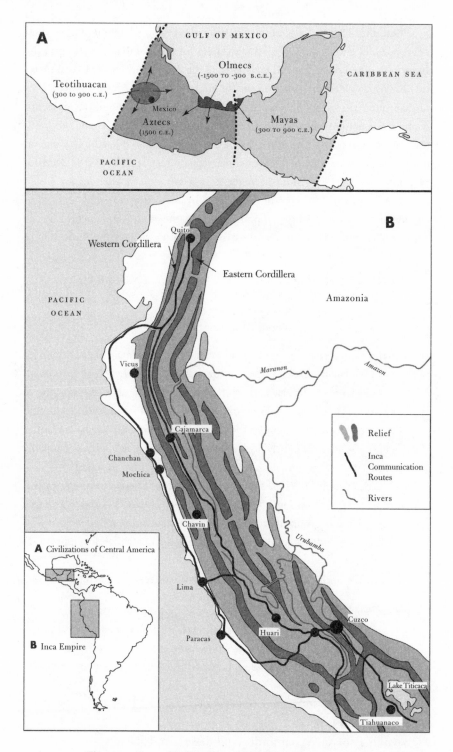

**A**

GULF OF MEXICO

Olmecs
(-1500 TO -300 B.C.E.)

CARIBBEAN SEA

Teotihuacan
(300 to 900 C.E.)

Mexico

Aztecs
(1500 C.E.)

Mayas
(300 to 900 C.E.)

PACIFIC
OCEAN

**B**

Quito

Western Cordillera

Eastern Cordillera

Amazonia

PACIFIC
OCEAN

Vicus

Maranon

Amazon

Cajamarca

Chanchan

Mochica

Chavin

Urubamba

Relief

Inca
Communication
Routes

Rivers

**A** Civilizations of Central America

Lima

**B** Inca Empire

Cuzco

Paracas

Huari

Lake Titicaca

Tiahuanaco

*Figure 5.1 Great Civilizations of Pre-Columbian America*

Then, after having defeated the Chanca military confederation in 1440, their neighbor and rival that struggled with them for control of a part of Peru, the Incas conquered the tribes and city-states of the Andes and Pacific Coast, thereby forming a vast empire, the empire "of Four Directions" (Tohuantsuyn), centered around the capital Cuzco (the navel). At the beginning of the sixteenth century, this empire extended to the north up to Quito in Ecuador, to the south up to Chile and to the Argentine pampas, to the west to the Pacific Ocean, and to the east to the borders of the great Amazonian forest. It thus covered a territory 4,000 kilometers long and 300 to 400 kilometers wide. It federated some seventy ethnic groups. This empire was unified by an economic, social, and political organization that reproduced on a large scale the model implemented by the Incas in the Sacred Valley and on the two neighboring sides of the Andes.

## 2. AGRICULTURAL PRODUCTION AND TRADE IN THE INCA EMPIRE

### Diverse Bioclimatic Zones

In Peru, the natural environment can be divided into three main areas, beginning in the west: the Pacific coastal plain, the Andes massif, and the Amazonian plain. The coastal plain is a desert, with scattered oases situated at the mouths of Andean streams. The presence from May to December of an anticyclone, formed from the rising of cold waters under the effect of the Humboldt marine current, is the origin of the simultaneously arid and foggy climate. The fog dissipates only from January to April, when the anticyclone moves away from the coast.

The Andes massif, which occupies one-quarter of Peruvian territory, is composed of two ranges running from northeast to southeast. These high ranges encompass a diverse group of cold and dry high plateaus, perched valleys with more or less steep sides, and high-altitude alluvial plains which form, properly speaking, the *altiplano*.

It is possible to distinguish several levels in the interior of this massif:

— the *quechua* level, which goes from the valley bottoms and their sides up to 3,600 meters in height, and the *suni* level, which extends from 3,600 to 4,200 meters altitude, two levels whose present natural cover, which is both sparse and highly degraded, is made up of little more than grassy formations with scattered shrubs

— the *puna* level, from 4,200 to 4,500 meters high, which is covered with meadows and steppes

— beyond 4,500 meters altitude, cold deserts and glaciers occupy the slopes and summits, which reach their highest point at 6,000 meters

In the Andes, a high-altitude, semiarid tropical climate predominates, which consists of a dry and cold off-season from May to September and a hotter and more humid season from November to March. The average temperatures, rather low, decrease with altitude (13°C at 3,000 meters, 5°C at 4,000 meters), while the period of sunshine and precipitation increases (600 millimeters in the valleys, 1,000 millimeters in the *puna*). The daily temperature variations are important. It freezes every night in June and July and occasional freezes can also occur in March and November. Beyond the freeze, the dryness constitutes a handicap for agriculture: only three to five months per year, depending on the place, is there adequate water. Moreover, the meteorological conditions change as a function of the topography and the direction of the slopes.

On the other side of the Andes, the Amazonian plain represents more than half of the total surface area of Peru and it is dominated by a dense equatorial forest. Descending from the eastern cordillera to the Amazonian plain, the slopes and piedmont are covered with a grassy vegetation at first, then with a more and more varied and denser bushlike, arboreal vegetation. The climate is constantly hot (the average temperatures exceed 23°C) and humid (the rainfall exceeds 1,500 millimeters per year almost everywhere). The surplus water that saturates the soil is the principal factor limiting its agricultural use.

Thus in Peru, whether it be on the desert coast, in the Andes, or in the Amazon, it is uncommon to find any area that combines all the conditions of temperature, water supply, and slope that would allow cultivation. As a result, agricultural territory is quite fragmented and dispersed amid vast uncultivated areas. The oases of the coast are separated from one another by many kilometers of desert. Cultivated, and possibly irrigated, middle and high valleys of the Andes are interspersed among immense arid or semiarid slopes, pastures, and high, cold deserts. On the eastern slope, the farmers' villages are dotted along rivers that disappear into the vast Amazonian forest. This is a discontinuous agrarian world, an archipelago composed of populated and cultivated islands and islets, dispersed along the coast and the Andean and Amazonian rivers. Lastly, these very heterogeneous agricultural lands are affected by mixed climates and unstable meteorological conditions, which make crop yields often very uncertain.

## Differentiated and Discontinuous Pre-Inca Agrarian Systems at Different Elevations

As early as the pre-Inca period, farming peoples who were scattered across this discontinuous world had adapted themselves to the particular difficulties of this type of environment. In order to reduce the risks of poor harvests or none at all, they multiplied the cultivated parcels in the most varied conditions and diversified the crops and varieties grown on the same parcel. The *ayllus*—population groups with an endogamous tendency, claiming a common ancestor, and composed of elementary families—exploited several lands situated in different ecological niches so that they could take advantage of the possibilities of complementary production. The leaders federated several *ayllus*, generally formed around a central Andean nucleus. They extended their cultivated lands, from the oases to the Amazonian clearings by way of the puna.[2] At the time of the first city-states along the Pacific Coast and in the Andes, differentiated agrarian systems already occupied different bioclimatic levels, namely:

— systems of irrigated cultivation based on maize, beans, and cotton in the oases of the coastal plain
— systems of irrigated cultivation based on maize, beans, lupine, and quinoa (a type of cereal of the chenopodiacea family, which the Spanish called "small rice") in the *quechua* area
— systems of cultivation based on the potato in the *suni* area
— systems of pastoral animal herding in the *puna* area
— systems of slash-and-burn cultivation based on manioc, maize, and coca on the forested Amazonian slope

These differentiated agrarian systems, situated at various elevations and scattered across vast spaces which were either sparsely populated or totally unpopulated, were linked together by trade in agricultural and mining products. This trade, which had begun to develop well ahead of the Inca conquest, already made it possible to exploit the complementarities existing between the different areas. These complementarities were then facilitated by the conquests, tribal confederations, and the first empires (Tiahuanaco, Chimú). The pre-Inca civilizations had also developed elaborate techniques for organizing, irrigating, and manuring the land. Long canals of several dozen kilometers supplied the coastal oases with water. The Andean valleys were arranged into terraces, some irrigated, others not, on very great heights. The collective organization of hydraulic and agricultural works, the management of water and trade, were already performed by a caste of priest or warrior origin, which appropriated a portion of peasant agricultural production for its own maintenance.

The Inca Empire largely depended on the heritage of these older civilizations and took over this organization of production and agricultural trade. From its first conquests, the Inca state had great advantages over its neighbors. It controlled the most extensive and the best organized of the high Andean valleys. From this valley, they could take advantage of several relatively close and complementary Andean levels: the valley bottom, irrigated and planted with maize, the

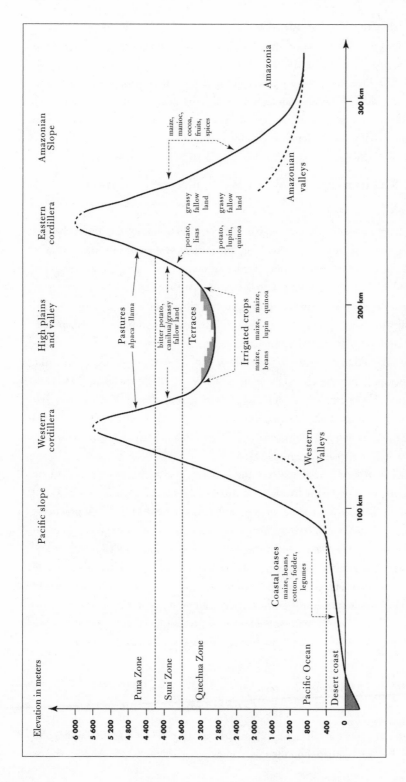

Figure 5.2 Schematic Section of the Cultivation and Animal Herding Systems of the Pacific Coast, the Andes, and the Amazonian Slope During the Inca Epoch

potato zone, the high-altitude pastoral area where herds of llamas and alpacas grazed and the Amazonian slope, which gave them access to coca, off-season maize (maize produced during the Andes winter ), and precious metals, especially gold. The Inca state had incomparable currencies of exchange using livestock, coca, and gold. All these advantages undoubtedly explain why this state could, better than the others, achieve the conquest and the economic and political unification of the vast legacy from pre-Inca societies.

## The Inca Agrarian System: Complimentary Subsystems at Different Elevation Levels

Controlling the totality of agro-ecological niches and monopolizing the exchanges between them, the Inca state could organize them systematically and push the specialization of each region much further, thereby reinforcing the interregional division of labor throughout the Inca Empire. At that time, however, llamas or humans were the sole means of transporting merchandise. Each territory retained the ability to be self-supporting in food products as much as possible and specialization could only be partial: it concerned only one or several products more particularly adapted to the area and from which only the *surplus* was exported to the other regions. Thus the agrarian system at each level, which was necessarily greatly diversified in order to satisfy local needs, was nevertheless relatively specialized and was part of the trade system. The agrarian systems belonging to each area formed a set of complementary local subsystems that participated in a much larger whole: the imperial Inca agrarian system.

The information available on the Inca agrarian society is certainly inadequate to describe with precision the agriculture of each zone. But the long survival of the principal traits of the agrarian geography and ancient agricultural practices, even masked, deformed, or transformed by colonization, furnish enough facts to supplement this information and make it possible to try to present a broad outline of the organization and functioning of this system. This singular, even original system, provides a kind of archetype of a system formed from complementary subsystems, unified by an all-powerful state ruling over undifferentiated peasant communities. Before presenting the social organization and role of the state, we will lay out each of the regional systems composing the Inca agrarian world.

### *The System of Irrigated Cultivation in the Oases of the Coastal Plain*

In the desert plain, only the oases organized and arranged for irrigation were cultivated. They were situated on the raised alluvial sediments beside Andean rivers, in neighboring depressions and at the end of canals descending from the cordillera. The rate of flow of the water carried by these canals was sometimes controlled by a system of reservoirs and locks and, once arrived on the plain, the water was distributed by a network of paved drains.[3] The principal food crops were maize and beans. The cultivation of long-fiber cotton was an original feature of the cultivation system in the oases, and the surplus was exported to the other regions of the empire. Manioc, groundnuts, squash, and peppers were also cultivated, as well as fodder legumes (*Desmodium, Centrosema*), which were, along with crop residues, the only means of feeding animals. Indeed, these oases were surrounded by the desert and, as in the Nile Valley, the presence of a fodder legume in the rotations was very old. The animals consisted of passing llama caravans, which ensured the transportation of merchandise between this region and the rest of the empire. The zone was poor in animal products, which were imported from the high Andes. Villages of fishers established on the coast also supplied dried fish, which usefully supplemented the dietary regime of the population of the oases, in exchange for agricultural products.

Fodder legumes and beans contributed to the nitrogenous fertility of the cultivated soil. But their contribution was not sufficient because, contrary to the silt-laden floodwaters carried by the Nile, the irrigation water, which came in part from melting snow and glaciers in the Andes, was low in mineral salts. Certainly, that protected the soil in the oases from salinization, but it also meant that this irrigation water was not fertilizing. Beyond that, the coastal *guano* deposits, riche in nitrates and phosphates, had long been exploited as mineral fertilizer mines. *Guano*, the decomposed and mineralized product of the feces and skeletons of millions of marine birds accumulated over the centuries on the Pacific Coast, was used as fertilizer in the oases and in the Andean valleys, to which it was carried by llama. It is possible to compare this use with the practice of exploiting the *tells* of the Nile Valley as fertilizer mines during the Pharaonic period. Tells were mounds that resulted from the accumulation of domestic wastes on the sites of ancient villages over thousands of years.

## The System of Irrigated Maize Growing and Associated Animal Breeding on the Quechua Level

The *quechua* level, which is composed of the valley bottoms and their sides up to 3,600 meters in altitude, concentrated the bulk of the empire's population. The ecosystem was composed of irrigated lands, arranged into terraces and planted with maize; non-irrigated, cultivated lands, possibly arranged into terraces also; spontaneous grassy and shrubby formations, exploited as pasture; and lands left wild.

The organization of irrigated valleys is among the great accomplishments of the Inca civilization. In general, gravity irrigation of a valley section was based on an arrangement that included, at least, a water intake located on the river bed upstream from the irrigable perimeter. Starting from this water intake, a primary diversion canal moved across the hillside with a small incline, so as to remain elevated enough and to dominate whole portions of the slope. The slope was arranged into terraces supported by stone walls. The steeper the terrain, the narrower the terraces and the higher the retaining walls. Terracing of the slopes made it possible to fight erosion and obtain deeper soil, which was arranged in flat, cultivable strips and enriched by fine ingredients (clays, organic matter, mineral elements) brought by oblique leaching of the lands located upstream. Secondary canals, added to the primary diversion canal, conducted the water from terrace to terrace. The water then circulated throughout the terraces, before being finally distributed to the cultivated parcels by a network of tertiary canals. The organization of a whole valley segment bounded by two narrow passes often included several basic irrigation networks each controlled by a water intake.

In the Inca installations, rather small water reservoirs (large reservoir dams did not exist) were located at the level of the water intakes or at the elevated parts of the system. They made it possible to store water for irrigation. The steepest and most impressive installations were close to high defensive sites and perched fortresses, like Machu Picchu, where the cut and fitted dry-stone walls, sometimes of cyclopean dimensions, reach heights of two to three meters. Such installations were designed to ensure a minimum of agricultural production while resisting a state of siege. The large and flat valleys, by contrast, required less terracing and masonry work. Such was the case for the Sacred Valley at the center of the Inca Empire, with its large terraces and its low walls. Moreover, dikes were built along the river in this valley.

All these large-scale installations were principally intended for the cultivation of maize. The surplus was exported to other areas. Maize was consumed boiled, grilled, as cornmeal, or even as a fermented drink, *chicha*. It was cultivated each year, possibly in association with beans, lupine, quinoa, or fodder legumes. On non-irrigated cultivated lands, tubers such as oca or potatoes, lupine, and quinoa were cultivated in rotation with grassy fallow land of moderate duration. Squash was also cultivated, but cotton, requiring heat, was not grown in high-altitude regions.

The undeveloped slopes and the heights were used as pastures for the local llamas and those passing through in caravans. However, these rather scanty pastures were inadequate. Hence fodder legumes were grown, the stalks and leaves from the maize plants were systematically harvested, and a portion of the herds was moved to summer pastures on the high-altitude prairies and steppes of the *puna* zone, all to supplement these pastures. As in the oases, the legume crops contributed to the reproduction of fertility. Moreover, apart from cultivation periods, the lamas, who grazed on the range during the day, were assembled at night onto the fallow lands. By means of their excrement, they transferred a portion of the organic matter from grazing on the open range onto the cultivated lands. In total, these fertilizing contributions were sufficient and, according to some Spanish chronicles, made it possible to obtain yields of irrigated maize on the order of 20 quintals per hectare (a maximum, according to our research).

Animal products from the *quechua* level came from the local herds of llamas, but also from raising chickens and guinea pigs, and from hunting. Fishing villages established on the edge of Lake Titicaca also had fresh or dried fish.

## Systems of Potato Cultivation with Associated Animal Breeding on the Suni Level

In the *suni* zone lying between 3,600 and 4,200 meters in altitude, both cooler and wetter, the ecosystem included non-irrigated cultivated lands, pastures, and lands left to the wild. The principal crop was the potato, transformed into *chuno*, in order to preserve it. It was dehydrated by alternately exposing it to nightly freezes and daily sunshine. Thus it could be preserved for two to three years and easily transported to other regions. Dozens of varieties of potato were domesticated in this region, which provided the rest of the world with original clones. The potato was placed at the beginning of diverse rotations that varied with the altitude. At the lowest elevations, rotations of potato-lupine-quinoa-grassy fallow land of several years duration, or potato-lisas (a tuber)-grassy fallow land were found. At the highest elevations, the rotation bitter potato-canihua (a cereal of the chenopodiaceae family)-grassy fallow land of several years duration was found.

The grassy fallow period performed a sanitary function: it eliminated the nematode cysts that live as parasites on the potato. Also, through its production of biomass, it contributed to the reproduction of fertility. In addition, the herds of llamas that consumed the spontaneous vegetation on the neighboring open ranges could manure this fallow land as well as the small fallow lands of the off-season. Once dried, llama dung also acted as a fuel in the high-altitude areas without wood. After seven years (or more) of lying fallow, the preparation of the soil for planting a new crop of potatoes required a true plowing, i.e., a turning over of the top layer of the soil in order to destroy and bury the spontaneous groundcover of the fallow period and to loosen and aerate the soil to allow the young potato plants to take root.

*Plowing with the Taclla.* The Andean farmers possessed neither spade nor large hoe, ard, or plow to clear groundcover. Instead, they made use of a type of well-developed digging stick, the *taclla.* It was used in an ingenious manner, making it possible to perform a true plowing. The working part of the *taclla* consists of a long, thick, narrow blade that, in the Inca period, was made of polished stone or, rarely, of bronze. This blade, today made of iron, is solidly tied onto a large shaft of wood, more than a meter long, provided with a handle for easy manipulation. The difference between the blade of a spade and that of a *taclla* is that the latter has no supports to provide a place for applying pressure with a foot in order to drive the tool into the soil. Instead, the shaft of the *taclla* is equipped at its base with a sort of small crosspiece made of wood, also solidly tied on, which provides a footrest. The blade of the *taclla* can then be pushed deeply into the soil by pressure from the foot. Since this blade is too narrow to permit large clods of earth to be lifted and turned, several workers use their *tacllas* together in a coordinated effort. They push the blades of their *tacllas* into the soil side-by-side so as to cut and lift a large clod, which is then turned over by a fourth person (most often a woman or child). This person buries and composts the spontaneous vegetation and surface organic matter. Further, this person breaks up the clods to loosen and aerate the soil, pulls up the weeds, and removes any stones that may be found. Plowing with the *taclla,* equipped with an iron blade, is still widely practiced today.

## Pastoral Systems on the Puna Level

Above the areas of potato cultivation and the nearby pastures associated with them, the meadows and steppes of the *puna* level, between 4,200 and 4,500 meters in elevation, were used by herds of llamas and alpacas. The *puna* level supplied other regions of the empire with pack llamas, wool, skins, and dried meat. The llama, or domesticated *guanaco,* is a camelid that made good use of the poor quality high-altitude pastures. Although its carrying capacity is weak (20 to 30 kilograms maximum per animal), its great stamina made it an excellent means of transport in difficult zones. The llama can go for several days without eating or drinking. Its meat was consumed, but not its milk, and its skin was worked and used for various things. The alpaca, or domesticated *vicuña,* is another camelid that supplied a long and delicate wool of very high quality.

## Systems of Forest Cultivation on the Amazonian Slopes

On the Amazonian side of the cordillera, the high forest sheltered villages of slash-and-burn farmers, who cultivated maize, manioc, and coca, a leaf for chewing or brewing. This zone exported off-season maize to the rest of the

*Figure 5.3 Scenes of Peruvian Indians Performing Agricultural Work According to a Christian Calendar from the Beginning of the Colonial Period*

empire, which formed an appreciable food supplement to meet consumption needs in the Andean zone, coca leaves, fruits, spices, and ornamental feathers. The Amazonian plain, forested and often swampy, was relatively unpopulated and unexploited.

## Inter-Level Exchanges

In summary, each zone of the empire had an autonomous food base: maize and beans in the oases of the coastal plain and in the irrigated valleys, potatoes and meat in the high-elevation zones, maize and manioc on the Amazonian slope. However, each region supplied the others with goods for the production of which it had bioclimatic advantages. Thus, the coastal oases supplied cotton and *guano*, the irrigated valleys of the *quechua* level of the Andes, the heart of the Inca agrarian system, provided maize to a large portion of the population in the cities and mines of the *altiplano*. The *suni* level supplied potatoes, in the form of *chuno*. The high-altitude pastures of the *puna* acted as a supply base to other regions for pack animals as well as animals for slaughter, wool, leather, and skins. The Amazon supplied products from gathering activities, coca leaves, and off-season maize.

## Tools and Labor Productivity

The agricultural tools that Inca society used were rudimentary and not very powerful: an improved digging stick (the *taclla*); a wooden sledgehammer to break up the clods of earth; small hoes for hoeing, opening up furrows or digging canals; a knife for harvesting; baskets for humans and packsaddles for llamas for carrying products; and different kinds of pottery. This equipment, in fact, corresponds to the end of the Neolithic epoch and beginnings of the Bronze Age. Inca society did not have the wheel, the harness, or iron. Given such a lack, the productivity of agricultural labor was not high. The cultivated area per worker was less than one hectare for pluvial cultivation and less than one-half hectare for irrigated cultivation. Yields did not exceed 20 quintals of cereal-equivalents for irrigated and manured cultivation and did not reach 10 quintals for pluvial cultivation. Furthermore, because of the poor means of transportation, peasant communities were forced to devote much time to transporting goods and supplies, as is still seen today on the roads, country paths, and in the cities of the Andes. For all these reasons, agricultural production could hardly exceed the needs of the producers and their families. The surplus that could be used by groups in the population that did not participate directly in agricultural production was therefore usually quite small.

## 3. SOCIAL ORGANIZATION AND THE ROLE
## OF THE STATE

## Social Classes

The immense majority of Inca society was made up of peasant communities (the *ayllus*), which had little or no social differentiation. At the summit of the society and state was the *Inca*, male heir of a patrilineal clan considered descendants of the Sun-God. The Inca was the oldest son of the preceding sovereign and the latter's own sister or half-sister. The Inca practiced polygamy on a large scale (the last Inca had, it is said, more than 700 spouses, born of other noble clans of Cuzco or the provinces) so as to extend his influence and his alliances and thereby preserve his hegemony. The nobility consisted of descendants of the Inca, his numerous living relatives in Cuzco, other noble lineages of the Inca tribe and provincial tribes, as well as lineages ennobled for services rendered. Members of the clergy and administration formed two privileged classes, exempted from manual labor and tribute. The upper nobility filled posts in the higher ranks of these two bodies, while the common people filled posts at lower ranks. From the high priest to the lowest official, from the Inca to the smallest local functionary, the administration and clergy were strictly centralized and hierarchical.

For administrative purposes, the population, counted using the decimal system, was divided into groups of 10, 50, 100, 500, 1,000, 10,000, and 40,000 families (an *ayllu* consisted of around 100 families), and the rank of a functionary, called *curaca*, depended on the size of the group for which he was responsible. The *curacas* closely organized and controlled the work of the population, attended to its well-being, rendered justice, and regularly informed the central authorities of the state about the things and people within their jurisdiction. Artisans formed another social class, consisting of different artisans (stonecutters, ceramists, metallurgists, etc.) exclusively in the service of the state. While the members of this class originated from the peasantry, from which they were taken and sent to Cuzco for education, they should not be confused with the mass of corvée laborers, momentarily put in state service by peasant communities to carry out tasks requiring few skills.

## *The Distribution of Land and Livestock*

All the empire's land formally belonged to the Inca, who had "eminent domain" over it. Cultivated lands were not, without exception, private property. To simplify, they were divided into three categories. First of all, peasant lands were distributed periodically among family groups as a function of the number of laborers and mouths to feed, which changed over time. Thus, each couple received as usufruct a *toupou*, that is, the area necessary for its subsistence, to which was added one *toupou* per son and one-half *toupou* per daughter. The size of the *toupou* varied between one-third of a hectare and one hectare depending on the quality of the land, such that a family cultivated, in general, one to two hectares, seldom more. On another part of the land, those belonging to the Sun-God, the harvests were reserved for the clergy. Finally, the yield of the Inca's land was reserved for the support of the emperor and his family, as well as the support of the rest of the nobility, administration, artisans, miners, caravan leaders, armies on campaign, and peasants carrying out corvée labor on building sites far from their homes. It was also used to ensure the food security of the population: maize and quinoa, *chuno* and dried meats were stocked in large warehouses and redistributed in case of local or general shortages.

The Inca was also the formal owner of all the herds in the empire, which were also divided into three categories: the Inca's herds, the clergy's herds, and the herds belonging to the peasant communities. While a part of the latter existed as collective property, the other part consisted of family herds. Each peasant family could possess, beyond its house and the patch of land around it, a few animals. Livestock, in fact, constituted the only possible form of accumulation of wealth for the peasantry. The size of private herds could be up to a hundred head and sometimes even more.

All of the lands were cultivated according to an agricultural calendar established by the administration and appropriate to each area. The fields of the Inca, of the Sun-God, and possibly those of the local *curacas* were cultivated by peasant corvee labor. Corvée laborers tended the herds of the Inca and clergy. In addition, corvée laborers or young girls spun and wove the wool in workshops serving the state or clergy.

## Collective Corvees

The tribute in labor imposed on the peasantry, the *mita*, was not limited to productive agricultural tasks. It was also used for large work projects organized by the administration during the agricultural off-season. A large part of this work was devoted to the construction and maintenance of hydraulic networks, terraces, roads, relay stations, and warehouses, work that is indirectly but clearly productive. Another part consisted of building military works and cities. Finally, some of this work was devoted to building extravagant palaces, temples, and tombs. These large work projects temporarily displaced the population, which was then fed from reserves stocked in the numerous state granaries. Other temporary services demanded of the peasants were: the army, transport of merchandise, labor on the lands of people who were ill, aged, infirm, or orphaned, and harvesting of cotton on the coast or of coca in the Amazon. Every corvée was imposed collectively on an entire peasant community, and it was the local *curacas* who managed the distribution of tasks.

In addition, peasant communities supplied lifetime servants to the administration, the *yanaconas*, a type of serf used as domestics, animal herders, or workers by the Inca, nobles, and some *curacas*. This form of servile labor was not extensive on the eve of colonization, but was much developed after that.[4]

## The Importance of Tribute in Labor

In view of the poor quality of the equipment, low level of labor productivity and the importance of the corvées, every able-bodied individual (men, women, children, the elderly) had to participate, in so far as they were able, in agricultural and domestic tasks. For that purpose, the male population was divided into ten age classes. It is possible to distinguish the following categories: babies in the cradle; children from 1 to 5 years of age, who played; children from 5 to 9 years of age, who were employed at tasks of secondary importance; children from 9 to 12, who were put in charge of chasing birds away from the fields; adolescents from 12 to 18, who supervised the llamas or were apprentices; youth from 18 to 25, who assisted their parents in all work; adults from 25 to 50, who worked and were subjected to the corvée and conscription; men from 50 to 60, who still rendered various services; "sluggish old men" older than 60 who still carried out small tasks and provided counsel.[5] A tenth category included the ill and infirm, who were incapable of working. The same division into age classes existed for women.

Moreover, because the lands granted to peasant families were just large enough to support them, the Inca tributary system, unlike the Pharaonic tributary system, did not rest on a tribute in kind, i.e., on taking part of the family harvest, save for an exceptional harvest that allowed the reserves to be filled. It rested completely on a tribute in labor, i.e., on the use of all the surplus labor force from peasant families by means of corvées.[6] In this system, the peasantry had no possibility of consuming beyond the satisfaction of its essential needs, nor any possibility of acquiring consumer durables of whatever kind. Peasant families and communities were kept in a state of privation from which neither enterprise nor significant equipment could emerge. All the means of investment were in the hands of the state, and agricultural production progressed in step with the development of the collective means of production, such as the hydraulic installations. Very little progress in agricultural production was due to advances in the instruments of individual production.

In this type of society, in order to increase the volume of the surplus intended for the consumption of the privileged classes, the state had to reserve a sufficient part of this surplus to expand the productive base of the system. Such an expansion occurred by conquering new territories and building new irrigation networks in those territories. That required the massive mobilization of peasant forces, in particular during the off-season, in order to do the least harm to directly productive agricultural labor. In order to accomplish that, the discipline of peasant labor and the competence, diligence, and honesty of the administration had to be firmly maintained and extravagant consumption by the privileged classes of the society could not be allowed to take away from the necessary resources for expansion, or indeed even the simple reproduction of the system.

## Role of the State

Beyond the political, military, and administrative functions that are usually incumbent on a state, the Inca state carried out, like other "hydraulic" states (Pharaonic, Sumerian, Chinese, Vietnamese, etc.), extended technical and economic functions. In particular, it directly organized hydraulic construction and trade among the different zones of the empire.

It is well-known how difficult it is to formulate a comprehensive plan for a river basin in order to extend irrigated lands as much as possible, taking into account the availability of water and irrigable lands, and minimizing works and terraces. In order to do that, it is necessary one way or another to devise an organizational plan that combines as well as possible the different hydraulic sections of the same basin. It is sometimes necessary to anticipate the renewal or reconstruction of older works and manage the construction of the whole structure, which can last for several years, indeed decades, through successive phases, making necessary adjustments to the very end of the process. Lastly, it is necessary to oversee the distribution of water in time and space, as a function of the needs of the irrigated lands from different parts of the basin, an all the more delicate operation given more fully developed land and more complete use of water resources. These problems are difficult and far from being correctly resolved in most modern developments, despite the improved methods and means of calculation available. Furthermore, when a valley is small, only a unified and experienced hydraulic authority can carry out the planning of installations, the regulation of labor and the management of water. As long as the authority all through the same valley remains divided, hydro-agricultural coordination of the whole is not possible.

A skillful administration, composed of architects, agronomists, hydraulic experts, and specialists in civil and military engineering, educated at the University of Cuzco, coordinated and managed all these tasks of designing and directing the hydraulic works. It also organized the long-distance trade between the different regions of the empire, thereby strengthening economic integration and political unity. To accomplish this, the administration used the Inca's large llama herds and a vast network of well-maintained paved roads, marked with relay stations that acted as warehouses for provisions, clothes, sandals, and arms. Two major roads, one coastal and the other on the crest, traversed the empire along its length, connected by multiple transversal roads that linked together all the levels of the Andes. Beyond trade organized by the state, goods were exchanged directly among the peasants in rural markets that enlivened local life. Neither merchants nor money existed. Small copper axes or shells were used to facilitate transactions.

In order to manage all these activities and control all this wealth, the administration used accountants who made a census of, counted, registered, and kept up-to-date lists of workers, stocks of merchandise, lands, and corvées. Inca society had no writing system, but it did use a system of counting consisting of cords with knots, the *kipu*. Different according to region and activity, the *kipus* were true accounting registers. They were dispatched from the provinces toward Cuzco and kept the central power constantly and precisely informed on the state of things and people in the entire empire. By forcing the trait a little, one could say that the Inca agrarian economy was, like the Pharaonic agrarian economy, if not planned, at least centrally administered. Despite the poor equipment and the great difficulties of exploiting the Andean environment, Inca society developed a powerful civilization under the leadership of the state and by relying on the heritage of pre-Inca societies.

Lastly, the function of the state also extended to cultural and religious domains. Thus, so long as the Inca state advanced militarily, it imposed *Quechua* as the official language of the administration and the clergy. Functionaries entrusted with teaching it were sent out to the most distant places and, at the time of the Spanish conquest, only a century after the beginning of the formation of this vast empire, around one-third of the population spoke *Quechua*. Despite a certain tolerance toward other religious beliefs, all the conquered tribes had to submit to the Inca cult of the Sun-God, a cult in which agricultural labor was considered sacred. The Inca himself and his family set the example by ceremoniously cultivating, in festival costume, the fields of the Sun-God.

In the conquered provinces, the local elites were subjected to the influence of the Inca nobles and functionaries who were put there in order better to control them. But local populations preserved some of their traditions, such that the empire displayed, despite everything, a great cultural diversity. However, resistance of conquered peoples was both common and vigorous. In case of revolt, populations could be transferred en masse hundreds or thousands of kilometers, and replaced by Inca or assimilated populations. Population displacements also aimed at better balancing the population with the resources and labor needs of different regions.

## 4. THE DESTRUCTION OF INCA SOCIETY

In 1527, when the Spanish conqueror Pizarro, financed by a rich merchant established in Mexico, disembarked for the first time with his troops in the north of the Inca Empire, his objective, like that of other conquerors, was the discovery and exploitation of the mineral wealth of the new territories, beginning with gold and silver, reportedly abundant. To that end, the Spanish Crown awarded to the conquerors an exclusive title (the *capitulation*) over the new countries that they had subjugated. At the time of his third expedition, in 1531, Pizarro, with his troop of 182 men, destroyed the Inca political and military organization in a few weeks. This rapid victory seemed so extraordinary that explanations just as extraordinary were put forward in an attempt to explain it: misunderstanding by the Incas, who had confused the Spaniards with the hypothetical messengers of the creator god of the world or surprise, hesitations and superiority complex of the inheritors of a dynasty that seemed invincible for over a century. It is certainly possible that circumstances of this nature had facilitated the Inca defeat.

It remains the case, however, that at the time, any society in America, Africa, or Asia could not resist the armored cavalry and firearms of the Europeans, which the former did not have. As opposed to the Spanish, the Incas possessed neither swords of tempered steel nor armor, horses, harquebuses, or cannons, which they had never seen or heard about. The easy conquest of Peru is explained first by the unstoppable superiority of Spanish armaments and it certainly would have happened regardless of the dominant political and moral circumstances in Inca society at that time.

But it is necessary to point out that the Spanish conquerors benefited from the complicity of some Indian tribes who had never accepted Inca domination as well as rivalries between the partisans of the legitimate heir of the Inca and partisans of his brother. Moreover, as the Spanish priest Bartolomé de Las Casas bore witness to in *The Devastation of the Indies: A Brief Account*, the conquerors showed unheard-of treachery and violence. Under cover of peaceful meetings, several ambushes were organized to massacre the nobility. The Inca himself was captured and executed, despite the payment of a considerable ransom in gold and silver objects by his people. Such massacres were sufficient to bring down the state. They were not, alas, unique to the Spanish colonization of Peru. It is remarkable, on the other hand, that the "black legend" denouncing the atrocities of this colonization, revealed by Las Casas and others, could be expressed with such force and be understood in the Spanish society of the time.

Other colonizations just as murderous did not give rise, in their time, to so many scruples. Lastly, one should note that, as in Egypt at the time of the Greek conquest, it sufficed for the Spanish invaders to strike and take the head of this "pyramidal" type of society in order to subjugate it for a long time.

## 5. INSTALLATION OF A SATELLITE
## COLONIAL ECONOMY

### Non-renewable Exploitation of the Colony

For the most part, the conquerors were adventurers who fled the worsening conditions and the poverty of their lives in Spain. Certainly, functionaries appointed by the royal crown and members of religious orders also participated in colonization, but they were in the minority compared to ruined nobles and their henchmen who, motivated by the lure of gain, borrowed from European financiers for their voyage and their settlement in the New World. Squeezed by debts, greedy, and devoid of scruples, the conquistadors began to pillage the treasures of Inca society and rob its tombs. Then they organized the colony in order to exploit the gold and silver mines of the high Andes.

The viceroyalty of New Castile was founded in 1535. In order to be in a better position to carry out its exports of mineral products and conduct its relations with Spain, the colonial power established its capital at Lima, along the coast, and not at Cuzco, the center of the ancient Inca economic system. The lands of the Inca were granted to the Spanish Crown, those of the Sun-God were allocated to the Church, while the Indian communities conserved in principal the usufruct of the lands they had exploited previously. The territory and population were divided into types of fiefs, the *encomiendas*, the control and exploitation of which were granted for one or two generations to the conquistadors or to Inca dignitaries who had rallied to the Spanish, the *encomenderos*. The other fiefs were governed directly by representatives of the Spanish royal power or by religious orders.

Reserving a good share of the lands for their own use, the encomenderos forced subject Indian populations onto marginal lands. Enjoying an absolute right of life and death over these populations, they exploited them without mercy, while hijacking the ancient Inca institutions for their own profit. Under the *mita*, the *encomenderos* demanded of the Indian peasants sizeable corvees in labor in order to exploit their lands and their factories, as well as long periods of forced labor in the mines. They requisitioned, moreover, a share of the agricultural produce from the Indian communities, in order to supply the urban centers, the mine workers, and factory workers.

The former *curacas* were charged by the *encomenderos* with distributing the appropriations in kind and corvées among the members of the Indian communities. While these were generally tolerable in certain ways, the conquest of the New World prolonged the reconquest over the Arabs in the Iberian Peninsula. This huge enterprise united nascent merchant capital, royal power, the Church, and a part of the marginalized nobility and its henchmen. The discovery of territories, wealth, and populations so unarmed that they were easy to conquer made it possible for them to reproduce in the colonies forms of servile exploitation that were in the process of disappearing in western Europe at the time.

The relics of the administration and other Inca institutions, even if they sometimes kept the same appearance, were used by the colonial system for its own ends, to the detriment of their former functions. Thus the maintenance of the hydraulic networks and other infrastructure was neglected, the security reserves were not renewed, and trade among the different zones was disorganized. As a result, agricultural production in the colony collapsed, scarcity and famine multiplied, the weakened population became prey to disease, in particular diseases imported from Europe (smallpox, measles, leprosy, etc.) against which it was not immune. The population of the former Inca Empire shrunk from 10 million inhabitants in 1530 to around 2.5 million in 1560.[7] In 1590, the population was no more than 1.4 million inhabitants and it practically remained at this level until the beginning of the nineteenth century. It is only at the beginning of the twentieth century that this population again achieved its pre-conquest level. Because this population collapse threatened the future of the colony, slaves were brought from Africa beginning in the sixteenth century. Undoubtedly, the massacres, ill treatment, and disease played their role in this hecatomb, but it was above all the result of the collapse of agricultural production, a collapse due to the disorganization of the Inca economic, social, and political system. This proves, by the negative, that this system, which aimed to increase the power of the Inca, also had to fulfill the functions necessary to maintain the development of production and of the whole population in order to be successful.

For moral reasons, as well as economic ones, members of religious orders, in the first place B. de Las Casas, and royal officials, such as Garci Diez, condemned the *encomienda* system in written reports intended for the Spanish Crown, some going so far as to advocate the restoration of some Inca institutions. To curb the exactions of the *encomenderos*, take back from them lands in the royal domain and reestablish its authority in the colony, the Crown had to conduct a veritable war of reconquest, which ended in the middle of the sixteenth century. Subsequently, the royal power tried to dismantle the *encomiendas* (the last ones disappeared at the beginning of the eighteenth century), and reproduce in the colony the agrarian structures of the metropolis.

## The Formation of Large Estates and the Marginalization of the Indian Peasantry

Beginning in 1570, the first *haciendas*, large estates whose legal status rested on private property in the soil and subsoil, were created. The formation of the haciendas was accompanied by the institution of a policy known as the *reductions*, which consisted in regrouping what remained of the Indian population into villages and limiting its cultivated lands. Also, the illegitimate appropriations by the *encomenderos* of lands belonging to Indian communities and to the royal estate were legalized in return for payment of a sum of money to the Royal Treasury.[8] The haciendas were thus most often formed from former *encomiendas* and neighboring lands, originally Indian or royal. The property owners were former *encomenderos*, functionaries, Spanish soldiers, former Indian *curacas*, as well as the Church, which benefited from numerous donations and became the primary property owner in the colony.

The haciendas were thus exploited by a servile Indian labor force, which was granted small plots of land for its subsistence. The rest of the Indian population lived on the reductions, compelled to pay a tribute in kind and a tribute in labor (in the mines). Little by little, the tribute in kind was converted into a tax in money to be sent to the Royal Treasury. Thus the Indians had to be employed as day or seasonal workers on large estates, in order to obtain the necessary money to pay the tax. But, for a number of them, indebtedness to the owners led to a new form of servitude.

## An Underequipped Agricultural Economy, Exporter of Primary Goods

The Spanish introduced new plants into Peru: broad beans, a food legume that became part of the rotations based on the potato; wheat, not labor intensive, which was substituted, in part, for the potato, but also barley, oats, rye, alfalfa, grapevines, olives, other fruit trees, and sugar cane. They also introduced animals of European origin, oxen, sheep, pigs, poultry, horses, donkeys and mules, and some tools: the sickle, hoe, spade, ard, and cart, as well as grain mills, sugarcane crushers, and spinning and weaving equipment. These new instruments made it possible, in part, to make up for the shortage of labor power and to increase the productivity of agricultural labor on the haciendas.

But since in the middle of the sixteenth century the medieval agricultural revolution had hardly touched the north of the Iberian Peninsula, the Spanish colonizers did not transfer the scythe or the plow. Even today the Andean peasantry often lacks this type of equipment, working most often with the hoe, the *taclla*, and the ard.

One after the other, mineral resources (gold, copper, zinc, lead, saltpeter, guano) and agricultural resources (sugarcane, cotton, cattle, tobacco, rubber, quinine, coffee) from different regions were mobilized for export. New regional specializations were formed: sheep raising in the Andes, sugar in the coastal oases of the south and cotton in the coastal oases of the north, tobacco in the Andean valleys, coca, rubber, quinine and coffee in the Amazon. In contrast to the regional agricultural specializations of the Inca system, which were only partial and were part of the food production equilibrium of the country, the new regional specializations were most often agro-exporting mono-productions, as extensive as possible, contributing to the food supplies of the developed countries of Europe and North America. Consequently, the space for Indian peasant subsistence was reduced.

## Independence and Economic Subjection

As in the other countries of Latin America, the large owners and beneficiaries of the import-export commerce formed a landowning, mercantile, and financial oligarchy. In Peru, this oligarcy obtained independence in 1821 and it took the reins of power with the support of the army.

But the riches of Peru again attracted the covetousness of Spain, which tried to reestablish itself in its former colony (1863), and later Chile (1880), which annexed the entire southern part of the country where the largest deposits of guano were found. This wealth was also the origin of a ruinous policy of modernization (excessive and for the most part unproductive public works) which led the country to become overindebted, mortgage its resources, and finally to submit to foreign financial control, a situation reminiscent of Egypt's at the end of the nineteenth century (see chapter 4). In 1890, the United Kingdom and the other creditor countries created the Peruvian Corporation, which was granted, for a seventy-year lease, the control of the Peruvian railroads, the right to exploit the guano mines and oil wells, and the free use of seven Peruvian ports, as reimbursement for a debt of 50 million pounds. Politically independent, at least formally, Peru lost its economic independence for several decades.

## Persistence of the Latifundia and Minifundia

In order to remedy the shortage of labor power resulting from the hecatomb of the Indian population, Peru, like most of the tropical American colonies, imported African slaves. But on the Pacific coast, far from the main movements of the African slave trade, these imports were never very important (100,000 slaves in three centuries). Formally abolished since Independence in 1821, slavery was not actually abolished in practice until the second half of the nineteenth century. The large sugar-growing and cotton-growing estates of the coastal oases then made use of low-cost labor power, principally of Chinese origin.

As for the large animal-raising estates of the Andes, they made use of a quasi-servile labor force, poorly paid or not paid at all, originating from peasant communities confined to insufficient lands. The haciendas, which had never stopped encroaching on the lands of Indian communities, continued to expand to the detriment of peasant property, after its establishment at the beginning of the nineteenth century. Thus in Peru, as in most of the Latin American countries, a distribution of property on the latifundia-minifundia model was again reinforced. On one side, a small number of latifundia, huge estates of thousands, tens of thousands, sometimes even hundreds of thousands of hectares, held three-fourths of the land. On the other side, millions of minifundia, peasant plots too small to occupy and make a living for a family, disposed of less than 10 percent of the land, not counting the more numerous landless peasants, always searching for always more precarious employment.

Furthermore, while peasant agriculture most often remained at the level of manual cultivation or a poorly developed animal-drawn cultivation, the large sugar- and cotton-growing estates of the coast profited from the favorable conjunctures of the 1950s and 1960s to achieve motorization and large-scale mechanization. By eliminating nine-tenths of the agricultural jobs, moto-mechanization made the situation of the landless "peasants" and the minifundists, which supplied the labor power in this sector up until that time, even more untenable. To a certain extent, Amazonia served as an outlet for these hard-pressed peasants who, at the cost of thousands of difficulties, went there to cultivate coca.

In the interstices of the latifundia system, a middle peasantry, possessing generally only light equipment and a few heads of cattle, began to form. But for a long time it was confronted with the narrowness of the internal market and competition from basic agricultural produce coming from better-equipped and more productive farms. With the urban explosion, it turned toward the production of fresh and perishable produce. Fruit growing, truck farming, and dairy zones were thus formed in the vicinity of cities and in valleys close to the best roads.

Not having left its peasantry the necessary space to flourish, Peru saw them sink into poverty, migrate to the cities, suffer unemployment, and undergo a demographic explosion, a usual corollary of mass poverty in the twentieth century. The agrarian reforms of the second half of the century came too late to curb this development.

## 6. CONCLUSION

At the beginning of the 1990s, according to the UN Food and Agriculture Organization, Peru was among the twenty-two developing countries whose food supplies provided fewer than 2,100 calories per person per day. More than a quarter of the working-age population is unemployed. While a tiny minority (2 percent) of the population has one-third of the national revenue, more than 10 million people out of a total of 23 million live in extreme poverty and are undernourished. The great majority of the poor (around 70 percent) live in the countryside, particularly in the Andes. Nevertheless, poverty is also present in the cities, above all in Lima, a megalopolis of 8 million inhabitants making up one-third of the population, in which there are hundreds of thousands of poor and tens of thousands of abandoned children.

As with most of the Latin American countries afflicted with the latifundia-minifundia system and unequal exchanges with the developed countries, Peru and the other Andean countries are in a economic and social impasse, which the usual politics of expediency cannot overcome.

What the "devestation of the West Indies" cost in misery for the Andean peoples, and what it still costs them each day in misery living in a society that is distorted and tormented by so many difficulties, is beyond measure. What humanity lost by this historic cataclysm, which deprived it of cultural riches that could have supported the development of an original civilization, is absolutely unimaginable. And what it will cost the Andean populations themselves and the world, in perceptiveness and political courage, time, work, and money to reconstruct a society with a human face in the Andes is again immeasurable. But it is certain that the collective costs of colonization and its consequences will, in the end, be infinitely greater than the unjustified benefits some derived from it.

# 6

# Agrarian Systems Based on Fallowing and Animal-Drawn Cultivation with the Ard in the Temperate Region: The Agricultural Revolution in Antiquity

So they will live pleasantly together; and a prudent fear of poverty or war will keep them from begetting children beyond their means. ... The community I have described seems to me the ideal one, in sound health as it were [limited to what is strictly necessary], but if you want to see one suffering from inflammation [populous and luxurious], there is nothing to hinder us. ... The country, too, which was large enough to support the original inhabitants, will now be too small. If we are to have enough pasture and plow land, we shall have to cut off a slice of our neighbors' territory; and if they too are not content with necessaries, but give themselves up to getting unlimited wealth, they will want a slice of ours. ... We need not say yet whether war does good or harm, but only that we have discovered its origin in desires which are the most fruitful source of evils both to individuals and to states.

—PLATO, *The Republic*

The agrarian systems based on fallowing and cultivation using the ard in the temperate regions are derived from systems of slash-and-burn cultivation that occupied the forested environments of these areas since the Neolithic period.[1] They first developed in the hot temperate countries of the Mediterranean region, then in the cold temperate countries of Europe as they were deforested. This deforestation progressively expanded from east to west and from south to north in the age of metals, from 2500 B.C.E. to the first centuries of the common era. The development of systems using fallowing was around 2,000 years after that of the hydro-agricultures of the arid regions (Mesopotamia, Nile, and Indus Valleys). In the hot temperate regions, the predominance of systems based on fallowing did not exclude the limited presence of hydro-agricultures.

Undertaken in environments receiving enough rain to make possible the rain-fed cultivation of cereals and deforested enough to allow the development of pastoral animal herding, these systems rest on the combination of these two activities. Cereal crops are concentrated on the most fertile arable lands (the *ager*) where they alternate with the natural growth of grass during a fallow period, thereby forming a rotation of short duration, generally biennial.[2] The livestock exploit the relatively extended peripheral pastures (the *saltus*) and play a role in fieldwork and in the reproduction of the fertility of the cultivated lands. They supply the necessary energy for pulling the ard and for transport with the packsaddle, instruments of labor characteristic of animal-drawn cultivation using an ard. Moreover, pasturing during the day on the saltus and during the night on the fallow lands, the livestock ensure a certain transfer of fertility from the pastures to the arable lands by means of their manure.

However, despite the decisive role played by the animals, the productivity of systems based on fallowing and cultivation using the ard remained limited due to the inherent weakness of the means of plowing and transport. Indeed, the ard, which scarifies the soil without turning it over, does not accomplish a true plowing. That must be carried out manually, with a spade or hoe. But manual plowing is long, hard work, with such low productivity that it cannot be extended to all of the fallow lands. As a result, the soil is generally poorly prepared before the sowing.

Beyond that, transport using pack animals does not make it possible to move large quantities of organic matter (fodder, manure) from the *saltus* to the *ager*. Since transfers of fertility by simply penning the livestock every night on fallow land are not very effective, the lands cultivated with cereals are not well fertilized.

The cultivated lands are not extended and are poorly prepared and manured. As a result, the yield and overall production are low. Moreover, as the area cultivated per worker is limited by the deficiencies of the tools, labor productivity is hardly adequate to cover the needs of the population. This poor performance is the origin of the chronic subsistence crisis of the Mediterranean and European societies of antiquity. This crisis appears to us as inseparable from the development of war, the formation of militarized city-states, colonization and slavery that characterized these societies up to the end of the first millennium of the common era.

In fact, it is only after the year 1000 that these inadequacies were remedied. In the cold temperate regions, cultivation using the ard was replaced by cultivation using the plow, properly speaking, and the wagon. At the same time, in the hot temperate regions, animal-drawn cultivation with the ard was perpetuated for centuries, while a whole series of improvements appropriate to these regions were implemented, such as terracing of slopes, irrigation, arboriculture, planting of associated crops, all of which were already practiced in antiquity. Even today, systems based on fallowing and cultivation with the ard persist in diverse forms in several regions of northern Africa, the Near East, Asia, and Latin America.

This chapter aims to explain the agricultural revolution of antiquity, i.e., to discover how and why, in the temperate regions, deforestation generally led to the development of systems based on fallowing and animal-drawn cultivation using the ard, with associated pasturage and animal breeding. It also has the objective of explaining the structure and functioning of these systems, that is, the type of equipment, productive practices and cultivated ecosystem that characterize them, as well as the resulting outcomes. Finally, it attempts to relate the agrarian and food crisis of societies in antiquity to certain traits in their social organization and politics.

## 1. THE ORIGIN OF THE SYSTEMS BASED ON FALLOWING IN THE TEMPERATE REGIONS

In the semi-arid regions close to the Fertile Crescent, deforestation and desertification goes back to the sixth millennium before the present. But in the Mediterranean and European regions with a temperate climate, the original denser and less fragile forests resisted the ax and fire for a much longer period of time. The slash-and-burn systems of cultivation lasted until much later. However, beginning with the Bronze Age, around 2500 B.C.E., deforestation was already well advanced on the eastern rivers of the Mediterranean and, during the 2000 years that followed, spread to the whole of the hot temperate regions of the Mediterranean from east to west. Subsequently, deforestation gradually extended to the cold temperate regions in the northern half of Europe, up to the first centuries of the common era. It is commonly accepted that systems of cultivation using the ard with a biannual fallow period and pastoral animal herding became predominant in the temperate regions beginning in antiquity. But little is known about the manner in which such systems gradually developed in a region that was undergoing deforestation.

It is precisely this movement from slash-and-burn systems of cultivation to systems of cultivation based on fallowing that we want to try to reconstruct here, by examining the following questions: How and why did the new constituent elements of the cultivated ecosystem (the *ager* and the *saltus*) acquire distinct identities? How did the new tools—the ard, the spade, the hoe—appear and why did they become widespread? Why did the biennial rotation of fallow land become predominant? We will treat first of all the case of the hot temperate regions, before examining the case of the cold temperate regions.

# The Case of the Hot Temperate Regions

The Mediterranean climate is a hot and dry temperate climate in summer, with a short, mild winter, and moderate rainfall concentrated primarily in autumn, although it also rains in winter and spring. Under this climate, the period of plant dormancy always occurs in summer, but a slowing down of vegetation also takes place in winter. The Mediterranean climax is a forest of average biomass consisting of three levels of vegetation. The wooded level is typically made up of oaks combined with other species such as pines and maples. The holm oak, which withstands dry conditions better, is adapted to different terrains, including chalky ones, while the cork-oak is more widespread on sandy terrains. The shrubby sublevel is composed of pistachios, carobs, laurels, and junipers and the undergrowth is made up of heather, lavender, rockrose, etc.

## The Formation of a New Post-Forest Cultivated Ecosystem

Once attacked by ax and fire, then subjected to cycles of frequently repeated cultivation, forests of this type evolve toward degraded plant formations, such as the maquis and garrigue, types of scrubland typical of the hot temperate regions. The maquis is a rather dense, closed formation, on sandy soil, composed of bushes and shrubs, while the garrigue is an open formation on chalky terrain. In fact, the garrigue is a species of discontinuous shrubby steppe where the vegetation occupies fragments of brown soil, red soil, and rather small amounts of rendzina, and where portions of the skeletal terrain deprived of vegetation show on the surface as large slabs or as piles of fallen rock. In areas that are not too hilly and subjected to erosion, these scrub formations become less and less favorable to cultivation. They are most often reserved for pasturing domestic herbivores and subjected to periodic burning, which, while favoring the regrowth of grass in the spring and autumn, makes the regeneration of trees difficult. These grassy and bushlike formations, set aside for grazing and subject to fire and erosion, formed what is called in Latin the *saltus*, the first constitutive element of the new post-forest cultivated ecosystem.

Next to this generally uneven and eroded *saltus*, valleys, depressions, basins, dolines—in sum all the hollows of the terrain—benefit from increased sedimentation from the hills. These lands with deeper soils, which are continually rejuvenated and enriched, are reserved for the cultivation of cereals. Since they are often not extended, the cultivations must be repeated all the more often as the population increases. As a result, each cereal crop ends up alternating with a grassy fallow period of short duration, lasting hardly more than one year, with which it forms a biennial rotation. This grassy fallow land grazed on by domestic animals and fertilized by their manure is also plowed.[3] The plowable cereal

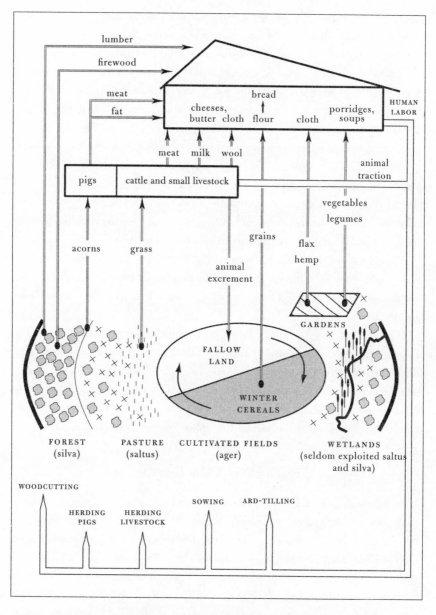

*Figure 6.1 Diagram of the Structure and Functioning of the Cultivated Ecosystem in Agrarian Systems Based on Fallowing and Cultivation Using the Ard with Associated Animal Breeding and Grazing*

lands, composed of a set of contiguous fields, formed what is called in Latin *ager*, the second element of the new cultivated ecosystem.

But there also exist wooded lands not propitious for cultivation, because they are too high, too hilly, rocky, permeable, wet, dense, or even quite simply too far from dwellings. On these lands, the forest was hardly subjected to slash-and-burn and persisted even if it were more or less degraded by wood harvesting. Thus, next to the *saltus* and the *ager*, some portions of the territory preserved a population of trees large and important enough to merit the name forest, woods, or grove, depending upon whether their surface areas were large, medium, or small. The generic Latin term *silva* is used to designate all of the lands that form the third element of the new cultivated ecosystem. *Silva* and *saltus* were not, moreover, always distinct from each other: the *saltus* remained cluttered with trees, and herds used the partially deforested *silva* as well.

Lastly, the garden-orchards formed the *hortus*, the fourth element of this ecosystem. They were the successors to the permanent crop enclosures adjoining the houses in a village. These enclosures already existed at the time of slash-and-burn cultivation.

The residual *silva*, the grazed *saltus*, the *ager* dedicated to cereal cultivation in rotation with a grassy fallow period of short duration, and the *hortus*, such were the four parts of the new cultivated ecosystem resulting from the process of deforestation in the temperate regions. However, in order for the cultivation of cereals to be possible in such conditions, it was necessary to resolve a problem: clearing the grassy fallow lands and renewing the fertility of lands planted in cereals.

## The Adoption of New Equipment

If the ax and fire are the appropriate means to clear a forest or wooded fallow land, they are ineffective for clearing the natural grass cover of plowable fallow land. That requires other tools. Therefore, the farmers of antiquity used manual tools, the spade and the hoe, and a tool pulled by animals, the ard. The spade and the hoe both make it possible to plow the soil, i.e., to turn the soil over and then bury and, in large measure, destroy the wild grasses of the fallow land. But this long and difficult labor could not be carried out on all of the fallow land. It was necessary to complete that work by using the ard. The ard, conceived originally to bury the seeds after a sowing, is a tool that uses animal traction (ox, donkey, mule) and is provided with a simple point, hardened by fire, or fitted with a metal tip, which scarifies the soil without turning it over and incompletely destroys the weeds. But as the ard-tilling is relatively quick, one can repeat it several times.

In fact, the ard, spade, and hoe were not invented as a response to the needs of the new systems based on fallowing: they were borrowed from the hydro-agricultures of Mesopotamia, where they were in use for a long time, by farmers of neighboring regions affected, in turn, by deforestation. The ard appeared, in fact, in lower Mesopotamia and it spread to the Near East in the fourth millennium B.C.E. Then it reached the Nile Valley, the Mediterranean region, and Europe, where its presence is attested from the third millennium in several regions by stone carvings, terra-cotta models, and traces of the tool exceptionally well preserved and dated, under burials for example.[5] The presence of the ard, which cannot be used in heavily wooded terrain, leads one to think that at least a portion of the cultivated lands were deforested and an embryonic *ager* already existed in this period, at least in some regions.

## A New Mode of Renewing Fertility

In contrast to wooded idling of long duration[2] and grassy idling of medium duration, both of which existed in the early years of deforestation, grassy fallowing of a little more than one year produces too little biomass to play an important role in renewing the fertility of cultivated lands. In a hot temperate climate, at least three years are required to reconstitute a relatively dense grassy groundcover.

The saltus, on the other hand, a type of extended and permanent idled land,[3] produces enough biomass to reproduce the fertility of cultivated lands, on condition that it has a means to transfer a portion of the biomass to the *ager*. This transfer is not easy if carts and wagons are uncommon, as in antiquity. It is essentially carried out, then, by herds of domestic herbivores, based on the appropriate management of cultivation, grazing and animal breeding. The animals are led early in the morning onto the *saltus* close to the village to graze there all day long. Then they are led onto the fallow lands in the evening where they remain during the night and deposit their manure. In this way, a part of the grazed biomass from the *saltus* is gathered (in the form of excrement) on the fallow lands, while the reverse transfers of biomass, from the fallow lands to the *saltus*, are sufficiently reduced.

But, in the Mediterranean region, the slow growth of grass during the summer limits the size of the herds, such that it is indeed difficult to have enough livestock to consume all the biomass produced in the autumn and spring. Failing that, the transfers of biomass and fertility from the *saltus* to the *ager* are inevitably limited. Diverse arrangements for managing herds and grazing make it possible, however, to maximize year-round the number of animals grazing on the nearby *saltus* and ensure transfers of fertility by penning them at night. One of these arrangements involves grouping the births at the end of winter and end of summer so as to increase the number of animals grazing during the seasons in which there is the strongest growth of grass, in autumn and spring. But the most

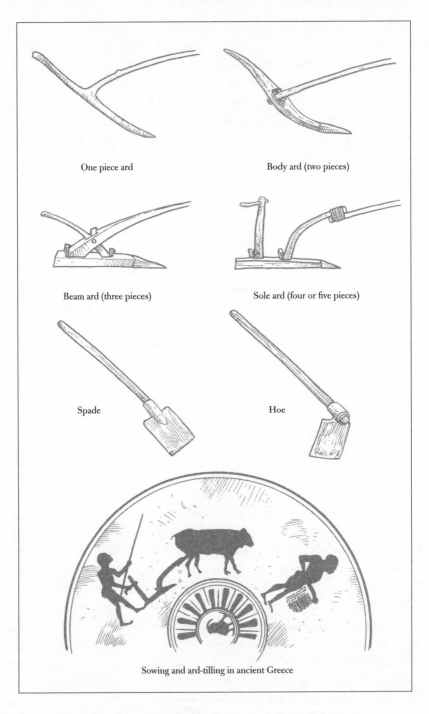

One piece ard

Body ard (two pieces)

Beam ard (three pieces)

Sole ard (four or five pieces)

Spade

Hoe

Sowing and ard-tilling in ancient Greece

*Figure 6.2 Tools for Tilling and Sowing in Agrarian Systems Based on Fallowing and Cultivation Using the Ard*

important arrangement is undoubtedly transhumance, which involves temporarily removing a surplus part of the herd to distant summer pastures (located higher up or farther north) as to have a sufficient number of livestock to consume the entire grass production from neighboring pastures during the rest of the year. Lastly, since the Mediterranean is dry enough during the summer to make it possible to preserve uncut, without too many losses, a portion of the overabundant grass from spring, it is also possible to make a portion of the pasture close to the village and cultivated lands off-limits for grazing in spring but available for grazing during the summer, when the grass is still consumable.

It remains the case, however, that penning animals at night on fallow land as a method of transfering fertility is altogether ineffective. It is necessary to have an extended *saltus* and numerous herds in order to succeed in fertilizing a very small area of *ager*, and even then it works rather poorly.

The fact that penning the animals at night was the mode of manuring characteristic of systems of cultivation using the ard does not mean people at that time did not know about the advantages of manure produced by animals in stables. These advantages were known since early antiquity but, lacking wagons and carts, the quantities of hay and manure that could be transported by hand or by animal were inevitably reduced, and the manure was often reserved for the gardens.[4] Nor were people at that time unaware of the advantages of rotations alternating cereals with legumes, but there again, as we will see, the obligation to practice fallowing was, in animal-drawn cultivation using the ard, practically inescapable.

## The Case of Cold Temperate Regions

In central European regions with a cold temperate climate, plant dormancy and the fall of leaves takes place in winter and a certain slowing down of vegetation growth occurs in summer. The climatic forest, composed of hardwoods, includes three levels of vegetation: the arboreal level of oaks, beeches, and hornbeams which can rise to thirty or forty meters; a shrubby sublevel made up of hazels, willows, hollies, dogwoods, etc.; and a bushy undergrowth of varying composition. The total biomass of such a forest, which can reach 400 tons of dry matter per hectare, is one of the highest there is. It is thus denser, stronger, more resistant to the ax and fire than the forest of the hot temperate regions.

However, the population growth at the end of the Neolithic period and beginning of the Bronze Age and, consequently, the more and more frequent repetition of slash-and-burn cultivations led to deforestation here too. In these regions, as in the Mediterranean area, a *silva*, a *saltus*, and an *ager* were formed, but their relative proportions varied from one region to another.

On the great silt-laden plains and in the large alluvial river valleys, with a rich, deep, and not very heavy soil, all land is potentially cultivable with the ard and other equipment associated with animal-drawn cultivation. However, it is necessary to preserve an adequate area of *silva* in order to support the wood needs of the population. The longer and more severe the winter, the larger this area must be. It is also necessary to devote an adequate area to the *saltus*, in order to feed herds that are large enough to fertilize the cereal lands of the *ager* adequately. Without effective means to harvest hay, the longer the cessation of growth during the winter the more extended was this area. Even when there are no lands unsuitable for cultivation, the new cultivated ecosystem must consist of one part *silva* and one part *saltus* proportional to the needs for wood and pasture.

In other regions, by contrast, some parts of the originally wooded land, cultivable by slash-and-burn methods, became unsuitable for cultivation after deforestation. This is particularly true of sandy lands low in fertility, which are covered by moors of heather and gorse, or indeed of the most skeletal and thinnest rendzina soils, on hard limestone, which are covered by meager prairies and calcicole heaths. These lands are exploited as *saltus*, while the *ager* must be concentrated on the deposits of silt-laden soil, on the colluvial deposits at the bottom of slopes, and on alluvial deposits in the river valleys.

In some regions the *silva* remained largely predominant, because the terrain is uncultivable using the ard and associated equipment. This was the case for the Nordic forests, high-altitude forests, and forests on hilly, rocky, permeable, humid, compact, etc., terrain. On the margins of regions settled because they were cultivable with the means of the time period, there persisted vast, heavily forested and little-populated mountains and "deserts," some of which were only put into cultivation in the Middle Ages, using the animal-drawn plow (see chapter 7).

## The Case of Non-Forested Temperate Regions

Originally, there also existed in the temperate zone grassy climatic formations, with some bushes, in which the forest could not develop. Some were not fertile, such as the moors, meadows, and high-altitude steppes situated above the conifer forest and the moors on the podzol soils of the wet Atlantic regions, on permeable, sandy soils or on skeletal soils, etc. Unsuited for cultivation, these grassy formations formed a sort of natural *saltus* from the beginning, exploitable by local herds of those that migrated with the seasons.

Other climatic grassy formations were, on the contrary, very fertile, such as the large prairies in regions with a continental climate (Danube Valley, Ukraine, etc.). In these regions, the heat and the dryness of summer caused a pronounced cessation in vegetation growth, which impeded the development

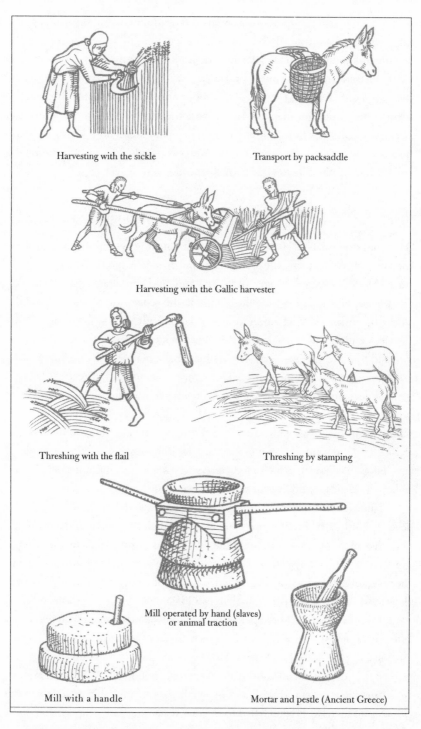

Harvesting with the sickle

Transport by packsaddle

Harvesting with the Gallic harvester

Threshing with the flail

Threshing by stamping

Mill operated by hand (slaves)
or animal traction

Mill with a handle

Mortar and pestle (Ancient Greece)

*Figure 6.5 Equipment for Harvesting, Transporting, and Processing Grains in Agrarian Systems Based on Fallowing and Cultivation Using the Ard*

of trees. These same conditions also favored evaporation and the capillary rising of the soil solution. By preventing the leaching of fine particles and the leaching of soluble mineral salts and the input of organic matter into the soil by the massive root systems of prarie vegetation, this mechanism lead to the formation of black soils (chernozem), among the richest soils in existence. For a long time, these continental prairies were the reserved domain of European pastoral societies or those coming from Asia. They were, however, more widely cultivated when larger populations had tools (spade, hoe, ard) for tilling the soil, making it possible to clear a thick, grassy groundcover.

## The Agricultural Revolution in Antiquity

The preceding analysis shows that the development of systems based on fallowing and cultivation using the ard was an appropriate response to the problems posed by deforestation in most of the hot and cold temperate regions. But this "response" is revealed to be quite complex: the separation of the *ager* and the *saltus*, the organization of the short-term rotation with a grassy fallow period, the development of new tools, herding the livestock onto the *saltus* and the fallow lands to transfer the most fertility possible for the benefit of the cereal-growing lands are many new organizational structures, means, and procedures. Their general development and particular adaptations in each locality took dozens of years. The development of systems based on fallowing and animal-drawn cultivation with the ard was not therefore the automatic and immediate result of deforestation but the product of a true agricultural revolution, the agricultural revolution of antiquity. What's more, it required a rather significant capitalization in farm implements and livestock, which necessarily took much time.

All indications are that the negative consequences of deforestation were felt for several centuries before the new systems developed. Beginning in the Neolithic era, erosion, the drying-out of lands, the difficulty of clearing lands that were more grass covered than forested, and low yields appear to have been instrumental in the abandonment of heavily deforested areas and the migration of entire populations to search for lands with enough forests to continue practicing slash-and-burn methods of cultivation. These phenomena are difficult to discern with precision, but they appear to be well attested. Thus many Mediterranean areas (Palestine, Anatolia, Cyprus, Malta) experienced a succession of periods of population, abandonment, and repopulation during the Neolithic period.[10] Since the first farmers in these regions practiced slash-and-burn cultivation, one can conclude that the population increase gradually led to the deforestation of these areas, then their abandonment. After the regeneration of a secondary forest, these same regions could once again be cultivated and colonized, then deforested again, and so on.

Furthermore, at the end of the Neolithic period and beginning of the Bronze Age, between 1800 and 1250 B.C.E., middle European areas were also affected by this type of exodus, in particular those that had been cultivated and deforested the earliest, i.e., the most fertile silt-laden plains and alluvial valleys.[11] These areas were then partially abandoned and people moved into areas of mid-elevation mountains, initially less favorable and which, as a result, were less populated and still had rather abundant cultivable forest reserves. At the end of the Bronze Age, around 1000 B.C.E., these fertile regions, after several centuries of abandonment, experienced a veritable agricultural revival, thanks no doubt to the agricultural revolution of antiquity. Once perfected, the systems based on fallowing and animal-drawn cultivation with the ard made regions that had formerly been exploited by slash-and-burn methods, leading to heavy deforestation, once again exploitable.

Finally, at the end of this long period of deforestation and transition, in the last centuries B.C.E., the systems based on fallowing and cultivation with the ard extended from North Africa to Scandinavia, and from the Atlantic to the Urals and the eastern shores of the Mediterranean. Certainly, slash-and-burn cultivation was still present in some areas that were still forested, pastoral systems occupied some deforested regions without plowable lands, and entire regions, forested or not, were still nearly deserted, because they were too cold, too hilly, rocky, marshy, in brief, inhospitable. It remains the case, however, that beginning in the Iron Age the new systems based on fallowing were predominant in this immense area, and for more than a millennium they supplied the basic essentials of subsistence for circum-Mediterranean and European societies. They set the tone for the agricultural economy of this part of the world.

## 2. STRUCTURE AND FUNCTIONING OF SYSTEMS BASED ON FALLOWING AND ANIMAL-DRAWN CULTIVATION USING THE ARD

Naturally, such extensive and long-lasting systems are not constant. From one region to another, even from one locality to another, as well as from one period to another, these systems take varying forms: the proportions and disposition of the *ager*, *saltus*, and *silva*, the species of cereals cultivated, the crop rotation, the form of the tools, and the agricultural calendar change. But whatever these variations may be, all these systems share characteristic common elements: the *ager*, *saltus*, and *silva*; the alternation between cereal growing and fallowing; the ard, the spade, the hoe, and the sickle; the clearing of lands for sowing by plowing, either with the spade or with the ard; and the transfers of fertility from the saltus to the ager through animal manure.

## Fields of Cereal Cultivation (the *Ager*)

The function of the *ager* is essentially to produce cereals, which supply more than three-quarters of the population's caloric ration. It is made up of fields in which the more and more frequent repetition of slashing, burning, and processes of working the soil have destroyed almost all the trees and shrubs. These fields are also stumped, that is, the last tree stumps and roots that clutter the soil are removed in order to be easily worked with the ard, spade, or hoe. Trees or shrubs can, however, be deliberately planted on the *ager*. They are either existing ones intentionally preserved or new ones planted because of their utility, e.g., trees bearing fruits consumed by humans or livestock, such as oaks, chestnuts, carobs, or olives; fodder trees such as the ash; or quite simply trees useful because of the shade or wood they provide and the organic matter they restore to the soil. These trees can be found in the open field, adequately spaced so as not to impede maneuvering the ard, or spread out along the border of the field.

### The Parcels

The fields that make up the *ager* are permanent, quadrangular, and contiguous, as opposed to the generally dispersed, multiform, and temporarily occupied parcels characteristic of slash-and-burn cultivation. On light, easy-to-work lands, two animals harnessed in front suffice to pull the ard. The harnessing is short enough to allow an easy turn at the end of the field. As a result, it is not necessary to stretch out the length of the fields in order to avoid difficult maneuvers. But since a second pass of the ard, at a right angle to the first one, is often practiced, the fields must not be too narrow. This is why cultivation with the ard accommodates fields that are rather small, not very long, or even quasi-square on soils that are easy to work. On the other hand, on heavy soil that requires a team of two pairs of oxen or more, which makes turning difficult, the fields must be larger.

### Arrangement of Village Lands and Distribution of Housing

In this system of animal-drawn cultivation, the poorly developed means of transport make it necessary to locate the houses as close as possible to the plowable lands. In areas that are not very hilly and of a relatively uniform fertility, the *ager* is thus conveniently grouped in a circle around each village, the peripheral lands being reserved for the *saltus* and the *silva*. In order to avoid too much transport and traveling, villages are no larger than a few hundred persons. When village territory is intersected by floodable lands that are steep or not very fertile, the *ager* is then made up of a fragmented set of plowable lands, in the middle of which the village is placed. When the *ager* is composed of small fragments, separated from one another by large uncultivable expanses, the settlement tends to be dispersed into small hamlets or even into isolated farms. Undoubtedly, the position of sources of water is also taken into account. But the spatial distribution of the houses is above all determined by the distribution of cultivable lands. Furthermore, plateaus without sources of water were indeed occupied by farms having an intake area (basin to collect rainwater) and a water reservoir.

In every agrarian system, the distribution of the housing tends, as a general rule, to be functional. This is why the concentration of dwellings in strategic sites or the forced relocations of populations by those in authority, in order to exercise more control over them, end up considerably increasing the time and energy devoted to traveling and transport. The productivity of labor is proportionately reduced, which can lead to a systemic crisis and famine. This type of relocation, sometimes ordered by the authorities for administrative convenience, has nothing to do with development.

## Rotations and Plot Allotments Between Cereal Cultivation and Fallow Land

In systems based on fallowing and animal-drawn cultivation with the ard, the biannual rotation, which is the most widespread, consists of a single cereal crop that alternates with one "year" of a grassy fallow land. Typically, this is a winter cereal crop, sown in autumn and harvested in summer of the following year. It occupies the land for around nine months. The winter cereal is often wheat on good lands, rye on less fertile lands, or a mixture of the two, which is called a mixed crop. It can also be winter barley or oats. After the harvest, the fifteen-month fallowing is begun, from the end of summer until autumn of the following year. This biennial rotation can be represented in the following manner:

| Two-Year Rotation | |
|:---:|:---:|
| YEAR 1 | YEAR 2 |
| August ..................... October | November .................... July |
| **fallow** | **winter cereal** |
| ‹—- 15 months —-› | ‹—- 9 months —-› |

From autumn until the middle of summer, the first plot is formed from the set of parcels occupied by the winter cereal, while the second plot is formed from the set of parcels occupied by the fallow land. The following year, the first plot will be occupied by the fallow land, while the second will be occupied by the winter cereals. From the end of the summer to the autumn sowing, both plots lie fallow.

One part of the cultivated plot can be set aside for spring cereals: spring barley, oats, millet, or even legume crops such as peas or lentils. These crops are also planted when it has not been possible to carry out the autumn sowing, or when the sowing has not been successful. Biennial rotation with a spring cereal that lasts only for three or four months includes a long fallowing of twenty to twenty-one months:

| Two-Year Rotation | |
| --- | --- |
| YEAR 1 | YEAR 2 |
| August....................March | April...................July |
| **fallow** | **spring cereal** |
| ‹—- 20 to 21 months —-› | ‹—- 3 to 4 months —-› |

In order to limit the risks of inadequate harvests, always a threat in these relatively unproductive systems, farmers from time immemorial simultaneously have used diverse species and varieties of cultivable cereals and have dispersed their cultivated parcels among different lands making up the *ager*.

Finally, we will see that there also exist rotations in which the fallowing, or more exactly the return of wild grasslands, lasts two years or even more, and others in which the fallowing lasts for less than one year, a spring cereal crop alternating each year with a short fallow period in autumn and winter. But in order to understand the reasons for these variations, we must now study more precisely the agricultural practices implemented in these systems and the functions of fallowing.

## Fallowing and its Functions

The fallow period is, as we have said, the condition of a cultivated land in rotation, not sown for several months, subjected to grazing by domestic animals and then, by definition, plowed.[12] It is important not to confuse the plowed fallow land in a rotation with forested idled land of medium or long duration cleared by slash-and-burn techniques or with a not yet plowed natural pasture in rotation.[5]

The fallow period begins in summer after the harvest and lasts until the next sowing. This unsown land is called uncultivated, vacant, waste, or even empty. However, far from being deserted, it is land where cereal stubble and residual weeds dominate during the months after the harvest, a land where a natural vegetation consisting of annual plants, which reproduce by seeds, and tenacious perennial plants, which reproduce by vegetative means, develop. The fallow land consists of a grassy fallow period of short duration, but this fallow land is not left to the wild. It is exploited as pasture, subjected to one or several ard-tillings, and, when possible, that is, once every two or three times, it is plowed deeply by hand with a spade or hoe.

Thus the widespread idea that fallowing is a period of "rest" for the soil, allowing it to "restore its strength" after a period of cultivation, is complete nonsense. The soil does not have strength, it does not tire, and it does not rest. And even if that were true, fallow land, invaded by weeds, subjected to grazing and multiple workings of the soil (ard-tillings, hoeing, spading) would hardly experience rest.

Another error is to consider that fallowing is designed to reproduce fertility in an analogous way to that of the long-term, forested idling of the slash-and-burn systems. Certainly, the biomass produced by fallowing is not negligible and, on fertile lands, it would be adequate to ensure acceptable yields, if it were allowed to develop fully. But, in a little more than one year, the vegetation of the fallow land does not have the time to root itself strongly because the ard-tillings repeatedly destroy its development. In any event, on relatively unfertile lands, this production of biomass would be inadequate to obtain satisfactory yields.

A third error consists in thinking that by simply having the livestock graze on the fallow land, the fertility of the cultivated lands would be improved. Nothing of the sort occurs, because as long as the animals do not eat elsewhere than on the fallow land, their excrement cannot contain organic matter and minerals other than those absorbed from the fallow land itself. This excrement even contains a little less organic matter, because the animals appropriate part of the matter necessary to their own growth. And, without a few elementary precautions, animal excrement can contain even much less organic matter and minerals if, after the animals have filled their bellies on the fallow lands, they are led away to wander elsewhere.

Another error consists in believing that in a Mediterranean climate, hot and dry in the summer, fallowing makes it possible to store a good portion of the rainwater that fell during the fallow year in the soil, to the benefit of the following year's cultivation. In fact, a large part of the rainwaters of autumn, winter, and spring is not reserved. It can, depending on the place, run off, stagnate on the surface and evaporate, or even percolate deeply into the soil. As for the water actually reserved in the soil and immediate subsoil, it is greatly reduced during the summer by direct evaporation from the soil surface and by the transpiration of the natural vegetation on the fallow land. It is indeed clear that a simple ard-tilling in the spring or summer does not sufficiently break the capillarity of the soil to prevent evaporation, and does not destroy enough of the grass on the fallow land to prevent transpiration. Consequently, under this hot and dry summer climate, the fallow lands are generally very dried out on the eve of the first rain and autumn sowing. Hence, the amount of water stored in the soil before summer to benefit the winter crop is greatly reduced.

The use of fallowing is completely different: the fallowing period is systematically used to carry out a whole series of operations that combine cultivation and animal herding and aim at restoring the soil to a condition where it can produce a new harvest. These operations have two functions: the first consists in renewing the fertility of the soil through the addition of organic matter. The second consists of clearing the weeds that tend to increase during the fallow period.

*Renewal of Fertility.* The transfer of fertility from the *saltus* to the *ager* by means of livestock excrement entails many losses, even when it is well organized. A part of the biomass consumed on the *saltus* returns to this same *saltus* or is lost along the roads, while a part of the biomass consumed on the fallow land undergoes a reverse transfer, from the fallow land to the *saltus*. In total, the net transfer of biomass benefiting the *ager* represents only a fraction of the biomass consumed daily on the *saltus* and, all things considered, the return from this mode of transferring fertility is low. Consequently, in order to obtain the most fertilization possible, the nightly penning of the animals must take place every day of the year and the long fallow period must last at least twelve months. It is also necessary to have an extended saltus and numerous livestock to fertilize a rather small *ager*, poorly. Finally, it is necessary to organize carefully the daily transits of the herds between the *saltus* and the *ager*.

Moreover, it should be noted that in the cold temperate regions, the availability of fodder in the winter is much lower than in the summer. Lacking the means to cut, transport, and store hay in sufficient quantity to feed the livestock in winter, as well as the ability to practice a winter transhumance to a milder climate in these regions, the size of the herd is quite limited and the largest part of the fodder production of spring and the beginning of summer is not consumed. In these conditions, transfers of biomass and fertility from the *saltus* to the *ager* are particularly insignificant. Thus, in the cold temperate regions, the surface area of *saltus* necessary to feed a herd and obtain enough fertilization through nightly penning of the animals for one hectare of *ager* is very high, higher than in the hot temperate regions where seasonal migration makes it possible to maximize the size of the herd and the transfers of fertility.

Sometimes in sheep-raising moors where a miniscule *ager* is lost in an immense *saltus*, the long fallow period of more than a year is not necessary. Six to seven months of fallowing is adequate to fertilize the plowable lands. This explains, for example, the old annual rotation in the moors of Gascogne, a rotation in which a spring cereal, such as millet, and later maize, alternated with a short fallow period in autumn and winter, in the following manner:

| One-Year Rotation | |
| --- | --- |
| September .................... March | April .................... August |
| **Short Fallow Period** | **Millet** |
| ‹—- 7 months —-› | ‹—- 5 months —-› |

But the fallowing should not last too long, otherwise the excrement deposited on the soil at the beginning of the fallow period, when subjected to alternate rainy and dry periods for too long, would be mineralized, then drained by percolating water or denitrified before the next cultivation. This is the reason why the long fallowing does not end up lasting more than fifteen months when it alternates with a winter cereal of nine months, and not more than twenty months when it alternates with a spring cereal of four months.

Sometimes there exist systems in which a grassy idled land period of several years alternates with one or sometimes two years of cereal. The vegetation of this idled land is not worked during the first years, and it has the time to firmly root itself. A natural grassland can then develop fully and accumulate biomass and fertility, while providing pasturage for livestock. Since ard-tilling would not suffice to clear it, a true plowing (manually or with a real plow, when one is available) is necessary in the year preceding its return to cultivation, when the following type of rotation is obtained:

| Five-Year Rotation | |
| --- | --- |
| Year 1    Year 2    Year 3    Year 4 | Year 5 |
| August ......./......./......./.......October / November .................... July | |
| **Natural Grass-Land** | **Winter Cereal** |
| ‹———— 4 years and 3 months ————› | ‹———— 9 months ————› |

Such systems—widespread in wet mountainous regions, which are more favorable to grasses than to cereals—can be considered as a variant of the agrarian systems based on plowed fallow lands. But when the return to natural vegetation during the idle period is prolonged to ten years or more, and the shrubby vegetation is developed to the point of requiring a clearing with ax and fire, then it is no longer a question of systems based on plowed fallow lands, but of slash-and-burn systems (see chapter 3).

*True Plowing and Ard-Tilling.* In a biannual rotation, even if the vegetation of the fallow land does not have the time to develop fully, it does have enough time to proliferate and "dirty" the lands planted with cereals to the point of eliminating all hope of a harvest. Indeed, when the season is hot and humid enough, the weed seeds, disseminated before the harvest or brought from the saltus to the fallow land via animal excrement, germinate and the hardy perennials resume their growth. In order to protect the fields planted with cereals from being overrun with weeds, it is necessary to limit the development of this wild vegetation and destroy it as much as possible before resuming cultivation. To this end, farmers using this biannual rotation system methodically graze their animals on the fallow lands and work the soil in several different ways.

The most effective of these processes is plowing. In the strict sense of the term, plowing consists of breaking the soil of the fallow land into clods or strips of land in rectangular sections and then turning them over. Thus the top layers of the soil are buried with what they contain, i.e., groundcover roots, and with what covers them, i.e., the exposed part of the vegetation, weed seeds, organic debris, and manure, if there is any.

In animal-drawn cultivation based on the ard, there is no equipment capable of carrying out a true plowing. The plow as such does not exist and, as we have seen, the ard does not turn over the soil. Plowing is carried out manually, either with a large hoe or with a spade. These two tools of "manual plowing" are made up of a blade around twenty centimeters long by ten centimeters wide and a shaft of wood more than a meter in length. In the best of cases, the blade is tempered steel and it is attached to the shaft by means of a metallic collar. In rocky soils, hoes and spades with two or three teeth are used. The spade, whose blade is in the same plane as its shaft, is buried in the soil by applying pressure with the foot, while the hoe, whose blade is perpendicular to the shaft, is balanced from the top downward and penetrates the soil by percussion. In both cases, the shaft is subjected only to light lateral pressure. The shaft, however, is subjected to much more force in lifting the clods of earth and turning them over, because it is used then as the arm of a lever. The collar attaching the blade to the shaft, also subjected to more force, must then be solidly soldered. To economize on metal, spades with wooden blades having metal edges were used for a long time on light soils. In this case, the shaft and the blade were made from a single piece of thick, hard wood, which made it possible to avoid the difficult problem of attaching the blade to the shaft.

A long shaft and a long and wide blade together make the spade and hoe real tools for plowing. These should not be confused with a whole series of quasi-hoes, such as the *dabas* used in the forest and savanna systems of West Africa, in which blade is attached to the often very short shaft by a simple ligature, a non-soldered collar or a metallic point directly affixed into the terminal bulge of the shaft. Depending upon their form, these *dabas* can be useful for superficially hoeing the soil and removing weeds, or even to gather the top layers of the soil into mounds, but they do not make it possible to plow, that is, to break, turn over, and compost a dense groundcover root network.

Following a manual plowing, the clods of earth are broken with the blade of a spade or head of a hoe. The roots and rhizomes are then freed from the earth, pulled up, shaken, dried, and, just to make sure, burned. The soil thus cleared, aerated, and broken up is ready to receive the plant or the seed. Manual plowing is long and exhausting work, much harder than harvesting or watching over the herds. It was once considered the archetype of hard work.[14] Since it was not possible to plow by hand all the lands for sowing, a much quicker method for working the soil had to be used: an ard-tilling, i.e., tilling with an ard, an unused term that we employ deliberately, in order to distinguish this process clearly from a so-called plowing properly.

The ard is a tool that, in its most elementary form, is composed of a large wooden fork. One branch is used as a pole for the harnessing of the draft animal, while the other branch, cut short and hardened with fire, is used to scarify the soil. The shaft acts as the handle for the operator of the instrument. More elaborate swing plows are made up of several pieces of wood: a horizontal working part, the frog, that ends in a reinforced point of flint or in a small, symmetrical plowshare made of iron that cuts through the soil. The frog can be directly attached to the base of the pole to which the draft animals are harnessed, but it can also be connected to the latter by a curved piece of wood, the beam. The handle, attached to the frog, is generally held with one hand. Sometimes a stanchion of hard wood can be used to reinforce the connection between the beam (or pole) and the frog.

Some ards include blades fastened in a V-shape onto the frog, which moves the dirt away from each side of the furrow as the plowing progresses. These ards carry out a quasi-plowing. The most sophisticated ards even make use of a front axle unit (like plows, see chapter 7), a small cart with two wheels on which the beam rests and which guides the instrument in heavy soils. Also, they sometimes have two handles. Finally, note that some ards have a vertical frog that can easily be extricated from the soil in case of an obstacle, which is necessary on rocky lands or lands that have been incompletely deforested and are encumbered with stumps and large roots. This type of ard that "jumps over the stumps," still widespread in northern and eastern Europe at the beginning of the twentieth century does not seem to have been used in early antiquity, where, however, it would have been quite useful.[6] It is clear that the ard is not as simple and easy an instrument to make as is often believed. Moreover, Hesiod, in *Works and Days*, underlines clearly the care required for selecting the proper species of wood to construct a good ard: laurel or elm for the pole, because these are "the woods which mildew the least." oak for the frog and holm oak for the beam, because this wood is "the most resistant."[16]

As the ard moves across the field, the point scarifies, opens and breaks up the soil. It lays bare the roots of vegetation, uprooting and destroying some of them. After an ard-tilling, it is possible to remove the weeds and some of the roots and rhizomes and then dry and burn them. Therefore, in order to prevent weeds from invading the fallow land, ard-tillings (or harrowings) are generally done at the end of summer or in spring to facilitate the removal of the weeds, which are then grazed to prevent their flowering and multiplying. The systematic alternation between ard-tilling and grazing the fallow lands is thus a method of fighting against weeds.

There is also one ard-tilling before the sowing, in order to prepare a relatively suitable seedbed by loosening and aerating the soil, then a second one after the sowing, to bury the seeds. At the same time, this ard-tilling prepares the soil for seeds from weeds that were not buried so deeply. Cereals sown on fallow land prepared with an ard, without any additional manual plowing, are often invaded with weeds that must be hoed by hand until the harvest. This is why cereals could, in the past, be considered hoed plants. Numerous Greek and Latin authors (Hesiod, Xenophon, Theophrastus, Varro, Columella) report the practice of multiple ard-plowings (generally three) and it still persists today in the Mediterranean region and Latin America.

In systems based on fallowing and animal-drawn cultivation with the ard, the *ager*, essentially devoted to cereals, functions to satisfy the basic dietary needs of the population in the form of biscuits, bread, semolina, porridge, etc. Of course, it is possible to replace cereals partly with other crops (legumes, textiles), but since the production of cereals is already hardly sufficient, this substitution cannot go very far. Furthermore, the ager cannot satisfy every need. These systems, crops other than cereals are frequently grouped in the garden-orchards close to the houses, which form the *hortus*.

## Gardens, Vineyards, and Orchards (the *Hortus*)

The garden is small, protected from animals by an enclosure, cultivated every year without clearing or fallowing and enriched by domestic waste, ashes from the hearth, and animal excrement. This type of garden is descended from the one that already existed in the forest systems. In antiquity and the Middle Ages, food legumes (peas, lentils), vegetables (onion, garlic, fennel, turnips, cabbage), textile plants (flax and later hemp), oil-producing plants (oil poppy, rapeseed), and fodder plants (clover, vetch, and later alfalfa) were cultivated in these gardens, either separately or in combination. There were also fruit trees and possibly vineyards.

## The *Saltus* and Other Pastures

In a few regions, as we have seen, the *saltus* is a direct result of an original grassland. But in most cases it grows out of a predominantly grassy, secondary formation resulting from the gradual deforestation of the primitive forest. If it were left to itself, this *saltus* would generally evolve toward a secondary forested formation. It survives as pasture only in so far as it is constantly exploited and maintained by grazing a sufficient number of livestock to prevent the overwhelming return of woody vegetation. Furthermore, shepherds clear by fire the bushy formations that tend to prevent access to grazing to keep the saltus open for their herds, eliminate dried vegetation, and encourage the growth of fresh grass. They also destroy with hoe and brush hook the plants that livestock refuse to consume and that develop to the detriment of useful vegetation.

Livestock do not graze only on the *saltus*. They also graze on fallow lands for all the reasons that we have indicated. When the fields are large enough, grazing on fallow lands can be implemented either by guarding the herds or by using permanent or mobile enclosures, made from bushes or wooden hurdles. But when the parcels are too small to be enclosed, grazing the fallow parcels, which are mixed with parcels cultivated with cereals, becomes difficult. Two solutions are nevertheless possible: an individual solution of attaching each animal with a short cord to a stake that is moved from time to time, and a collective solution of organizing the plot allotment at the level of the whole village and establishing common grazing land on consolidated fallow lands.

### *Regulated Plot Allotment and Common Grazing Land*

In order to implement the collective solution, it is necessary to divide all of the village's cultivable lands into two large, equal plots. Each farmer's land is divided in half between these two plots and each is required to plant their cereals at the same time on the same plot and leave all their other lands in fallow on the other plot, and vice versa the following year. This is the basic principle of a regulated plot allotment. In this way, once past the harvest and gleaning time, the fallow plot can be opened to everyone's herds without harm. Such is the principle of common grazing land, compatible with small, intermingled, and open parcels. A regulated plot allotment and common grazing land system coincides most often with a system of open fields. But it must be emphasized that this coordinated management of cultivation and animal herding has nothing in common with a cooperative or collective agriculture. Regulated plot allotment and common grazing ground do not prevent each farmer from exploiting his "own" fields and animals for his own profit. Beyond that, in Europe from the end of antiquity up to the elimination of fallowing in the eighteenth and nineteenth centuries, landowners and owners of large herds were forced to respect common grazing grounds for their own benefit, sometimes even with the support of peasants who had little livestock and found a way in this system to fertilize their lands with the animals of others.

It is sometimes asked how all the tenants of a village could work together to organize such a system. It should be remembered that in Europe this way of carrying out cultivation and animal herding existed earlier on the large estates of antiquity and on the seigniorial estates of the Middle Ages. It was quite naturally imposed on bound serfs at the time the large estates broke up at the end of antiquity, just as it was imposed on tenants at the time that plots from the manorial lands at the end of the Middle Ages were sold or rented. Also, remember that villagers who practice slash-and-burn cultivation in a forested environment are capable of adopting a multiannual regulated plot allotment that is much more complex than biannual plot allotment (see chapter 3).

The practice of common grazing land can be seen as a survival of the old slash-and-burn cultivation systems in which, once the harvest was over, parcels that were abandoned to the wild fell into the common domain. Each person could then gather, hunt, and allow their livestock to wander in that area. In the same way, with common grazing land, fallow land is open to others' gleaning and livestock. It is temporarily united into an undivided area open to all. The *saltus* and the *silva* were also part of this, at least originally.

Lastly, in addition to the fallow lands and the *saltus*, pasturage is also extended to the *silva*. Herbivores find additional food there when there is a shortage of grass because of dryness in summer or cold and snow in winter. The *silva* is also frequented in autumn by herds of pigs that fatten on acorns, beechnuts, and chestnuts.

## The Forest (the *Silva*)

During antiquity and the early Middle Ages, the forest still occupied an important place in the life of humanity. Beyond the more or less degraded remnants of original forest, saved from destruction because they were unsuited to cultivation or preserved as a source of wood or as game reserves, secondary forests existed that were reconstituted after excessive deforestation during heavily populated periods.

Whether residual or secondary, the forest played an important role, providing firewood for cooking, warmth, and baking bread, and providing lumber in order to make tools and agricultural equipment, but also shoes, stakes, casks, chests, and other furniture, as well as houses. Beyond wood, many other products were taken from the forest: hunting still supplied a not insignificant part of the meat supply; the gathering of berries, fruits, roots, mushrooms, honey, etc., supplemented and broke the monotony of a diet essentially made up of cereals. The forest was also, as we have seen, occasionally an important place for grazing.

When agrarian systems based on fallowing first began, standing timber, still overabundant in relation to the needs of the population, was an unappropriated common resource, which each person could use without restriction. The forest was also commonly exploited by harvesting dead wood as well as selective cutting. Trees of various species and in varying diameters were cut as a function of different needs, leaving aside trees that were too small, too large, or of an inappropriate species. With this type of forest use, trees were most often renewed by shoots coming from freshly cut stumps, but occasionally by seed as well. In a temperate climate, at least twenty years are required for the regeneration of firewood several centimeters in diameter, and fifty years or more to obtain small lumber trees of 15 to 20 centimeters. As for mature trees, several dozen centimeters in diameter, they are one hundred years old or more. Difficult to cut down and chop up, they are only cut in response to exceptional needs for lumber, such as building river- and canal-boats.

Exploitation by selective cutting is convenient and without any drawbacks as long as the needs of a rather small population do not outweigh the possibilities for renewing the forest. However, severe disadvantages appear in the opposite case. In fact, to the extent that the population grows, the *ager* and the *saltus* are extended and the residual forest dwindles while the need for wood increases. The overexploitation of wood begins when more wood is cut each year than the forest produces. Trees are cut at increasingly younger ages and are increasingly smaller in size. Then reserves of large trees come to be cut down and, finally, the forest is no more than an increasingly stunted thicket, producing less and less lumber. This phenomenon first appears in the most populated fertile regions, in areas close to dwellings and along access roads. At the edges of the villages, the boundary between the still wooded *saltus* and the increasingly deforested *silva* became utterly indistinct. To fight against this tendency, the forests were placed in a protective status: wood cutting and grazing were prohibited or strictly regulated. Thus the forest changed from an inexhaustible and freely exploitable natural resource to one that was limited and exploited in a controlled, planned and renewable manner (see chapter 7).

## The Performance and Limits of Systems Based on Fallowing and Cultivation with the Ard

### *The Yields*

Ancient documents relative to cereal production in antiquity are uncertain, heterogeneous, and sometimes even fanciful. Some authors go so far as to claim yields several hundred times greater than the seeds sown. A. Jardé, who has performed a detailed analysis of sources, has shown it is very difficult to form reasonable estimates concerning yields, size of production, relative proportions of *ager* and *saltus*, and population density.[17]

In an attempt to formulate a judgment, Jardé refers particularly to yields posited by Mathieu de Dombasle for France at the beginning of the nineteenth century, which are a bit higher than 10 quintals per hectare. He concludes from that figure that yields in ancient Greece could have averaged around 7.5 quintals per hectare for wheat and 13.5 quintals for barley, which is certainly an exaggeration. This conclusion overlooks the fact that between antiquity and the nineteenth century, the agricultural revolution of the Middle Ages (see chapter 7) made it possible to double, at the very least, cereal yields in the northern half of Europe. Moreover, G. Duby, in *L'Économie rurale et la Vie des campagnes dans l'Occident médiéval*, estimates that around the year 1000, i.e., before this agricultural revolution, cereal yields in Europe were not even half those maintained by Jardé.[18]

Several authors, moreover, have revised Jardé's estimates downward, particularly P. Garnsey, who maintains that grass yields for the Athenian countryside were on the order of 7 quintals of wheat and 9 quintals of barley per hectare, which still seems exaggerated to us. In the nineteenth century, the average cereal yields in southern Europe, where systems of fallowing and animal-drawn cultivation with the ard still existed, did not reach this level.[19] Observations of several systems of this type, still extant in the 1960s without the use of mineral fertilizers, showed that the grass yield in grain hardly surpassed 5 quintals per hectare, which corresponds to a net consumable yield (seeds and losses after the harvest deducted) of around 3 quintals per sown hectare.[20] It is then reasonable to conclude that the average net yield of cereals in antiquity did not surpass this level, and it is that figure that we will retain in the following calculations. Of course, this is not to forget that, from one region to another and one year to the next, yields could vary by one to two times and even more.

## Productivity of Labor

Let's assume that the average net yield (seeds and losses deducted) of cereals in these systems is on the order of 3 quintals per hectare and that, using the relatively ineffective tools associated with this type of animal-drawn cultivation (spade, hoe, ard, pack-saddle, sickle), an agricultural worker and family helpers can cultivate 6 to 7 hectares of the *ager* or, in a biannual rotation, 3 to 3.5 hectares of cereals. In these conditions, the net cereal productivity reaches just 9 to 11 quintals per worker, or hardly enough to feed a family of five. Certainly, there exist territories providing better yields, but there also exist many that provide less. Lastly, because of overpopulation or inequality in land distribution, many small farmers do not possess enough land, equipment and livestock to obtain such results. Consequently, it is clear that it is generally difficult, in systems of this type, to retain a surplus that makes it possible to feed a nonagricultural population of any importance.

## The System's Production Capacity and Population Density

Let's assume that, in order to produce the 10 quintals of grain required to support the needs of a family of five, it is necessary to make use of at least 6 hectares of the *ager*. In addition, in order to obtain a grass cereal yield of 5 quintals per hectare (3 quintals net) in a system in which the livestock is penned up at night, one large animal (bovine) or, what amounts to the same thing, five to six heads of small livestock (sheep, goats), are required in order to fertilize one hectare of sown land. In fact, one head of large livestock consumes around 6 tons of dry matter each year and produces 15 tons of excrement, of which around one-third is usefully transferred to the fallow lands in such a system. Now, in the Mediterranean region with a hot temperate climate and moderate rainfall, nearly three hectares of nearby *saltus* are required in order to feed one head of large livestock and thus to fertilize one hectare of fallow land. In order to fertilize the 3 hectares of cereals necessary for a family (or 6 hectares of the *ager*), it is necessary to make use of around 9 hectares of nearby *saltus*. Lastly, one can assume that, in regions of this type, it is necessary to count 0.2 hectare of forest per person in order to satisfy the wood needs of the population. In total, in order to support the needs of five people, it is necessary to make use of 16 hectares (6 hectares of *ager*, 9 hectares of *saltus*, 1 hectare of *silva*), which corresponds to a population density of thirty inhabitants per square kilometer. On the basis of this estimate, in antiquity the Athenian countryside could feed on its own (without permanent imports) a population of around 70,000 (2,400 km² x 30 inhabitants/km² = 72,000 inhabitants). By hypothesizing higher yields, P. Garnsey estimates a much larger self-supporting population of 132,000 people.[21]

In a much drier Mediterranean climate, where the yields from pastures and forests are less than half, in order to provide for the needs of five people, it is necessary to have access to a surface area of 26 hectares (6 hectares of *ager*, 18 hectares of *saltus* and 2 hectares of *silva*), which corresponds to a population density of twenty inhabitants per square kilometer.

In a more northern region with a cold temperate climate, the yields from pastures and forests are undoubtedly higher than in the first case, but this apparent advantage is canceled by the harshness and duration of the winter. Since the need for wood for heating is much greater, it is necessary to have available around 0.7 hectare of forest per person. Since fodder production is very low in winter, if the hay is not harvested, 8 hectares of *saltus* are required to support one head of large livestock. In total, in order to provide for the needs of five persons, it is necessary to have available 33.5 hectares (6 hectares of *ager*, 24 hectares of *saltus*, 3.5 hectares of *silva*), which corresponds to a population density of around fifteen inhabitants per square kilometer.

In regions with a clearly colder temperate climate, such as those situated at higher altitudes and in northern Europe where the yields from pastures and forests are less than half those of the preceding case, the surface areas of *saltus* and of *silva* must be twice as large. In order to provide for the needs of five people, it is necessary to have access to 61 hectares (6 hectares of *ager*, 48 hectares of *saltus*, and 7 hectares of *silva*), which corresponds to a population density on the order of eight inhabitants per square kilometer. Finally, as we have said, most cold and relatively infertile regions were unexploitable using the equipment found in this type of animal-drawn cultivation.

Thus, until the year 1000, French territory within its current boundaries undoubtedly did not have more than 15 million hectares of cultivable lands or, in a biannual rotation, 7.5 million hectares planted with cereals. With a net yield on the order of 3 quintals per hectare, these 7.5 million hectares made it possible to obtain around 22.5 million quintals of grain, which provided for the basic needs of slightly more than 10 million inhabitants.

## The Limits of Systems Based on Fallowing and Animal-Drawn Cultivation with the Ard

These calculations, as approximate as they are, show that the production capacity of the systems based on fallowing and ard-cultivation is quite limited, maybe even more limited than that of the slash-and-burn cultivation systems that preceded them. In the best of cases, the "average" hot temperate climate of the immediate Mediterranean region, these systems can barely support population densities greater than thirty inhabitants per square kilometer. Everywhere else, the farther one moves toward the south with increasingly drier climates or toward the north with increasingly colder temperate climates, the performance of the system increasingly diminishes and the maximum population densities fall rather quickly below twenty inhabitants per square kilometer. Beyond a certain level of dryness or cold, these systems even become impracticable, thereby setting limits to their expansion. Thus, in the most southern regions of North Africa, cereal-growing systems with fallowing give way to pastoral systems and oasis cultivation systems, while cold regions in the mountains and in northern Europe, uncultivable with the ard, remained forested until the agricultural revolution of the Middle Ages.

From the preceding analysis, it is possible to conclude that systems based on fallowing and animal-drawn cultivation using the ard were hardly more productive than the slash-and-burn systems that they replaced. Therefore, they did not put an end to the crisis that began at the end of the Neolithic epoch in the temperate regions undergoing deforestation. This crisis was continually apparent throughout antiquity in a chronic lack of lands and provisions. It was also apparent in the constant difficulty of extracting the surplus necessary to feed the nonagricultural population and supply the nascent cities. The crisis formed the backdrop to the agrarian and food question during the whole of antiquity.

## 3. THE AGRARIAN AND FOOD QUESTIONS IN ANTIQUITY

### Permanent War and the Formation of Militarized City-States

Deforestation and the development of systems based on fallowing began in the Near East some 2,000 years before the common era, expanding from east to west and from south to north, into the Mediterranean region and Europe. It is striking to note that, in this part of the world, palaces, cities, states, and empires developed parallel to vast agro-ecological upheaval. The first palaces in Crete (Knossos) and the Peloponnesian peninsula (Mycenae) and the first city-states of Asia Minor (Hattusas in Anatolia) appeared between 2000 and 1500 B.C.E. Between 1000 and 500 B.C.E. the Phoenician cities (Tyre, Sidon) and Greek cities (Athens, Sparta) were formed, as well as the cities of their western colonies: the Phoenician colonies of North Africa (Carthage), the Greek colonies of Sicily and southern Italy (Syracuse, Tarente), and Etruscan colonies of central Italy (Volsinia, Populonia, Volterra). Between 500 B.C.E. and the beginning of the Christian era, Rome experienced a rapid expansion and formed a vast Mediterranean empire. Finally, from the fifth century the Germanic, Slavic, and Scandinavian kingdoms and empires formed farther north.

## Colonization

In all these societies, the crisis of deforestation, the lack of cultivable lands, and the lack of food were strongly felt. They gave rise to migrations of peoples searching for new lands to colonize, expeditions of pillage and quasi-permanent wars, which entailed the growing militarization of the Mediterranean and European societies of antiquity. Ramparts and citadels were built on natural defensive sites, in which the rural population found refuge in case of invasion. During the Archaic period in Greece, for example, wars between lineages and tribes increased. Harvests were pillaged, lands appropriated, and neighboring populations reduced to servitude. This led the most powerful chiefs to make themselves into an aristocracy that concentrated in its hands the largest portion of the lands, costly and effective metallic arms, horses, and war chariots. This landowning military aristocracy was part of the origin of the fortified city and nascent state. The most powerful militarized city-states could then prolong their pillaging expeditions to neighboring cities by colonizing them and thus resolve their food-supply problems either by imposing a tribute on them or by occupying and exploiting their lands. As P. Garnsey writes:

> The Romans fed their starving thanks to the harvest of their neighbors and culti-
> vated the lands ceded by their conquered enemies. The vanquished were also obli-
> gated to furnish provisions (and labor) in order to make possible subsequent
> stages of the conquest. Over time, the Romans pillaged and exploited the
> resources of countries overseas. Roman soldiers and nonproductive civilians were
> fed from the surplus appropriated from subject states.[22]

But from the moment that these permanent and relatively large city-states were formed, a not insignificant fraction of the population (nobles, warriors, magistrates, artisans, merchants, servants) was removed from agricultural work. As we have seen, agricultural yield at that time was generally hardly sufficient to feed the farmers and their families. From that moment on, the dominant and growing ancient city not only needed colonies, always more colonies, but also slaves in order to feed itself.

## "Necessary" Slavery?

In fact, as Claude Meillassoux emphasizes in *Anthropologie de l'esclavage*, the slave is generally prohibited from reproducing and thus does not have a family to support. Their needs are reduced to nothing more than their own minimum daily requirement and, in these conditions, the slave working in agricultural production can provide a "surplus" as opposed to a free laborer with a family to support.[23] Of course, this "surplus" is only apparent, because there are peripheral societies that have provided this captured and enslaved labor force, societies that have, in effect, had their labor force pillaged. For the slaveholding city, the renewal cost of the slave is limited first to the cost of his/her capture and sale, the capture being that much easier the greater the military superiority of the conquering city, and subsequently to the cost of support, which is reduced to food and guarding the slave.

This analysis is very different from the one sometimes attributed to Friedrich Engels (*The Origin of the Family, Private Property and the State*), according to which slavery developed historically from the moment that the output of an active laborer became greater than his/her own needs. It then became advantageous to keep captives of war as slaves rather than exterminate them as before. This point of view is hardly tenable. In reality, in order for any society to be able to reproduce itself through its own means, it is absolutely necessary that the output of a laborer be greater than his/her own needs, if only to feed the children, the ill, the temporarily disabled, etc. (see chapter 1, point 4). This rule is valid for every society, including those that existed prior to the development of slavery.

According to us, the development of ancient slavery in the West and its perpetuation for more than a millennium is explained otherwise. What made slavery "necessary," from the moment the ancient city appeared, derived from the fact that agricultural productivity of the period was inadequate to both ensure the reproduction of future generations and obtain a surplus able to support the city. Moreover, beyond the military superiority of the slaveholding city, what made slavery possible was the existence at the periphery of less powerful peoples forming a vast reserve of labor power. Such was indeed the viewpoint of the Ancients on the question:

> The use made of slaves, too, departs but little from that made of other animals; for both slaves and tame animals contribute to the necessities of life with the aid of their bodies. ... For this reason, the art of war, too, would be by nature an art of acquisition in some sense. For the art [of acquisition] includes the art of hunting, which should be used against brutes and those men who, born by nature to be ruled, refuse to do so.... According to nature, then, one species of the art of acquisition is a part of the art of household management; and this [part] should either exist or be secured [by the household manager]. It is concerned with the accumulation of goods which are necessary for and useful to life and contribute to the political association or to the household.[24]

Besides, slavery for indebtedness often preceded the development of slavery through war. From the moment that the ancient city and unproductive social groups were formed and taxation acquired a certain importance, many farmers who previously had barely succeeded in supporting their own and their families' needs were forced to fall into the web of a growing indebtedness. This led a number of them to lose both their goods and their independence. The mechanism of this servitude for debt is well-known: a barely self-sufficient peasant, who must sell a large part of the harvest in order to pay taxes, is obliged to go into debt in order to buy back some expensive grain in order to "bridge the gap" before the following harvest. In order to pay back this debt, a portion of this harvest must be sold at a low price. That amounts to borrowing money for several months at a high interest rate. From year to year, still impoverished by the interest on the debt, the peasant must go further and further into debt by mortgaging an increasingly larger part of the land and future labor, both the peasant's own and that of the peasant's family. Arriving at the point where the value of the annual harvest becomes less than the value of what is owed, the peasant is forced to deliver to the creditor all the mortgaged means of production, including the peasant and peasant's family. Thus the peasant is reduced to a state of servitude for indebtedness to the advantage of the creditor, who becomes owner of the farmer's lands, person, and descendents.

In ancient societies, the magnitude of this mechanism, the expansion of different forms of servitude and the later development of slave-hunting war demonstrate that, in the conditions of the time period, the slave became, as Aristotle said, a "natural" necessity for supplying the needs of both the family and the state.

## The Case of Greece

In the "occidental" societies of the Mediterranean region and Europe, cultivable lands were not, as in the "oriental" hydraulic societies (Egypt, Mesopotamia, the Indus Valley), the result of large installations constructed under the aegis of an all-powerful sovereign who grants lands to particular individuals or to relatively undifferentiated communities in the vicinity. In the Occident, the cultivable lands of the *ager* were more the object of a generally unequal appropriation or private usufruct.

### Colonization and Servitude

Beginning in the eighth century B.C.E., the concentration of lands in the hands of a minority of large property owners grew, above all in fertile regions of Greece. Farmer victims of this concentration, but also perhaps of overpopulation, were confined to lots that were too small or driven into the most unprofitable areas.

Increasing numbers of farmers who found themselves unable to pay the taxes were obliged to go into debt and finally had no other choice than servitude for indebtedness or emigration. Greek colonization toward the west (southern Italy, Sicily) then toward the east (Asia Minor, Black Sea) and the south (north Africa) was organized. The aristocracy, artisans, merchants, and ruined peasants, either mercenaries or in servitude, participated in this colonization, which was largely agrarian. It occurred generally on plains that were more extensive, more fertile, and less overpopulated than those of Greece. To a large extent, it was based on the exploitation of local labor power or on that of immigrants subjected to various forms of servitude. It produced a surplus that, in turn, supplied the metropolis.

## Agrarian Reform and Democracy

The ruin and bondage of a part of the peasantry impoverished the countryside, reduced the demand for artisanal products, and weakened general economic activity. The constant aggravation of inequalities encouraged revolutionary movements that consistently demanded, throughout antiquity, both the abolition of debts and redistribution of lands. These movements carried to power either democratically elected reformist legislators or tyrants who were imposed by violence.

In Athens, at the beginning of the sixth century B.C.E., the legislator Solon exempted the enslaved peasants from their heavy obligations and prohibited servitude for indebtedness and the sale of children as slaves. Every Athenian citizen was thus considered free in the eyes of the state. Solon also took a whole series of measures to distribute the undivided lands of the *saltus* and apportion more justly the taxes and obligations of the different categories of citizens as a function of their wealth. But these reforms displeased both the oligarchy, who found them too radical, and the peasantry, who called for the redistribution of lands. In 524 B.C.E., Pisistratus, representing an aristocratic faction that had taken leadership of the Diacrian movement (expropriated peasants forced to the infertile mountains on the periphery of Attica, notably the plateau of Diacria), seized power in Athens. He then imposed radical reforms by extensively distributing estates confiscated from the aristocracy and uncultivated lands. As a result, a strong class of small and middle peasants was formed. He encouraged them, through state credits at low interest rates, to invest in planting vines and fruit trees. Different from Solon, who represented the people of the cities, Pisistratus relied on the support of the dispossessed and repressed peasantry and, by redistributing lands to their benefit, carried out one of the first agrarian reforms of history. However, if the reforms of Solon and Pisistratus, and afterward those of Cleisthenes, established democracy and protected Athenian citizens from servitude, they did not, for all that, abolish slavery for foreigners, in the metropolis or the colonies.[25]

## The Question of Supplying the City

Following the reforms of the sixth century B.C.E., the rural economy flourished again in Attica. This was the golden age of the independent small and medium property owners, who produced their own grain, sold the produce from their vines and orchards, worked with their families and a few slaves, lived frugally and limited their descendents to one or two children. Xenophon, in *The Economist*, exalts this ideal of life. Nevertheless, this agrarian system produced very little surplus, and various signs show that the chronic shortage of grain continued. Restriction of births was the rule; abortion and infanticide were frequent. Newborn children, especially girls, were "exposed," on the public road, thereby abandoning them to slave hunters and most assuredly to death. The dietary regime remained frugal. Food shortages were frequent, as were epidemics (plague) and endemic diseases (malaria, tuberculosis).

Certainly, from the sixth to the fourth century B.C.E., the Athenian population, in both the city and country, more than doubled, going from 100,000 inhabitants to more than 200,000. But it is also important to note that in the fifth century, the Athenian city, master of the seas, imported at least half of its wheat, above all from the Black Sea regions but also, later, from Sicily, southern Italy, Egypt, and Thrace.[26] In the fourth century, on the other hand, having lost its maritime dominance, supplying the city with necessities became a constant preoccupation of the government. The law prohibited exports of wheat, under pain of death. It also prohibited all residents from financing any vessels not transporting wheat to Athens and required merchants to deliver to the city at least two-thirds of their wheat cargoes. The authorities fixed the prices of grain, flour, and bread. Purchases of cereals by merchants and their profits were limited in order to avoid monopoly, speculation and scarcity. Magistrates (the *sitophylakes*) were specially appointed to verify that all these rules were indeed applied. But, even in Athens, these legal arrangements lasted only a short period of time. And in most of the nondominant Greek cities the authorities did not take responsibility for the food security of the population. That responsibility was left to the ostentatious charity of the wealthy, which, even if it temporarily assisted the famished, could not resolve the basic problem.

## The Crisis and the Fall of Athens

From the end of the fifth century B.C.E., the situation in the Athenian countryside deteriorated again. Because of the effects of land distribution through inheritance, peasant farms became smaller. The size of most of them was between 2 and 4 hectares, much less than the minimum necessary to feed a family. Indebtedness and ruin spread leading to the formation of indigent masses

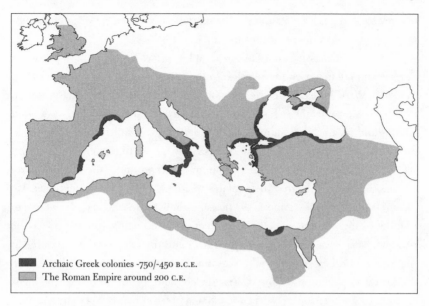

Archaic Greek colonies -750/-450 B.C.E.
The Roman Empire around 200 C.E.

*Figure 6.4 Map of Greek Colonies and the Roman Empire*

inclined to hire themselves out as mercenaries, even to the enemy, or to the development of large estates. Since the government refused reforms, civil wars returned that continued until the Macedonian conquests. The colonization of territories to the east, under the rule of Alexander of Macedon, made possible a temporary return to prosperity. Then Athens became depopulated and entered into a period of grave crisis until the end of the second century B.C.E., a crisis that only worsened until the Roman conquest.

## The Case of Italy: Colonization

Rome undertook an increasingly larger colonization effort essentially motivated by pillage, confiscation of cultivable lands, and the capture of slave labor. At first limited to Italy, Roman colonization expanded to the whole Mediterranean region and to south central and northwest Europe, after the victory over its Carthaginian rival. These conquests led to an enormous transfer of wealth from the conquered regions to the Italian peninsula and to Rome in particular. Agricultural goods, various commodities, silver, and slaves at low prices came in great quantities. This was essentially plunder, tribute paid by areas that became Roman provinces, products from the exploitation of the *ager publicus* (totality of landed property, mines, quarries, forests, salt-works, etc., confiscated by the Roman state in the conquered regions) and profits from companies and private persons who exploited the resources of the provinces. The revenues of the state became so large that, in 167 B.C.E.,, the republican government decided to suppress all direct taxation of citizens in Italy.

However, the massive influx of cereals at low prices from the colonies caused a significant fall in agricultural prices. In the two years that followed the end of the Punic Wars, for example, the price of wheat in Rome fell by a factor of four, then by a factor of eight.[27] Later, competition from colonial products extended to wine and olive oil, but animals, fresh fruits. and vegetables, difficult to transport, remained largely protected from this competition. In regions bordering on the sea and affected by competition from imported products, the large property owners generally converted their estates to animal raising and sometimes also toward fruit plantations. Among the peasants, only those who had capital due to plunder brought home from war could convert their lands. In these areas, the rural landscape was transformed: the *saltus* and the *hortus* were expanded at the expense of the *ager*. For poor peasants, the only outcome was often to sell their lands and rejoin the more or less idle common people of Rome (the plebs of Rome). Landed property was concentrated in fewer hands. Very large estates, the *latifundia*, were formed, cultivated by slaves who arrived in such numbers and at such low prices that they replaced the wage laborers and free tenant farmers. However, in fertile regions such as the Po plain, where cereal growing was productive, and in interior regions that imported cereal convoys did not reach, cereal growing diminished very little.

## The Agrarian Laws

This development, which aggravated the food dependence of Rome on its colonies and increased the number of Roman plebians, made some senators uneasy. At the beginning of the second century B.C.E., the Senate decided to confer portions of land taken from the *ager publicus* on Roman citizens and its deprived Latin allies, so as to reinforce the declining class of small and medium property owners. But this measure was hardly applied, because it came up against the interests of the large property owners, publicans (charged by the state with ensuring diverse administrative functions and management of public goods and funds) and the senators themselves, who possessed estates derived more from large portions of the *ager publicus* than from any other property.

Nevertheless, the idea was not dead and, in 133 B.C.E., Tiberius Sempronius Gracchus, people's tribune, made the Assembly vote for an agrarian law. The objective of this law was to reconstruct a stratum of numerous family farms, likely to restore the economy of the Italian countryside and restore to the state a larger social base. To this end, the law first limited what a family could possess to 125 hectares for the head of the family, plus 62.5 hectares per child, from the *ager publicus*. Beyond this ceiling, public lands had to be ceded back to the state. In return, the agrarian law accorded to the occupants of the *ager publicus* the complete ownership of the lands that they retained. Finally, the state had to redistribute the lands thus recovered to the greatest number of deprived citizens in plots of 7.5 hectares, plots considered inalienable and subjected to the payment of a regular tax, thereby ensuring that these lands were actually cultivated.

This law provoked strong opposition from the senatorial nobility, which used all of its power to try to suspend it, then to hold up its application and limit it to marginal regions. Tiberius Gracchus was assassinated the same year that the law was voted on, but the torch of reform was taken up by his brother Caius, who was also assassinated several years later.

Despite this opposition, the agrarian law, an expression of the popular will, was partially applied afterward though subjected to numerous modifications: the state granted financial compensation to concessionaries who ceded back part of the *ager publicus*; lands were redistributed in plots of 50 hectares rather than 7.5; the new farmers were no longer obliged to pay a tax; and finally, the law was above all applied in the conquered provinces, where it resulted in the founding of veritable Roman agrarian colonies. But the principal change came from the fact that, little by little, plots were awarded only to war veterans. It was only under the consulship of Caesar (59 B.C.E.) that grants of land to poor citizens resumed. Despite these successive modifications, the "reform of the Gracchi," from the name of its inspirers, reconstituted, at the end of a century, a class of peasant small and medium property owners who farmed a few dozen hectares, resided on their lands, and experienced relative prosperity thanks to grapevines and olives. But in the end these reforms were of limited significance: not only did large estates not disappear, they were even enlarged from that part of the *ager publicus* given to latifundia owners as sole owners, in compensation for that part of the *ager publicus* taken back by the tate.

## Wheat Distribution Laws

The application of the agrarian laws was inadequate to curb the rural exodus and the growth in the number of Roman plebians. Up to the middle of the second century B.C.E., the latter were still relatively small in number and the meals given by the important Roman families to their supporters sufficed to feed them. Beyond that, the state itself gave large public banquets. But the number of plebians grew and the extravagant expenditures of the wealthy no longer sufficed to feed the poor. In 123 B.C.E., Caius Sempronius Gracchus caused a "wheat distribution law" to be passed according to which the state would have to sell to citizens a certain quantity of cereals at very low prices. This law was then modified on several occasions, sometimes in a more restrictive sense (fewer beneficiaries, higher prices) or, at other times, in a more liberal sense, according to the relationship of force between the Senate and the Roman people. It reached a maximum of liberality with the Claudian law, in 58 B.C.E., which extended the public distribution of cereals to poor citizens. The number of beneficiaries exceeded 300,000, which leads to the conclusion that out of the million (approximately) persons who lived in Rome, more than half lived from these distributions. Under the dictatorship of Caesar, this number was reduced by half.

## Military and Economic Crisis

However, neither the agrarian laws nor the wheat distribution laws, which aimed at reducing the social crisis in country and city by reconstituting a middle peasantry and feeding the people of Rome, could resolve the military, and therefore economic, impasse into which the empire had sunk. From triumph to triumph, the empire had reached impassable limits. Because of the distance and lengthening of its front lines and the proliferation and increasing strength of the peoples that it fought, the Roman Empire could no longer expand and conquer cheaply the wealth, new lands, and new people upon which the Roman state and economy continually lived. Assailed from all sides, by partially subjugated peoples, slave revolts, and urban people from within the empire, the Roman state found itself overwhelmed by the growing cost of war, maintaining order and public policies, while at the same time its revenues were increasingly limited. In order to overcome its budget deficit, the state resorted to modifying the money: for the same stated value, the weight of the pieces and their content in precious metals were lowered.

The military and budgetary crisis of the state partly explains the economic crisis. War retrieved fewer and fewer slaves. Labor power became more expensive and in short supply. The growing troubles accentuated the decline of agricultural and artisanal production. Deprived of low-priced slave labor, the agriculture of the Italian peninsula descended into crisis. Estates remained uncultivated and the supply of cereals depended increasingly on imports from the eastern empires. Shortages were common. Confronted with this disastrous development, some emperors attempted to slow it down. Domitian prohibited the planting of new vines in Italy and required half of the existing vines in the provinces to be uprooted. Under Trajan, agricultural credit funds, mixing public and private sources, made loans at low interest rates to farmers who wanted to invest. Hadrian granted increasingly liberal conditions to those who agreed to develop a part, even a small part, of the imperial property. Privileges were granted to guilds of artisans who contributed toward feeding Rome (bakers, butchers, maritime carriers of wheat, etc.) in exchange for services rendered.

These disparate and unevenly implemented measures were far from curbing the fall in agricultural production and the increase in social tensions. From the end of the early empire, the first barbarian invasions, combined with shortages and civil wars, spread terror, epidemics, desolation, and death throughout the empire. There followed a dramatic fall in production, population, and fiscal receipts, while the state needed supplementary resources to repulse the barbarians and attempt to maintain internal order. Runaway inflation resulted both from insufficient production and the abundant creation of devalued money. Frustrated by what was, at the time, a new phenomenon, the state tried to intervene directly in the economy. In 301, the emperor Diocletian signed the "edict of the maximum," which fixed a maximum price for a thousand commodities and allowed imposition of the death penalty on anyone who paid or demanded a higher price, as well as any speculator who concealed reserves. This attempt to control prices in such an extensive territory as the Roman Empire ended in total failure. In fact, this edict, which did not foresee necessary regional variations in price resulting from transport costs, led to concealment of products and a rise in their prices. From 304, the state bought commodities from Egypt paying ten times more than the maximum price fixed by this edict. Between 294 and 344, the price of Egyptian wheat increased by 6,700 percent! The same type of edict, signed in 362 by the emperor Julian, experienced the same failure. Indeed, up to the end of the Empire, stability of prices was never truly reestablished.[28] Formerly, the Roman state made war, maintained order, constructed necessary infrastructure for the army and commerce, and generally practiced a noninterventionist political economy in relation to private agents: farmers, artisans, merchants, associations of tax-gatherers, etc. However, through agrarian and wheat distribution laws, it had already intervened in the agricultural economy and distribution of food products. But, at the end of the late empire, in an attempt to remedy increasingly numerous and serious shortages, the Roman state intervened even more heavily in trade and even took a growing part of production directly in hand. Direct state control, state monopolies, compulsory deliveries, various levies, fixed-price orders, etc., supplanted every other economic form.

## The Emergence of Serfdom

In order to compensate for the labor shortage, the state attempted to encourage the diffusion of more productive technical means (the Gallic harvester and water mill, for example), and "serfdom" became the law. In fact, being a colonist ceased to be, as in the early empire, a contract freely concluded and broken. Henceforth, the colonists were juridically bound to the land that they farmed, indeed even bound to the owner of this land by a bond of personal dependence specific to serfdom (in the modern sense of the word).

In the declining empire, the large property owners increasingly took refuge in their country villas, protected from the urban mobs who always appeared to them to be more demanding and more threatening. Organizing the defense of their estates against attacks from disbanded legions, barbarian bands, and pillagers, they gradually escaped from the authority of the declining central power and established their own laws. Slaves and fleeing peasants placed themselves under the authority of these landowners, both in order to find a means of living and to benefit from the protection provided on the estates. The owner granted to each family a plot of land that it could farm in exchange for both a tax on a part of the harvest and large corvées to cultivate lands reserved to the master of the estate. In so far as they could no longer escape from their new master, these dependent peasants were hardly distinct from the former bound slaves, that is, the serfs.

With serfdom, buying men and women pillaged from neighboring peoples was no longer the preferred method for renewing the labor force of the large estates. It was ensured by the serf families who produced and raised children, children who were born serfs and remained so, just as their descendents did.

Rome and the other cities of the empire declined when the Roman state, supplier of lands, slaves, subsistence, and other wealth, fell apart. A military and landed nobility from many origins (Roman, Germanic, Gallic) began to organize the production and protection of subsistence and people on its own "fiefs." But this new political, economic, and social order in the countryside, based on private estates of the nobility and peasant tenures, whether serf or free, took centuries to be imposed on the Occident. These were centuries during which armed bands of all sorts continued to traverse Europe, pillaging, destroying, decimating livestock and people, even if the formation of Germanic and Nordic kingdoms (barbarian kingdoms) and the renaissance of a Christian Empire of the Occident (the Carolingian Empire) temporarily instituted a sort of order. These were centuries during which slavery (captured and sold slaves) certainly experienced ups and downs, but persisted all the same.

Slavery ceased to exist in the Occident from the moment when war no longer made it possible to renew stocks of slaves (sold captives) and serfs (bound and with family responsibilities) by capturing them, on which the ancient economy had rested.[29] That would essentially explain the gradual exhaustion of this reserve, further depleted by the emancipations that became more numerous after the year 1000. Can it be concluded from that that war to obtain slaves then became in itself nonprofitable for the Occident? That is difficult to say.

We believe for our part that, in the agrarian economy of the Early Middle Ages, the raising of children by serf families did not suffice to renew the active agricultural population entirely, because the productivity of systems based on fallowing and cultivation with the ard was undoubtedly inadequate to ensure the reproduction of the peasant family and to support a large tribute in kind (part of the harvest) and in labor (corvées on private estates). Beginning in the year 1000, however, as we will see in the following chapter, vast productive investments—new tools, livestock, clearings, mills—became possible, which finally brought the agrarian economy of the Occident out of its ancient poverty. With the agricultural revolution of the Middle Ages, production and population finally developed rapidly. The productivity of agricultural labor markedly increased, which made the ancient forms of servitude (slavery and serfdom) increasingly "unnecessary," in the Occident at least. It appears to us then plausible to conclude that wars to obtain slaves, which were in themselves perhaps less and less profitable, in any case became less profitable than the new productive investments.

# 7

# Agrarian Systems Based on Fallowing and Animal-Drawn Cultivation with the Plow in the Cold Temperate Regions: The Agricultural Revolution of the Middle Ages in Northwestern Europe

Thus, although cathedral art was urban art, it relied on the nearby countryside for the major factor in its growth, and it was the efforts of countless pioneers, clearers of land, planters of vinestocks, diggers of ditches, and builders of dikes, all flushed with the successes of a flourishing agriculture, that brought cathedral art to its fulfillment. The towers of Laon rose against a backdrop of new harvests and young vineyards; the image of the oxen used in plowing, carved in stone, crowned those towers; vineshoots appeared on the capitals of all the cathedrals. The façades of the cathedrals in Amiens and Paris showed the turning of the seasons by depicting different types of peasant labor. It was only right to honor them in this way, for it was the work of the harvester sharpening his scythe, of the vine-grower pruning or layering his vines or spading about them, that had made the edifice rise little by little. The cathedral was the fruit of the system of manor lords—in other words, of the peasants' labor.

—GEORGES DUBY, *The Age of the Cathedrals*

The agrarian systems based on fallowing and cultivation with the animal-drawn plow derive from earlier systems also based on fallowing, but that used the ard for cultivation.[1] Like the ard-based system, these systems are based on the combination of rainfed cereal growing and animal raising. The cereals occupy the arable lands where they alternate with a fallow period forming a short-term

259

rotation, while the livestock draw their subsistence from peripheral natural pastures and play a central role in fieldwork and in renewing the fertility of the cereal-growing lands. But the new systems are clearly distinguished from their predecessors by the use of more powerful means of transport and equipment for working the soil. Wheeled carts take the place of the packsaddle, and the plow, contrary to the ard that it replaces, makes possible true plowing.

In the cold temperate regions, this new equipment made it possible to expand the previously limited practices of cultivation and animal raising by using hay, stabling livestock during the dead season, and using manure. The development of these practices gave rise to a new cultivated ecosystem, which from then on included hay meadows and extended arable lands. The latter were better manured, better plowed, and generally cultivated in a triennial rotation. Thus a new agrarian system appeared which, despite the high cost of the necessary equipment, spread widely throughout the cold, temperate regions, where it facilitated a considerable growth in production and agricultural productivity. In the Mediterranean regions, where the lack of fodder during the winter is not an important limiting factor, use of the plow was much less profitable. These regions adopted other, more appropriate ways of improving agricultural production, such as arboriculture, terracing of slopes, and irrigation.

The scythe, wheeled carts, the plow, hay, stabling, manure, and a whole series of complementary methods and practices were familiar in the West since antiquity or the early Middle Ages. But it is only with the central Middle Ages, from the eleventh to the thirteenth century, that the agrarian systems characterized by the combination of fallowing and cultivation with the animal-drawn plow were widely developed in the northern half of Europe. Subsequently, they were transferred through European colonization into the temperate regions of the two Americas, South Africa, Australia, and New Zealand.

The agricultural revolution of the Middle Ages carried the rural economy in the West to the threshold of modern times. This agricultural revolution sustained an unprecedented demographic, economic, and urban expansion over three centuries, an expansion that ended with the horrifying crisis of the fourteenth century, during which more than half of the European population perished. After a century of crisis and upheaval, reconstruction began at the end of the fifteenth and into the sixteenth century. After that, the crisis began again and lasted until a new agricultural revolution, based on cultivating the fallow lands, developed during the seventeenth, eighteenth, and nineteenth centuries. Despite these developments, the practice of cultivation using the animal-drawn plow existed well after the Middle Ages. In fact, the use of wheeled carts and animal-drawn plows, whether combined with fallowing or not, lasted in the West until motorization in the twentieth century.

Even today, on condition of being used advisedly, cultivation using the animal-drawn plow can be of great service in many regions of Africa, Asia, and Latin America, where relatively unproductive forms of manual cultivation and cultivation using the ard continue to exist. That is why in this chapter we focus on presenting the origin of the equipment characteristic of this new system of cultivation. We analyze the conditions of development of the farming and corresponding animal breeding practices, as well as the cultivated ecosystem that originated from them. We also attempt to explain the structure, the functioning and the results of the new systems, to understand the consequences of their development as well as their historical and geographical limits.

## 1. THE GENESIS OF CULTIVATION USING THE ANIMAL-DRAWN PLOW

### The Inadequacies of Agrarian Systems Based on Fallowing and Cultivation Using the Ard

As we saw in the preceding chapter, the performance of the systems based on fallowing and cultivation with the ard is limited by the poor means available for plowing and transport. Plowing by hand, with spade, or hoe demands so much time and effort that it cannot be extended to all of the fallow lands, while use of the ard results only in a very imperfect quasi-plowing. As a result, the soil is poorly prepared before the sowing. Furthermore, transportation using pack-saddles does not permit the transfer of large quantities of organic matter (fodder, litter, and manure) from the pastures to the cultivated areas. Transferring fertility by simply penning the animals at night is not effective, because a large part of the animal manure is lost on lands and roads along the way to and from the pastures, while the manure that is actually deposited on the fallow lands is poorly buried in the soil. In the end, the maintenance of fertility on cultivated lands is not assured.

Beyond that, in the cold temperate regions, the size of the herds is strongly limited by the availability of fodder during the winter. Each autumn, most of the young born in the spring and the unfit animals have to be slaughtered in order to retain only a small number of reproductive animals during the winter. Nevertheless, extended pastures are necessary in order to feed even such a reduced herd, so much so that during the growing season the largest part of the grass produced on these large pastures is effectively lost and cannot contribute to the reproduction of the fertility of the cultivated lands. In the final analysis, in this type of system, cereal crops cannot be cultivated widely, are poorly fertilized, poorly prepared, and produce small yields.

## The Innovations of Antiquity and the Early Middle Ages

### *The Scythe and Hay*

In order to make up for the lack of fodder in the winter, and thus increase the number of livestock and the consequent transfers of fertility, it was common, beginning in antiquity, to harvest a part of the excess grass in summer, dry it in the sun to obtain hay and then save it to give to the animals during the winter. Originally, farmers were equipped only with a grain sickle to cut the grass, yielding poor results. As a result, the use of hay was limited until the invention and diffusion of the large scythe, used with two hands.

The scythe appeared in Gaul in the last century B.C.E. and its use was gradually extended to the northern half of Europe during the first millennium of the common era. Since the scythe was not yet used during this period to cut cereals, it can be concluded that hay making and the use of hay in the winter were widespread. In fact, the productivity of the scythe is much greater than that of the sickle, but it remained a rare and expensive instrument until around the year 1000, since its manufacture required a mastery of working with iron. In the central Middle Ages, on the other hand, the progress of metallurgy and the rural craft industry made possible the scythe's more widespread use.

However, use of the scythe alone does not suffice to expand the use of hay. It is also necessary to protect the grasses reserved for this purpose from the livestock until the hay-making season. During antiquity and the early Middle Ages, meadows set aside for mowing, which would have been enclosed and thereby not available for common pasturage on the *saltus*, were uncommon. Thus it was often necessary to harvest hay in distant, open grassy clearings in the middle of the forest, "hay shelters" protected from crossing herds. Lacking effective means of transport and without any buildings for shelter, this hay was stored in place, piled around poles in conical stacks down which rainwater would flow without getting the hay too wet. During the coming winter, the livestock were led to these clearings to feed on the fodder.

As a result of the increased use of the scythe, the supplies of hay were increased and the herds of livestock grew, as did the animal excrement transferred to the fallow lands. However, in order for this transfer of fertility to take place, the livestock, which had passed the day in the hay clearings, had to be returned to the fallow lands at night. Such was, undoubtedly, the reason for the ban on "passing the night" pointed out by G. and C. Bertrand in *L'Histoire de la France rurale*, which stipulated that the herds should not spend the night in the clearings or in the neighboring woods.[2] The daily comings and goings of the livestock between the fallow lands and the hay reserves were nevertheless long and difficult, due to the distance of the clearings and the winter weather. Thus, much time, energy, and animal excrement were lost en route, to such an extent that this impractical system had an altogether limited potential.

## Heavy Transport, Stabling, and Manure

In order to avoid the difficulties involved in daily displacements of livestock during the winter, buildings designed to shelter the animals (cowsheds, stables, sheep folds) and hay reserves (barns or haylofts) were constructed close to houses. As a result, livestock could spend the entire winter in stalls, thereby making it possible to collect all of the excrement, both day and night. Since this excrement was wet and difficult to handle, it was mixed with a litter composed of brushwood, leaves, or straw from the grain. Thus a sort of compost was obtained, easily handled with a pitchfork and transportable.

The use of manure is a far more effective means of transferring fertility from the pastures to the cultivated lands than penning the animals at night. Not only is all the animal excrement, from both day and night, collected during the winter (while, in the earlier system, penning of the animals took place only at night), but it is enriched with plant matter taken from the forest or other uncultivated lands, which acts as bedding for the livestock and contributes to the transfer of fertility to the cultivated lands in the same way as animal excrement does. Note that straw from grain used as litter adds nothing to these transfers of fertility when and if it originates from the cereal-growing lands themselves. Beyond that, stored manure with bedding has the distinct advantage of being amenable to preservation and thus can be spread at the most favorable moment.

But in order for the practice of stabling to develop, it was also necessary to resolve the problems of transporting hay, litter, and manure. During a long winter, one large head of livestock (cow or horse) or five or six small heads of livestock (goat or sheep), which comes to the same thing, consume several tons of hay and produces several tons of excrement. The longer the period of winter stabling, the higher the amount of hay consumed and excrement produced. In total, for one head of large livestock spending the winter in the stable, it is necessary to transport 8 to 16 tons of diverse materials over several kilometers: 2 to 4 tons of hay, 1 to 2 tons of litter, and 5 to 10 tons of manure.

The solution to the problems of transporting heavy and bulky materials came from using wheeled carts for work in the fields, pulled by oxen, horses, mules, or donkeys. Used in Mesopotamia in the fourth millenium B.C.E., wheeled vehicles of Oriental origin began to become widespread in the Near East and Europe in the third millenium B.C.E.. But during all of antiquity and the early Middle

Ages, this equipment remained costly and, even if one can point to some rare agricultural uses, its use was limited principally to chariots for war or parade, and carts for transporting people and high-priced commodities.

Moreover, Roman methods of harnessing, which lasted in Europe until the end of the early Middle Ages, were not very effective. The breast harness for horses and the withers yoke for cattle "choked" the animals to the point that it was necessary to use four horses to pull a war chariot and a pair, or even several pairs, of oxen to pull a transport cart or a swing-plow. Not only were such harnesses difficult to handle, they were expensive. For all these reasons, the use of carts in agriculture remained very limited in this epoch. No one would know better than Hesiod (*Works and Days*) how to make quite clear the difficulties involved in obtaining this equipment: "The man full of illusions speaks of building a chariot. The poor fool! He does not know that there are a hundred pieces in a chariot and that it is first necessary to make a point of gathering them all together in his house."[3]

### The Plow, the Harrow, and the Roller

The manure obtained by stabling the animals and feeding them hay during the winter must be carefully buried over the whole surface of the land being sown in order to gain the most from its use. Neither the ard, which does not turn the soil over, nor manual plowing, which can be accomplished only on a small part of the fallow lands, makes it possible to carry out this work completely in the required time. In order to resolve this problem, it was necessary to make use of new equipment, the plow, which is capable of accomplishing a true plowing and doing so quickly enough to bury huge amounts of manure on all of the fallow lands each year.

The plow appeared in several places in the northern half of Europe independently of one another, all at the beginning of the Christian era. It had different names in different places: *carruca* in Gaul, *Pflug* in Germany. The plow is a complex instrument composed of several tools. The *colter* cuts the soil vertically while the *plowshare*, triangular and asymmetric, cuts the soil horizontally. These two tools are assembled in such a way that together they cut a continuous strip of land in a rectangular shape as the machine advances. The *moldboard* extends the plowshare and turns over the strip of cut earth into the open furrow created by an earlier crossing of the plow. While the colter and the plowshare are always made from iron, the moldboard can be made from a simple plank of wood. In more recent, improved plows, the moldboard is iron and curved toward the outside in order to turn over the soil more effectively. The plow turns over the soil from one side alone. It is an asymmetric instrument, difficult to keep in a straight line because of the lateral forces exerted on the moldboard. While a single handle generally suffices to steer an ard, it is usually necessary to have two, firmly held in the hands, to maintain a straight line with a plow. Vertical pressures are equally exerted on the moldboard, which sometimes push the plowshare down and sometimes pull it up. Thus the handles are also necessary to maintain a constant depth in the plowing.

But on soils that are even a little heavy or rocky, the handles are not adequate to control the plow. It is then necessary to use a *forecarriage*. A classic forecarriage is a small, two-wheeled cart on which the pole (or beam) of the plow rests. One wheel moves in the previous furrow, which guides the forward movement of the machine, and the other on the tillage, i.e., on the earth that remains to be plowed. Hence it is important to open up the first furrow in a straight line. The beam of the plow rests on a mobile crosspiece, which can be adjusted vertically to set the depth of the plowing. It is placed between two pegs that can be adjusted laterally to set the width of the plowing. After making these rough adjustments, the farmer must adapt the operation of the plow to the nature of the soil and its variations by using the handles. On soils that are easier to work, the two wheels can be replaced by a single wheel or by a wooden shoe or runner that, while gliding in the bottom of the previous furrow, acts as a guide for the plow. On very light sandy or silty soils, it is possible to do without the forecarriage and sometimes also the colter.

Undoubtedly, hundreds of repeated trials and errors occurred before all these pieces could be coherently assembled on the same machine. There are, moreover, old representations of incomplete or deformed plows. Perhaps such plows existed, but it is also possible that at the beginning, this new, revolutionary instrument—not widespread, relatively unknown, and a bit mythical—gave rise to some chimerical representations.

However, even if such plowing were relatively fast, it is not as thorough as manual plowing with a spade or hoe. In fact, it generally leaves the ground encumbered with clods of earth that are too large and weeds that are either poorly uprooted or poorly buried. This is why the work of plowing must be completed by manually breaking up the clods of earth and weeding, either by crisscross runs with an ard or, more effectively, by using a new instrument, the *harrow*. Pulled by animals, the harrow consists of a wooden frame in which long points or teeth are fixed. It scarifies, breaks up, and loosens the soil, while uprooting residual weeds during its forward movement. It is used before sowing in order to prepare the seedbed and after that in order to cover the seeds with soil, which then must be packed down with a roller. Both the harrow and the roller, each in its own way, complete the plow's work. They are integral parts of the technology used in the new system of cultivation.

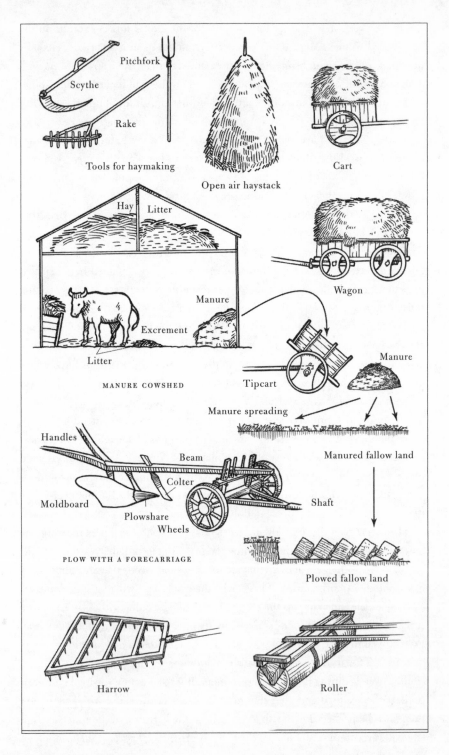

*Figure 7.1 Technology Associated with the Cultivation System Based on the Plow*

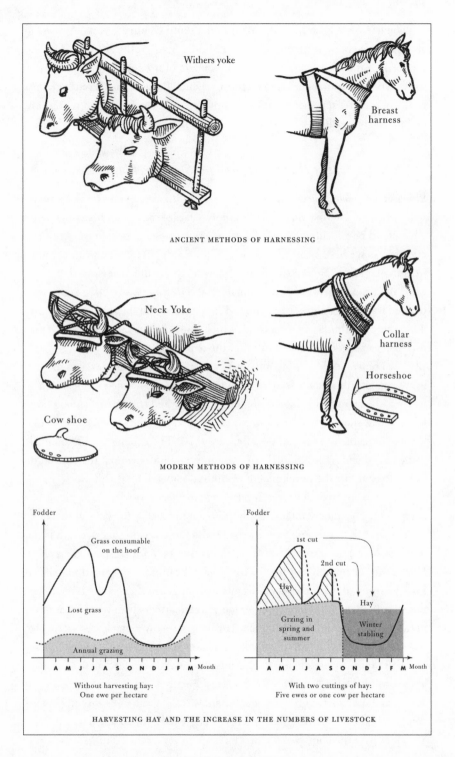

Withers yoke

Breast harness

ANCIENT METHODS OF HARNESSING

Neck Yoke

Collar harness

Horseshoe

Cow shoe

MODERN METHODS OF HARNESSING

Fodder

Grass consumable
on the hoof

Lost grass

Annual grazing

A M J J A S O N D J F M Month

Without harvesting hay:
One ewe per hectare

Fodder

1st cut

2nd cut

Hay

Grzing in
spring and
summer

Hay

Winter
stabling

A M J J A S O N D J F M Month

With two cuttings of hay:
Five ewes or one cow per hectare

HARVESTING HAY AND THE INCREASE IN THE NUMBERS OF LIVESTOCK

*Figure 7.1 (continued) Technology Associated with Cultivation System Based on the Plow*

The harrow appeared in the West in the ninth century and its use spread, along with the plow, through the central Middle Ages. At the end of this period, the teeth of the harrows were more often made of iron, while previously they were made of wood. The Bayeux Tapestry, which dates to the eleventh century, is one of the first representations of a mule and a horse, harnessed with shoulder collars, one pulling a plow and the other a harrow.

## New Methods of Harnessing and the Shoeing of Draft Animals

Plows and wheeled carts required a strong pulling force in order to be pulled with a full load across fields or on barely visible roads, much stronger than what could be provided by older methods of harnessing, which were not very effective, as we have seen. The growth of the new system of cultivation consequently depended on the diffusion of new methods of harnessing that multiplied the tractive power of the animals: the *collar harness*, with a rigid, well-padded framework, for horses, donkeys, and mules, and the *neck yoke* for cows. These new methods of harnessing, which appeared in Europe in the eighth century, became widespread only after the tenth century.[4]

The contribution of draft animals to agricultural work increases immensely in systems of cultivation based on the plow. Day after day, year-round, and on all terrains, the animals are at work, pulling the plow and harrow, or hauling heavy loads of hay, sheaves, manure, and wood. While doing such work, the hooves of horses and cattle wear out, unless they are shod. The shoeing of draft animals, using nailed shoes, began in Europe around the ninth century. It removed the last factor limiting the development of cultivation based on the plow.

The breeding of draft horses developed in connection with the use of this new equipment. The tractive force of the horse is approximately the same as that of the ox, but since it moves one and one-half times more quickly, its strength is that much greater. Furthermore, it can work two hours more each day than an ox. Hence the raising of draft horses took on great importance in some parts of Europe during the Middle Ages, despite their high price, generally three to four times more than oxen.[5] However, in most areas, farmers continued to use oxen because of their lower cost and their hardiness. Moreover, in cleared areas, they were more suitable for working a soil still cluttered with stumps. On small farms, the draft animals were often cows because they were even less costly than oxen and, in addition, supplied milk and calves. Finally, donkeys and mules, widespread in the southern regions where the ard was still used, were also increasingly common in the northern regions where use of the plow predominated.

Thus, over the course of the first millennium, all of the equipment associated with the use of a true animal-drawn plow began to be employed in the agriculture of the northern half of Europe, whether that equipment had long been in existence, such as the scythe, the cart, and the wagon, or was relatively new, such as the dumpcart (a type of cart in which the bed tips up, well suited for dumping manure), the plow, the harrow, the roller, the collar harness, the neck yoke, and the shoeing of draft animals. This equipment made possible the development of previously limited agricultural and animal-raising practices (hay making, winter stabling, production and use of manure, plowing, harrowing, rolling) and remedied many of the serious inadequacies of systems using the ard in the cold temperate regions (weak carrying capacity of the animals, mediocre manure, poor preparation of the soil). Systems of fallowing and cultivation using the plow emerged toward the end of the first millennium based on the coordinated use of these new instruments of labor and the combined development of these practices. These systems subsequently spread into most areas of northern Europe during the central Middle Ages, from the eleventh to the thirteenth centuries.

Such systems took on various forms from one region to another and from one century to another, forms that will never be completely known. The proportions and arrangement of pastures, hay meadows, and arable lands, both fallow and cereal growing, vary, as do the forms of the plows and carts and the dates of plowing and manure spreading. But regardless of these variations, these systems retain structural (equipment, cultivated ecosystem) and functional (method of managing crops, pastures and animal raising, method of renewing fertility and method of clearing) characteristics that clearly distinguish them from the preceding systems based on fallowing and cultivation with an ard and the systems without fallowing, which will replace them.

## 2. STRUCTURE AND FUNCTIONING OF SYSTEMS BASED ON FALLOWING AND ANIMAL-DRAWN CULTIVATION USING A PLOW

Let us look more closely at the structural and functional characteristics, as well as the performance and limits of these new systems.

## New Equipment

The introduction and use of each new piece of equipment associated with these systems of cultivation make it possible to loosen a constraint limiting the development of more effective agricultural and animal-raising practices. As long as an implement is used in isolation, it has a reduced impact. In fact, as soon as a constraint is lifted, another constraint appears which, in turn, blocks development unless it too is lifted by a new innovation. The scythe allows the expansion of hay making, but without carts, winter stabling, increase in herd size, and production of manure remain limited. Using the scythe and cart, large quantities of manure can be produced, but without the plow, this manure cannot be buried over large areas in the required time, and without the harrow, the preparation of the soil cannot be completed. Lastly, without improved harnessing and the shoeing of draft animals, there could be no truly effective cultivation using a plow. Thus it is only when all the new means were assembled and articulated into a new, coherent technical system that the new practices could develop fully and fulfill all their potential.

One or two scythes, a cart, a plow, a harrow, a roller, and relatively large farm buildings to shelter the hay, litter, and increased numbers of livestock are, essentially, the working capital of the new farmer of the thirteenth century, not counting the small tools, sickles, hoes, and spades that from then on have working parts made of iron. All of that represents, in the end, ten times the value of the equipment, buildings, and livestock of its much smaller homologue of the tenth century, which hardly possessed more than an ard, a packsaddle, small tools, often entirely made of wood, a simple house for the farmer and family, and far fewer animals.

It is then quite improbable that a relatively unproductive farm practicing cultivation with an ard could all at once increase its working capital tenfold by acquiring the whole set of means for implementing the new system based on the plow. Even on the largest estates, this costly accumulation of equipment had to be gradual. Among the peasants in villages undergoing transformation, mutual aid operated for a long time between those who owned a plow and those who owned a cart or a harrow. Undoubtedly, it necessarily took several generations for the majority of farms in a region to be outfitted with a nearly complete set of equipment. Moreover, the generalization of the new practices of cultivation and animal raising and the emergence of a new cultivated ecosystem did not require all farms of a village to be provided with the equipment for practicing the new system. Those who did have such equipment did the plowing and hauling for small farmers who lacked the necessary equipment in exchange for day labor. Thus, until the nineteenth century, 10 to 30 percent of the peasants in most of the villages of northern Europe owned only manual tools.

## The New Cultivated Ecosystem

The new cultivated ecosystem can be characterized in broad terms. Because of the more widespread use of the scythe and carts and the growth of hay making, hay meadows, along with pastures, henceforth occupied a large part of the old *saltus*. The development of animal raising, stabling, manure production, and the new plow gave rise to more extensive arable lands that were both better manured and better prepared. Furthermore, a triennial rotation tended to replace a biennial rotation. Finally, we will see that with more extensive and productive cereal-growing lands, the population increased, which means that gardens, orchards, and forests have to meet larger needs.

### Extended Hay Meadows and a Large Growth in Livestock

In cold, temperate climates, once the problems of mowing, transporting, and storing hay are resolved, a large part of the grass produced in the spring and early summer can be harvested and stored. Because this harvest largely exceeds the needs of the livestock at this time of year, part of it is available to feed the animals in the stables during the winter. However, in order to create such a surplus, a portion of the natural pastures must be protected and thereby transformed into hay meadows, which subsequently form a new and important element of the cultivated ecosystem. The rest of the grasslands remain as pastures to feed the livestock during the summer months. In regions where the summer is long enough, the hay meadows can be protected during the summer and a second crop of hay can be mowed at the end of September or beginning of October. But if the first mowing is adequate, the second crop is given over to common grazing land.

With the storing of hay, the availability of fodder in winter is larger and the herd can grow. But, at the same time, because the herd is larger, its needs increase during the summer. Yet the surface area set aside for pastures is now reduced in order to create hay meadows. Consequently, the size of the herd increases up to the point where the proportion between the pastures and the meadows is such that all of the fodder production is used. Depending upon the duration of the winter and the length of the stabling, which varies from three to eight months, hay meadows thus occupy between one-quarter and two-thirds of the natural pastures. The number of livestock can therefore be four, five, or six times greater, or even more, in relation to what it was in the system based on cultivation with an ard.

To facilitate hay making, and in particular the use of the large scythe, hay meadows are preferably established on relatively productive pastures that are not very hilly, have no stones or have had the stones removed, and are cleared of all shrubby vegetation. Hay meadows can be accommodated on lands that are not very favorable to grazing, however, such as those in cold areas where the vegetation gets started late in the spring, or even on wetlands where livestock would tend to sink into the ground and contract diseases. As a result, hay meadows are often situated in the wettest, low-lying areas, with a predominantly clay soil. These mead-

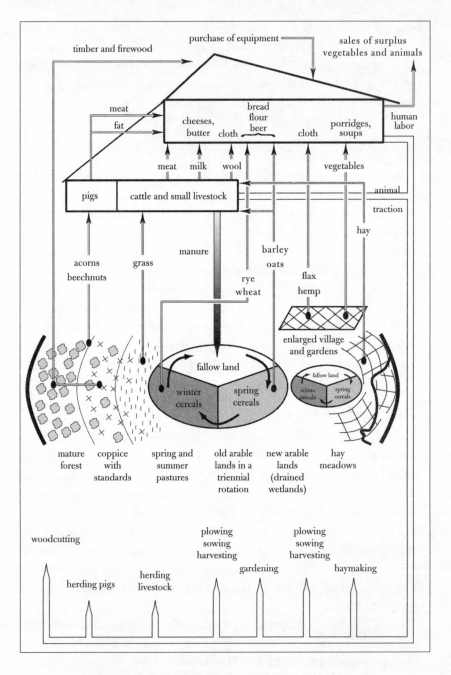

*Figure 7.2 Diagram of the Structure and Functioning of the Cultivated Ecosystem in Agrarian Systems Based on Fallowing and Cultivation Using the Plow*

ows are generally divided into independent parcels and protected from the live-stock by enclosures: dry stone walls on stony lands, fences in forested areas, quick-set hedges everywhere else. The meadows are simply guarded in situations where any sort of enclosure is lacking. Conversely, pastures most often remain undivided and subject to common grazing. They tend to occupy lands that easily withstand the constant tramping of livestock. These lands can be relatively unproductive, hilly, stony, indeed even rocky, and can include trees, shrubs, or scrubby moors. Thus, compared to the hay meadows, which are established on carefully selected and managed lands, the pastures continue to resemble the old saltus.

## More Abundant Manure and More Extended Arable Lands

The growth in the size of herds and the development of stabling led to an enor-mous increase in the availability of organic fertilizer, compared to what was available in systems of cultivation based on the ard. In the summer months, the animals always graze during the day and, at night, are either penned on the fal-low lands or returned to the stable. In either case, their nighttime excrement is collected, so that transfers of fertility increase in proportion to the increase in livestock. But during the winter months, these transfers are twice as high because permanent stabling makes it possible to collect all of the animal excre-ment, both day and night. Moreover, as is well-known, this excrement is mixed with vegetable matter (brushwood, leaves) coming from, in part, pastures and forests. Thus, in an area where the permanent stabling period lasts six months and where the use of hay permits the number of animals to quintuple per hectare of pasture, transfers of fertility become, at the very least, five times greater in summer and ten times greater in winter. In total, for the year, these transfers are thus 7.5 times higher in the system of cultivation based on the plow than they were in the system of cultivation using the ard.

A hectare of natural pasture in cold, temperate regions that is properly divid-ed between hay meadow and pasture can feed between one-half head and one head of large livestock, thus producing between 5 and 10 tons of manure. Let us consider the case of one hectare of pasture, from which 7.5 tons of manure can be obtained. If this manure is applied at the rate of 15 tons per hectare of fallow land, it is then necessary to have two hectares of pasture to manure one hectare of fallow land or, in a biennial rotation, two hectares of pasture to manure two hectares of arable lands or, in a triennial rotation, two hectares of pasture to manure three hectares of arable land. In a system based on cultivation with the plow, arable land can then occupy an area equal to or greater than the pasture, while in a system based on cultivation with the ard, the area of arable lands remains necessarily much less than that of the pastureland (see chapter 6).

*Triennial Rotation.* The growth in the availability of manure is an incentive to replace biennial rotation with triennial rotation. A large use of manure on fallow lands leads to a strong increase in the yield of cereals cultivated right after this fallow period. But using manure from the stable has a prolonged effect because the mineralization of its organic matter in a cold, temperate climate is far from complete at the end of one year. A second cereal crop can then draw on part of the remainder of this fertility, which would otherwise be lost to drainage and denitrification during a long fallow period. Triennial rotation thus becomes not only possible but desirable as soon as the supply of manure is large enough to make it worthwhile to cultivate a second cereal crop.

In a triennial rotation, the winter cereal crop, which lasts 9 months, is followed by a short fallow period of 8 months, followed by a spring cereal of 4 months (or 3 months).[6] Finally, a long fallow period of 15 months completes the rotation, which can be presented as follows:

| Triennial Rotation | | | |
|---|---|---|---|
| August ........ October | November ........ July | August ........ March | April ........ July |
| *long fallowing* | *winter cereal* | *short fallowing* | *spring cereal* |
| < 15 months > | < 9 months > | < 8 months > | < 4 months > |

The plot allotment corresponding to this new rotation can be represented in the following manner:

| Plot Allotment | | | |
|---|---|---|---|
| Rotation | Plot 1 | Plot 2 | Plot 3 |
| 1st year | long fallowing | winter cereal | spring cereal |
| 2nd year | winter cereal | spring cereal | long fallowing |
| 3rd year | spring cereal | long fallowing | winter cereal |

Triennial rotation includes, then, like a biennial rotation, a long fallowing that lasts more than 12 months and leaves time to carry out at least three plowings: the first in autumn, after the harvest; the second in spring, when the manure collected during the winter is buried; and the third in the following autumn, before sowing. The last plowing is completed with two runs using the harrow, one before and the other after the sowing. During the short fallow period, which lasts only 8 to 9 months, there is generally only one plowing. By means of all these operations, the soil is more thoroughly cleared, cleaned of weeds, broken up, and aerated than it was when cultivated with the ard. Triennial rotation also presents the advantage of lightening two very burdensome workload peaks. First, it allows the sowing to be divided into two periods rather than one and then it makes possible the lengthening of the harvest period by several days. The harvesting of spring cereals generally begins a little later than that of the winter cereals. Finally, another advantage is that the risks of a bad harvest are divided between two cultivation seasons rather than just one.

Triennial rotation, though long familiar, developed very slowly. Its use began to spread only in the thirteenth century, though cultivation using the plow had been expanding since the eleventh century. In the fourteenth century, biennial rotation was still the most widespread, and in the seventeenth century it was still used in several regions in the northern half of Europe. In order to explain this late development of triennial rotation, note first that such a rotation system was not necessary as long as a certain population density did not require a further expansion of cereal growing. But in regions with open fields and regulated plot allotment, the movement from biennial to triennial rotation required the entire rearrangement of arable lands. It was necessary to divide each segment (block of contiguous arable lands) into three plots instead of two, redivide each plot into parcels, and redistribute these among all the farmers so that each received as much land as before, equally divided among three plots. Thus there was a complicated process of land consolidation, which necessarily took a long time to become widespread.

We believe, however, that the deeper reason for this delay lies elsewhere. As long as hay meadows and larger herds were not yet widely developed, the amount of manure spread before the first cereal crop remained small, and the remaining fertility available the following year was inadequate for a second cereal crop. In order for triennial rotation to be more productive than biennial rotation, it is necessary that the yield of the second cereal crop be greater by one-half than the yield of the first: $r_1$ and $r_2$ being the yields of the first and second cereals, it is necessary that $(r_1 + r_2)/3 > r_1/2$, that is, that $r_2 > r_1/2$.

In the new systems that developed in the Middle Ages, the winter cereals were always, as in antiquity, wheat, rye, and winter barley. Among the spring cereals, the sowing of barley declined in comparison to oats, which was used as food for horses but also for humans. Sometimes, food legumes such as peas, broad beans, or lentils replaced the spring cereals. The associated crops, vetch and oats, were also planted, because they formed a fodder of excellent quality.

*Fields in Strips.* Finally, the form of the parcels had to be modified due to use of the plow. The plow is a long, heavy, and cumbersome machine, which turns with difficulty at the end of a field. To facilitate its use, the small, quasi-square parcels, which worked quite well for cultivation with an ard, had to be replaced by parcels that were as long as possible. As a result, many fields became narrow. Occasionally, they had only several furrows that were lengthened to more than a hundred meters, sometimes more. These fields in strips were so narrow and long that they could not be effectively enclosed. In order to exploit them easily, they were necessarily subject to a joint system of regulated plot allotment and common grazing.

In sum, with more extended and productive cereal-growing lands, the population grew, villages expanded and multiplied, and the need for other products, from gardens, orchards, and forests, grew in proportion.

## Expanded Gardens, Vineyards, and Orchards

Since the arable lands were generally reserved for cereal crops, all the other crops were planted in gardens or in small enclosures close to houses. These gardens were cleared, made free of weeds and rocks, enriched and abundantly manured. They were the most human-made part of the cultivated ecosystem and took over more land as the population increased, thereby slightly reducing the area devoted to cereal growing. Food legumes and other vegetables were commonly planted there and consumed in soups or eaten as an accompaniment to bread. These gardens were often referred to as "kitchen gardens."[7] Some aromatic, medicinal, or even ornamental plants were found there as well. Textile plants (flax, hemp), oil-producing plants, and tinctorial plants (pastel, madder, orchil, weld) were also cultivated, which took on great importance in some regions during the Middle Ages. These plants were included in the cereal-growing plot allotments only much later, when agrarian systems without fallowing were developed (see chapter 8).

Under the influence of the Roman conquest and the Catholic Church, vineyards took over land in all of Europe during antiquity. But expansion of vineyards was considerably larger during the Middle Ages. They were found in all villages where viticulture was possible, including the northern part of Europe, up to southern Great Britain and central Germania. Vines were sometimes found in a climbing form, planted in association with fruit trees in gardens and orchards. But they were also planted on sloping rocky hills, well positioned to produce good wine. In certain periods, the vine, more profitable than cereals, was extended even to wheat-growing plains. Political authorities tried to oppose these efforts, with more or less success, in order to preserve food security.

## Smaller but Planned Forests

The forest generally occupied lands that were the least propitious for cultivation and animal raising. These forests, as we have seen, were either residual—that is, more or less degraded relicts of the original forest, or secondary forest, spontaneously—or humanly reconstituted on lands formerly cleared for cultivation and animal raising.

In the central Middle Ages, the forest suffered terrible ravages. As a result of the demographic explosion, the taking of firewood and lumber increased in all regions. What is more, the construction of new agricultural buildings, the development of cities, metallurgy's use of wood and, later, the growth of naval construction gave rise to new demands for wood. In certain areas, there was from the twelfth century a shortage of wood, which led to the use of coal as an energy source. The powerful people in medieval society had the means to take precautions against the lack of wood, even if they were not particularly affected by it. The lords, whether lay or ecclesiastical, began to scale back the rights of the population to use the common forest closest to the villages, thereby limiting the overexploitation of wood for all purposes. Prohibitions were made against villagers hunting there or cutting trees of large diameter. A portion of the pole timber (trees of 10 to 20 centimeters in diameter at the base) and the saplings (coppice shoots remaining after the cut) had to be preserved in order to renew a forest of mature trees. In order to prevent uncontrolled cutting, wood cuttings were authorized only on a clearly defined section of the forest each year. Thus the neighboring forest, browsed in the past, exploited bit by bit to the liking of each, evolved toward a coppice with standards,[8] arranged into so many sections that were periodically exploited, each in turn, every twenty to thirty years. Each household thus had a right to a limited and verifiable cut on the section set aside for exploitation that year.

Beyond that, the lords attempted to incorporate forests that were the most distant from houses, and thus the least exploited, into the reserves of their private estates. They then reserved for themselves exclusive rights to hunt and cut mature trees in these areas. This distant forest was ultimately managed as a regulated high forest, exploited by thinning and by rotating the cutting, every hundred to two hundred years. Undoubtedly these restrictions and plans contributed to ensuring the renewal of wood resources and the possibility of pursuing their exploitation over the long term. However, these restrictions also led to the increased deprivation of the general population and encouraged expansion of the sometimes abusive privileges of the powerful. Since the population continued to increase, the authorities could not prevent the forest from becoming smaller and overexploited in the thirteenth century. Wood then became quite expensive, as it was in later periods with growing populations, notably from the sixteenth to the nineteenth century. Conversely, when the rural population diminished, as in the fourteenth century because of famines, wars, and the plague, the forest expanded again.

Thus, beginning in the Middle Ages, the shortage of wood led to important and reasonable strategies for the controlled management and renewable exploitation of wood, implemented both by lords and through collective efforts. As a result, coppice forests, coppice forests with standards, and regulated high forests developed. All the same, foraging for small pieces of wood in small forest parcels that are exploited individually has lasted up to the present. In some mountain forests that are distant from inhabited areas, the gradual harvesting of quality wood (wood used by string-instrument makers, for example) evolved toward forms of careful "gardening" of the forest population. So long as forests are carefully managed, a coppice selection system and a high forest selection system are modes of exploitation that are as reasonable and renewable as the preceding ones. Undoubtedly they are even more subtly ecological, even if their rationality is less apparent.

### Reinforcing the Association of Cultivation and Animal Breeding

The cultivation and animal breeding practices that developed in the central Middle Ages ultimately led to the organization of a cultivated ecosystem quite different from former ones. It was composed of more extended arable lands that were better manured and better prepared; smaller natural pastures that were appropriately balanced between pasturage and hay meadows and supported an increased number of better-nourished livestock; and a larger population that had to be fed with the increased yield from gardens and orchards and required more wood from the forests.

The reproduction of this ecosystem was assured by new and characteristic functional modalities: a mode of supervising herds not only based on grazing but also on the harvesting of hay and winter stabling; a mode of renewing the fertility of cereal-growing lands no longer based on penning the animals at night but on the use of manure; and finally a mode of clearing the fallow lands no longer based on using the ard but on a true plowing using the plow and the harrow.

Plowing, remember, is designed to fight against weeds, bury surface organic matter, loosen the soil in order to facilitate water circulation and the penetration of roots, and aerate the soil in order to favor the mineralization of organic matter. Increased plowings with the plow (and harrowings as well) then result in increased cereal yields, at least in the short term. In fact, the accelerated mineralization of soil organic matter liberates a large quantity of nutrients, and the cereals, freed from the competition of weeds, absorb those elements even more, resulting in increased yields. But a portion of the liberated mineral elements are also lost through drainage and denitrification, while, because of the accelerated mineralization of the humus, the content of the soil in organic matter is diminished over the long term. Consequently, the quantity of liberated minerals also diminishes each year. After several years, that inevitably leads to a lowering of crop yields. All increases in the number of plowings in an attempt to maintain these yields will, in the end, lower them even more. In other words, the increase in plowings and harrowings undoubtedly makes possible short-term gains in yields, but also leads just as assuredly, in the long term, to the impoverishment of the cultivated lands, unless the losses in organic matter and supplementary minerals are compensated for by an equivalent addition of manure.

Ultimately, the lasting growth in cereal yields in systems based on cultivation with the plow comes from an enlarged use of manure, that is, a clearly enhanced transfer of a fertility from pastures to arable lands. The plow and the harrow are only then able to take full advantage of this increased fertility. This is why, as soon as plows were available, the lack of meadows, hay, livestock, and manure was a cause of great concern throughout the Middle Ages. But it is absurd to conclude from such expressions of concern, as is so often done, that the higher cereal yields in this period owed nothing to the development of manure use and came exclusively from use of the plow and secondarily the harrow.

The deeper meaning of the move from cultivation using the ard to cultivation using the plow is that the close combination of cultivation with animal breeding is decisively strengthened. Animals henceforth became major participants in agricultural work, from carting the hay, litter, and manure to pulling the plow and harrow. In return, due to hay making, stabling, and the care they receive during the winter, animals benefit from a portion of the fruits of human agricultural labor. The total quantity of labor invested (by humans and animals) in the maintenance and exploitation of the fertility of the cultivated ecosystem is considerably increased. In the end, both the production capacity of the cultivated ecosystem (i.e., the volume of vegetable and animal products consumable by humans produced per surface unit on a long-term basis) and the productivity of human labor are greatly increased due to the use of new and stronger harnessing equipment and the massive intervention of animal labor. It is not then surprising, with these conditions, that the agricultural calendar is filled with new tasks.

## A Full Agricultural Calendar

Since cereals are always at the heart of an agricultural system, plowing, sowing, and harvesting continued to give rhythm to work in the fields. However, with triennial rotation, the preparation of the land and the sowing were divided between two seasons, autumn and spring, and cereal cultivation was expanded. The harvest, on the other hand, remained concentrated in the middle of the summer. Certainly, as soon as it was sheaved, the harvest was carried to the barns to be threshed later, which saved a lot of time in the fields. But since the grain ear was still cut with the relatively inefficient sickle, there ended up being less time for the harvest in those areas where triennial rotation was developing. That is why the scythe, more effective than the sickle, began to be used at the end of the Middle Ages to cut cereals. The scythe is fitted with a sort of comb or rake, composed of three to five long wooden teeth attached parallel to the blade. This configuration makes it possible to lift up the stalks and ears and then set them on the ground together in a properly laid-out pattern in order to cut them with one blow of the blade. While the grains harvested with the sickle were cut high under the ear, grains harvested with the scythe are cut full-length. In this way, after threshing in the barn, the straw, which will provide litter for bedding the livestock, is conveniently piled up close to the animal stables.

Harvesting with the scythe, threshing in the barn, and use of straw as litter are thus complementary practices that came to complete, in some way, the technical system associated with cultivation based on the plow. But these practices would develop later, when the pastures and meadows had been completely exploited and the woods no longer sufficed to provide litter for an increased number of livestock because they had been placed off-limits. In many areas, that would happen only in the nineteenth century, when human-made meadows, more productive than plowed fallow lands, had taken over much of the land (see chapter 8).

Whatever the case may be, with hay making and the stabling of livestock, there are two new seasons of work that come to be inserted between the plowings, sowings, and harvestings. Hay making takes place before the grain harvest, in June and at the beginning of July. As far as winter stabling of livestock is concerned, it can last for three to eight months, during which it is necessary to clean out the manure, put in clean litter, and provide fodder and water for the animals twice a day. Moreover, during the summer, herds of herbivores must always be guarded, and in the autumn, herds of pigs must be led into the forest to fatten on acorns and beechnuts. Hence the agricultural calendar becomes quite full. However, in its interstices, it is still necessary to carry out a whole series of tasks: in autumn, harvest grapes and make wine; in winter, prune vines and fruit trees, repair fences, clean out ditches and streams, make firewood and lumber; in spring and summer, garden, gather, hunt or poach, and also repair roofs, buildings and tools, spin, weave, grind the grain, bake bread, salt or smoke meats, curdle the milk, cook, etc.

The important work of the agricultural calendar is frequently represented in the sculptures, bas-reliefs, mosaics, and stained-glass windows that ornament Romanesque and Gothic cathedrals; in miniatures; in illuminations which illustrate the works of copyists; as well as in paintings and frescoes. In his book *Calendriers et Techniques Agricoles*, Perrine Mane studies one hundred twenty-seven calendars dating from the twelfth and thirteenth centuries in France and Italy. It is possible to see from his work that the calendars from the southern regions of France and Italy devote a lot of illustration space to haymaking and to new, heavy equipment such as wheeled carts and plows, while those from the more northern regions do not allude to such items.[9] In the fifteenth century, in their work *Les Très Riches Heures du duc de Berry*, the Limbourg friars illustrated with precision hay making with a scythe, grain harvesting with a sickle, plowing with a small-wheeled plow pulled by two yoked oxen, the loading of sheaves onto large, four-wheeled carts, the transporting of the grape harvest onto two-wheeled carts, broadcast sowing, and harrowing with the assistance of a horse fitted with a collar. But it is undoubtedly in *Les Heures de la Vierge*, a Flemish calendar from the fifteenth century, that the work, tools, harnesses, as well as the buildings, building frames, and even the furnishing and clothing of the peasants are illustrated with the most precision. This calendar clearly shows that the conditions of work and life that will be those of the north European peasantry until the beginning of the twentieth century were already well in place at this time.

## The Performance and Limits of the New Systems

As we've seen, a system based on cultivation using the plow requires large investments in equipment, buildings, livestock, and labor. Such a system can develop only on condition of leading to gains in productivity, which allow a return on these investments and gains in production, which make it possible to feed larger numbers of livestock and people. As we will demonstrate, this double condition, which was fulfilled in the cold temperate regions of middle Europe, could not be met in the very cold Nordic regions or in the hot temperate regions.

### Yields and Productivity

It is generally accepted that the agricultural revolution of the Middle Ages resulted in a near doubling of cereal yields.[10] For our part, we have estimated that, with cultivation using the ard, the average yield of cereals in a biennial rotation was on the order of 5 quintals gross per hectare, or 3 quintals net after deducting seeds and losses (see chapter 6). In order to remain consistent with this estimate, we maintain that in systems based on cultivation using the plow, the yield of cereals in a biennial rotation may be on the order of 8 quintals gross per hectare, or around 6 quintals net. In a triennial rotation, we consider that the yield of the first cereal remains unchanged, 6 quintals net, while that of the second cereal falls to 4 quintals. In these conditions, in order to produce the minimum 10 quintals necessary to support the basic needs of a family of five, it is sufficient, in a triennial rotation using the plow, to have 3 hectares of arable lands, while it is necessary to have 6 to 7 hectares in a biennial rotation using the ard (see chapter 6). With the equipment involved in this new system, a farmer and his family assistants can exploit up to 6 hectares of arable lands in a triennial rotation, which corresponds to a productivity per principal laborer on the order of 20 quintals (2 hectares x 6 quintals/hectare + 2 hectares x 4 quintals/hectare = 20 quintals), or twice the needs of such a family. Thus cultivation systems that use the plow can potentially extract a large surplus, unlike systems that use the ard, in which the yield was hardly sufficient to support the needs of a single peasant family.

## Production Capacity of the New Systems and Population

But in order to obtain such yields (8 quintals gross for the first cereal and 6 quintals for the second), it is necessary to apply, before the first cereal, fifteen tons of manure per hectare of fallow lands. In cold temperate regions, one head of large livestock (3,000 french feed units) passing six months of winter in a stable can produce ten tons of manure, on condition of having available around 1.5 hectares of pasturage for fodder (each hectare producing some 2,000 feed units) divided in half between hay meadows and pastures. In these conditions, in order to support the needs of five persons, it is then necessary in principle to have available 3 hectares of arable lands, 2.25 hectares of natural pasturage, and 3.5 hectares of forest (0.7 hectare per person), or a total of 9 hectares. This corresponds to a population density on the order of 55 inhabitants per useful square kilometer, or more than triple the maximum supportable population density in ard-based systems of cultivation under the same regional conditions (see chapter 6).

Naturally, the maximum population density attainable in a plow-based system of cultivation varies with soil and climatic conditions. In colder northern regions, the necessity for increased availability of wood and lower yields due to heavily leached soil, means that population density can fall to fewer than 30 inhabitants per square kilometer. On the other hand, under a milder climate and on fertile soils, such as loess, the population density can attain 80 inhabitants per square kilometer.

At the end of the agricultural revolution of the Middle Ages, France (within its present boundaries) counted some 18 million hectares of arable lands, equally divided between cultivation using the ard in the south and cultivation using the plow in the north. With some 9 million hectares in a biennial rotation, southern France could produce each year a little more than 13 million quintals net of cereals (4.5 million hectares of grains x 3 quintals). With some 9 million hectares in a triennial rotation, northern France could produce 30 million quintals net (3 million hectares of winter grains x 6 quintals/hectare + 3 million hectares of spring cereals x 4 quintals/hectare = 30 million quintals). In total, with 43 million quintals net of cereals, France could feed 20 million inhabitants. That is quite close to the population estimates advanced for the thirteenth and seventeenth centuries by several authors.[11]

## The Expansion of Cultivation Based on the Plow

Due to the use of hay and stabling of livestock during the winter, systems based on cultivation with the plow could move to cold regions in the north and at higher altitudes situated well beyond the limits of expansion of cultivation using the ard. As a result of the transfer of fertility by the use of manure, which is much more effective than simply penning the animals at night on fallow lands, plow-based systems could also be extended to shallow, sandy, permeable soils that are not very fertile. Finally, the development of the plow made it possible to cultivate heavy soils, until then impracticable. The potential expansion area of cultivation using the plow extended well beyond that of ard-based cultivation.

However, there were areas in which agrarian systems based on fallowing and cultivation with the plow was not practicable. This is the case for very cold regions occupied by conifer forests on highly leached and infertile podzol soil, located in far northern Europe or at very high elevations, where the winter requirements of wood and hay were enormous and where the cereal yields were low and uncertain. And it goes without saying that these systems ceased to be at all practicable in the tundra and in the arid steppes of Central Asia.

But in southern Europe, the Middle East, and North Africa, under a hot temperate climate, plow-based systems of cultivation were also problematic, but for different reasons. These regions did not experience a sufficient shortage of fodder during the winter to justify and make profitable the enormous investments required by the plow. In the valleys and plains, the summer shortage of fodder was compensated for by reserves of fodder "on the hoof" (well preserved in this climate) found in pastures, scrub, and bush protected in the spring. If need be, part of the herds spent the summer in the mountains (ascending migration) or in regions located much farther north. For example, herds from lower Provence climbed into the southern Alps and herds from Andalusia climbed up to the Pyrenees. Conversely, during the winter, a portion of the herds from medium-elevation mountain areas traveled down toward the low valleys and coastal plains, to a milder climate (descending migration).

In the high Mediterranean mountains, the shortage of winter fodder was, however, so large that it was necessary to store hay. Since at one time carts were not always available, this hay was generally stored in barns located at the hay-making sites, where the livestock went to spend a portion of the winter.

In some of these regions, the slope of the land was used quite ingeniously to resolve transport problems, and is sometimes still used for that purpose. Buildings were constructed at mid-slope, below the pastures where the hay making took place, and above the cultivated lands. In this way, large bundles of hay wrapped in netting could be dragged down the slope to the buildings and the manure could be brought down on simple sleds to the cultivated lands. In certain high Alpine valleys, such as Abondance and Illiez, amazing "sled-carts" were used, fitted with two wheels in the rear and two turned-up runners in front. They were brought down fully loaded by allowing them to slide and then braked at the appropriate moment. The draft animals brought the empty "sled-carts" back up the slopes using the wheels. Since there was no plow available, an ard was used. This made it possible to bury the manure on condition that the ard was steered by following the contour lines while holding it at an angle in order to turn over the earth toward the lower part of the slope.

## Improvements in Systems of Cultivation Based on the Ard in the Hot Temperate Regions

In the hot temperate regions, cultivation using the plow was most often not profitable. In order to increase the fertility of the cultivated ecosystem, other more appropriate means had to be utilized. Since antiquity, arable lands had been expanded by constructing walls that follow the contour lines of the land, laid out in tiers along the slopes. Behind these walls the earth accumulated, forming cultivable terraces with deep soil that were continually enriched by runoff water and organic matter eroded away from the *saltus* up above. These terraces, still visible in many hilly Mediterranean regions, are today often abandoned to the wild.

To make up for the lack of water in summer, perennial plantations of nutritive or fodder-producing shrubs and trees (vines, figs, olives, almonds, apricots, chestnuts, carobs, ashes, oaks) were also developed in these regions, which because of their own reserves of water better tolerated the summer dryness than annual plants and, due to their deep roots, could reach reserves of water out of reach for annual crops. Moreover, these plantations provided wood and produced litter that contributed to renewing the fertility of both the cultivated lands and the *saltus*. Olives, carobs, cork oaks, and chestnuts were able to form arboreal canopies above cereal-growing lands, but they could also be planted on the border of a parcel or in complex association with diverse annual crops to form highly productive terraced garden-orchards.[12] In the Minho in northwest Portugal, for example, a field of one hectare was planted with large, regularly pruned, fodder-producing ash trees that dominated rows of various fruit trees (peaches, almonds) that, in turn, supported climbing vines and between which were cultivated, in alternating rows, maize, beans, and other vegetables, all of which fed a milk cow, a pig, some poultry, and a small family. There are still some examples of this today.

Certainly, the most effective, but also the most costly, means of correcting the summer dryness of the Mediterranean climate was watering or irrigation. When the groundwater aquifer was not too deep, water was drawn from wells with the assistance of different machines (beams, pulleys, cranks, noria). In areas dominated by streams and rivers with an adequate flow, water was led onto terraces, into the bottoms of valleys, and onto the plains by diversion canals coming from water intakes located at higher elevations. Finally, in hilly areas lacking water-courses, water could be brought to the hillsides by means of gently sloping horizontal shafts which descended to groundwater located under the mountain.

In the Mediterranean area, terracing of the slopes made it possible to expand cultivated lands and increase crop yields. Perennial plantations made it possible to enlarge fodder and food resources and irrigation made possible the growing of crops in the middle of the summer, such as maize or sugar beets and even, in the hottest areas, tropical crops such as rice, cotton, sugar-cane, and citrus fruits. On the other hand, in the cold temperate regions, arrangements of this type were, in general, less effective and less profitable. That is why they occurred less frequently and often under simpler and less costly forms. Thus, on slopes in the mountainous regions of northern Europe there were quasi-terraces formed simply by the accumulation of soil above quickset hedges laid out on the contour lines. There were plantations of apple trees, ash trees for fodder, and chestnuts. There was also summer irrigation of pastures. The drainage network for the meadows—which are too wet at the end of winter—could effectively be adapted with less expense to carry out irrigation during the summer. Finally, sometimes there were steep hillsides in the north that were arranged into high terraced vineyards (Swiss vineyards, for example) and terraced garden-orchards (in numerous European villages) set high up in the mountains.

Sometimes in the north there were types of agricultural investments characteristic of the hot temperate regions and, conversely, equipment associated with plow-based cultivation were found in some southern regions. But, in essence, the agricultural revolution of the Middle Ages appeared quite different in northern and southern Europe.

## 3. THE AGRICULTURAL REVOLUTION
## OF THE MIDDLE AGES

The inventories of large estates, agricultural calendars, the works of historians, all indicate that the equipment and practices associated with cultivation using the animal-drawn plow were generalized in most of northern Europe in the eleventh, twelfth, and thirteenth centuries. However, there is insufficient information to reconstruct precisely how the new systems emerged or to follow their progress from year to year and region to region. Such systems already existed in the Carolingian epoch on some royal and monastic estates and began to develop from the year 950 in the regions between the Loire and the Rhine.[13]

In the tenth century, cultivation using the ard was largely predominant in western Europe. At this time, the population, which had fallen drastically in the late stages of the Roman Empire at the time of the great invasions (Germans, Huns, Arabs, Vikings), was restored. The clearings were resumed, cultivation and pasturage regained land lost during the periods of falling population and regions that had been entirely abandoned were recovered.

However, around the year 1000, there were increasing signs that Europe was beginning to be overpopulated in relation to the production capacities of the current agricultural system. Cereal prices increased, scarcity and unrest became more frequent and, in many areas, land tenures, which were subdivided for each succeeding generation, became smaller. The peasantry's conditions of existence, whether serf or free, deteriorated, and even the nobility and clergy began to face difficulties.

Although this tension lasted until the eleventh century, it does not seem to have led to any sort of major crisis. On the contrary, population and agricultural production continued to increase, slowly but surely. This phenomenon appears to be paradoxical, but is easily explained if one takes into consideration that the still dominant system of cultivation with the ard could no longer make progress, while the means of cultivation associated with the plow, which had developed during the preceding centuries and were already present in many places, were able to develop. At the turning point of the year 1000, Europe, overexploited and overpopulated in relation to the production capacities of the old ard-based system of cultivation, was underexploited and underpopulated in relation to those of the new plow-based system of cultivation.

In the northern half of Europe, the potential of the new system of cultivation was immense. In already populated regions, the transition from the ard to the plow made possible the doubling or tripling of production and population. Moreover, use of the plow could also develop in vast areas that had remained until then unexploited because they could not be cultivated under the old ard system. The new areas included forests and moors that existed either on permeable and leached soils that were not fertile for cultivation without manure or on soils that were too heavy to be cultivated without the plow. Other areas were coastal marshes, freshwater marshes, and wetlands in the interior that were difficult to drain and cultivate without heavy equipment. Finally, there were particularly cold regions in which hay and stabling was indispensable in order for livestock to be able to live through the winter, such as the hills and high plateaus of central Europe, the valleys and plateaus of the Alps, Jura, and Carpathians at an altitude of between 500 and 1,500 meters, and the northern areas of Scandinavia, Poland, and the Baltic countries. All these regions were thus relatively or totally uninhabited. Moreover, they were called "deserts" even if hunters, slash-and-burn farmers, shepherds, fugitives, and brigands were sometimes encountered there. These were relatively insecure regions, and roads suitable for wheeled vehicles often made large detours in order to avoid them.

Colonization of these vast underexploited or unexploited areas, far from preexisting centers of population, was a difficult undertaking, as difficult as was the transition from cultivation with the ard to cultivation with the plow in the already overpopulated inhabited areas.

It was much easier to use the new equipment to exploit the forests, moors, or swamps located close to preexisting villages. That is why the new system of cultivation began to develop in populated areas with nearby lands that had been difficult to exploit with the equipment associated with the old system.

## Clearing of Nearby Lands and New Villages

The clearing of nearby areas began in the tenth century. At the beginning, such additional clearings were generally made by peasants who did not own any land in the village. Alone or in small groups, equipped with axes, scythes, carts, and plows, they cleared neighboring unfarmed lands that were relatively inaccessible or unfertile, or whose soil was too heavy, in order to farm on a long-term basis with the new tools. It was no longer a question, as in the past, of planting some temporary crops after cutting and burning a wooded parcel or after burning the vegetation on a portion of a moor. Rather, it was a question of moving as quickly as possible to establish hay meadows, pastures, and arable lands that were cleared, stumped, and drained for long-term use, and making profitable use of the recently acquired new equipment associated with the animal-drawn plow. Naturally, this clearing of nearby lands hardly passed unnoticed by the local nobility. They quickly recognized the additional revenues that they could draw from the cleared lands and encouraged such clearing by imposing relatively small taxes on those performing the work. Thus, little by little, the unexploited lands around each village disappeared.

In areas where the already-populated lands bordered extensive quasi-desert zones, the lords themselves began to organize large clearings, which led to the creation of new villages. When the new lands overlapped old lands, the new villages came to be inserted into the preexisting network of villages. But when these new lands were clearly isolated, the new villages were established on a sort of pioneer front that gradually advanced onto new colonized lands. In this way, lords, abbots, and other entrepreneurs of clearing were learning methods that they then made the most of in launching larger and more distant colonization enterprises.

## Large Clearings of Distant Virgin Lands

Most of the distant and poorly controlled virgin lands were not without masters. High plateaus, hills, high valleys in central Europe, forested plains in northern Europe, moors, and freshwater and saltwater marshes were under the authority of powerful lords, princes, dukes, and counts. They became increasingly aware of the enormous revenues they could extract from these territories if they were populated and exploited using the new methods of cultivation and animal raising. But the colonization of these "deserts" demanded financial resources and organizational capabilities that quite often surpassed those of their masters. In order to undertake such enterprises successfully, these lords, as powerful as they were, had to seek out partners among those who were able to contribute to the financing and implementation of the necessary work. Thus there developed contracts of feudal property between two lords, or between a lord and a religious establishment. In overpopulated areas, wealthy religious establishments with branches spread over large areas were well placed to take charge. They launched campaigns of information and recruitment targeting possible candidates among the peasantry for colonization enterprises. These establishments were also in a position to finance the journey to and settlement on the lands to be cleared. All these efforts were organized and directed by entrepreneurs, who, for the most part, were bourgeois of the cities, or even the youngest sons of noble families, wealthy farmers, or servants who were confided this task by their masters. In exchange for their services and possible advances in money, these entrepreneurs received part of the profits from the operation, either in the form of lands to exploit on their own account or in the form of a fraction of the taxes due from the newly settled peasants.

### *The Management of Coastal and Freshwater Marshes*

Next to the major clearings, the conquest of lands from the sea, along the coasts of the North and Baltic Seas, was among the most spectacular projects completed in the Middle Ages. To complete a project of this nature, it was necessary first to build a dike facing the sea, which protected the areas to be dried out from the tides and encircle that area with an earthen levee surrounded by a ditch to protect it from waters coming from the interior. Then an organized network of canals to drain the excess surface waters at low tide had to be excavated. These drainage canals had to be closed off in order to stop rising saltwater at high tide and regulate the groundwater level. In addition, dikes and levees had to be continually repaired and canals cleaned and dug out again. Finally, a system for collectively managing water had to be put in place so that users from the same area could coordinate their activities. The organization and exploitation of coastal marshes thus demanded considerable investments and a large mobilization of interested social forces.

The construction of the first large polders in Flanders was, in this respect, exemplary. The lower valleys of the Rhine, the Yser, and the Aa were overpopulated and frequently submerged by marine encroachments. In the eleventh century, responding to pressures from the local populations and lords, the counts of Flanders, ultimate masters of this "low country," undertook to dry it out.[14] They had large dikes constructed and entrusted the exploitation of the contained lands to the monasteries. At first, the dried-out but still saline marshes were transformed into meadows for sheep, then into meadows for cows, with scattered sheep pens and cowsheds. In the twelfth century, when the lands were sufficiently desalinated, plowing began and cereals were planted. Villages of farmers were then established. In the thirteenth century, the maintenance of the installations and the management of water were taken over by local associations of users, the *draining syndicates* (or *wateringues*), which operated under the control of agents of the counts of Flanders. In two centuries, the Netherlands became a prosperous agricultural country and the Flemish accumulated a considerable expertise in the building of polders, an expertise that was called on by most of the countries bordering the Atlantic, the North Sea, and the Baltic Sea. In the interior, the drying out of freshwater marshes, the construction of dikes to guard against floods, and the exploitation of valleys susceptible to inundation also made progress.

## The Military Conquest and Agricultural Colonization of Sparsely Populated Countries

Lands marked for colonization were not all virgin, however. The great plains of northeastern Europe, for example, still largely covered with a mixed forest of broadleaf and conifer trees, were occupied by relatively sparse Slavic or Baltic populations who still practiced slash-and-burn agriculture. The colonization of these regions took place after their military conquest and the conqueror's consolidation of power. These preliminary tasks were entrusted by German princes to orders that were both military and religious, such as the order of the Teutonic Knights (*Chevaliers Teutoniques*) who conquered Eastern Prussia and the Baltic countries or the order of the Knights of the Sword (*Chevaliers Porte-Glaive*), who besieged Courland. These expeditions, presented as crusades aimed at evangelizing the pagan populations of the East, frequently ended up by subjugating them or even exterminating and replacing them with German colonists. The latter were attracted by the favorable conditions of settlement promised by the entrepreneurs. In the end, the exploitation of these regions with the powerful means provided by the equipment associated with the new agrarian system based on the plow led to the formation of a new and vast cereal-growing basin, well served by a network of rivers flowing into the Baltic. Over the centuries, the grain production of this basin was collected by the large merchants of the Hanseatic cities and exported to Scandinavia, England, the Netherlands, etc.

Generally, in newly cleared regions, agricultural productivity was relatively high because there was an abundance of land and the farms were large enough to make use of all the potential provided by the new system of cultivation. These regions provided a large marketable surplus that, despite their distance from markets, allowed them to make a profit from agricultural prices that were kept high enough by demographic and urban growth. That is why these new territories increasingly attracted holders of power and money, who reserved part of the cleared lands for themselves, which they exploited directly by using salaried workers. A little later, beginning in the thirteenth century, these estates were also sometimes rented out to tenant farmers or sharecroppers.

But these territories also attracted masses of peasants who were fleeing serfdom, abuses of power, lack of land and poverty, all of which were rife in overpopulated regions dominated by the older ard system of cultivation. During the whole period of the clearings, the powerful had to meet the needs of these peasants and assist them by providing seeds, equipment, and livestock. They also had to allow them a share in the profits of the operation by granting them perpetual title to a rather large tenure through payment of a moderate fixed tax called "quitrent" (*le cens*). Otherwise, free to come and go, these peasants were going to offer their services on other clearing sites where the conditions offered were more favorable.

Thus, at the periphery of the ancient world in which diverse forms of servitude still existed, a new world began to be formed. This world included independent peasants, whether quitrent farmers, tenant farmers, or sharecroppers, as well as entrepreneurs and wage earners—a modern world, in fact.

## The Agricultural Revolution in Overpopulated Regions

In previously occupied and overpopulated areas the agricultural revolution encountered many difficulties. Most of the peasants were much too poor to acquire the new equipment, while the lords, even if they had the means, had little interest in doing so as long as the mass of corvée laborers remained numerous and docile enough to cultivate their lands for free. The general reorganization of communal and village territories, necessary to set up hay meadows and extend arable lands, was not an easy undertaking. That is why systems of cultivation based on the plow developed rather slowly in areas that did not have any unexploited lands nearby.

### Competition with New Agricultural Territories

Eventually, the land clearings spread and the deliveries from the new agricultural areas to market of grain, animals, and other products increased, while population emigration toward these new areas grew even more. Agriculture in the old territories confronted competition on both the commodities market and the emerging labor market and had to conform to both the new methods of cultivation and the social conditions prevalent in the new lands. Thus the clearing of some cold, forested plateaus in eastern France (the plateau of Langres, for example) was undertaken from valleys populated long before. But, while the exploitation of these plateaus rested from the beginning on a combination of cultivation using the plow, wage labor, and settlement of free peasants, it was only much later that the valleys, relieved of their excess population, were converted to the new system of cultivation and abolished serfdom.

### The Transformation of Social Relations

Throughout the agricultural revolution, social relations underwent profound transformations. These changes differed from one region to another and were often confused and sometimes contradictory. Nevertheless, it is possible to outline a general trend.

The diffusion of new agricultural equipment had, first of all, a direct effect on the organization and labor conditions of the peasantry. The largely unproductive manual corvées declined and, in many regions of France and Germany, were replaced by high taxes. Conversely, corvées using the plow, harrow, and carts increased for the well-equipped farmers. If necessary, the lords began to employ underequipped small tenants as wage laborers.

The increase in production and the gains in yield resulting from the development of cultivation with the plow also entailed a strong growth in marketable surplus and income for the estates, even while taxes of all types, in kind and in money, continued to be levied by the lords on dependent peasants. Beginning in the eleventh century, new taxes abounded and tended to increase. Some had an economic character, such as the taxes that villagers paid when forced to use the mill, oven, or press built by the lord of the area, who reserved for himself the right to build these types of installations. But the lords also profited from the reduction in pasturage and forests resulting from the extension of meadows and arable lands by increasing taxes for grazing and woodcutting.

Other fees resulted from the exercise of public power, such as fines imposed by lords who had the power to render justice, taxes paid in return for defense of territory and public order assured by the powerful, tolls and taxes on commerce, and *tallages*, arbitrary and irregular taxes instituted in case of need. Many lords, however, did not have the power to levy these new taxes. In England, most of these profits already went to a strong royal power, and in France and Germany they were still largely in the hands of an ordinary nobility, i.e., the high aristocracy of dukes, counts, and princes. To protect themselves from the arbitrariness of the nobility, the population demanded and often obtained the conversion of all the taxes into a single tax agreed to by contract and payable annually in money.

Serfs and free peasants tended to draw closer together in juridical status. This happened because everyday obligations applied to everyone, free or not. Also, many serfs participated in the clearings and received lands like other peasants, which resulted in reduced responsibilities. Finally, in the twelfth century, the emancipation of the *serfs de corps*—the personal and hereditary property of the lord and master—increased, notably in France. But if the differences in juridical status became less marked, the economic disparities within the peasantry were accentuated. In the thirteenth century, when agricultural expansion came to an end and overpopulation appeared again, a stratum of wealthy farmers attempted to be lawmakers in the villages, while the landless peasants and casual laborers, deprived of any agricultural equipment, became more numerous and were sometimes even excluded from using common grazing grounds. The taxes that fell on small peasants and wealthy alike, as well as the indebtedness secured by tenures, played a determining role in the increase in landless peasants.[15] Due to the agricultural revolution of the Middle Ages, the old agrarian regime of the large estate, which was supported by the tenures of serfs subject to corvée labor, gradually gave way to a new rural society in northwestern Europe. This new society was made up of wealthy farmers and poor peasants, whether quitrent farmers, tenant farmers, or sharecroppers, as well as landless agricultural workers. There were also agricultural entrepreneurs, both bourgeois and noble, artisans, merchants, and lords, both lay and ecclesiastic. The latter group monopolized upstream industries (mines and iron metallurgy) and downstream industries (mills, presses, and ovens).

## 4. CAUSES AND CONSEQUENCES
## OF THE AGRICULTURAL REVOLUTION:
## DEMOGRAPHIC, ECONOMIC, URBAN,
## AND CULTURAL GROWTH

From the eleventh to the thirteenth century, the agricultural revolution was characterized both by an increase in production, which made possible the growth in population, and by an increase in yield, which made possible an improvement in diet and the extraction of an increased surplus. This surplus exercised a strong influence on the development of nonagricultural activities, i.e., handicraft, industrial, commercial, military, intellectual, and artistic activities, while, in return, industry and handicrafts supplied agriculture with new and more effective means of production. The growing demand for agricultural products coming from these other sectors of activity stimulated the development of agricultural production.

### The Demographic Explosion

E. Perroy estimates in *Le Moyen Âge* that the population of western Europe increased by three or four times in the central Middle Ages.[16] There is no doubt that the improvement in the dietary regime greatly contributed to this rapid demographic growth. Deadly famines became less frequent, almost disappearing, and local shortages were attenuated because of the development of trade in grains. Food was more abundant and of better quality. Bread remained the basic food, rye bread for most of the people and wheat bread for the wealthy. The consumption of bread was accompanied most often by a mixture of legumes (peas, lentils, broad beans), milk products (butter, cheese), eggs, fish or meat, especially among the well-off sections of the population. Since the population was better fed, it became more resistant to diseases. The death rate, especially among children, diminished. Malthusian practices (celibacy, late marriages, abortions, infanticides), utilized in the earlier period of overpopulation, in the tenth century, also declined.

### The Artisanal and Industrial Revolution

#### A New Rural Crafts Industry

The development of cultivation based on the plow went hand in hand with the development of a new generation of artisans. From then on, each village had to have a cartwright to make and maintain carts, wagons, plows, harrows, and

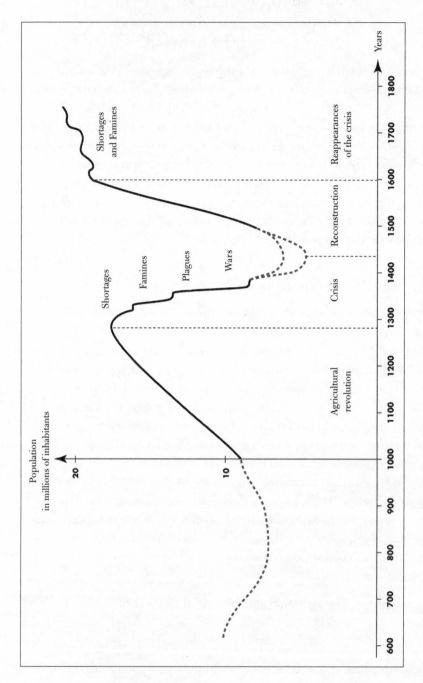

*Figure 7.2 Pace of Population Growth in France (within its current borders) from 1000 to 1750*

yokes, a blacksmith to make plowshares, colters, and other iron tools and also to shoe draft animals, not to mention saddlers to make collars and harnesses, and masons and carpenters to build stables, cowsheds, barns, and granaries. Initially, specialized servants of the castles and religious establishments made the new equipment, which was intended solely for the farms of the nobility. Later, some was sold to the peasantry, when the nobility became aware of the demand. With agricultural expansion, however, the demand for new equipment became such that some of these servants settled as artisans in the villages, with the authorization of their masters and in exchange for payment of a fee. Often they ended up purchasing their personal liberty and working on their own account. This network of rural artisans became both more extensive and intensive in proportion to agricultural growth.

Agricultural and artisanal growth also entailed an increase in the demand for iron in the countryside. Undoubtedly, the metallic tools of a farmer (scythe, sickle, hoe, spade, ax, plowshare, colter, possibly moldboard and various ironworks) weighed much less than the armor and armament of a knight. But it is likely that the wear on this set of agricultural tools was much harder than on the equipment of a warrior and thus the consumption of iron by a farmer was from then on much higher than that by a knight. Furthermore, the new rural artisans themselves accumulated in their workshops a panoply of tools that required iron and steel.

## The Iron and Steel Industry

These considerations lead us to conclude that the need for iron increased tenfold in the countryside, to which the demand from growing cities should be added. Stimulated by this demand, iron and steel production increased, affecting agricultural and artisanal growth. This expansion in the iron and steel industry was so strong that, beginning in the twelfth century, forest reserves around the iron factories, which used charcoal as fuel, began to run out. Furthermore, the extraction of the ore in simple open pits was no longer adequate. Mines with shafts and galleries equipped with hydraulic winches began to be excavated. At that time, the procedures for manufacturing iron were still generally ineffective, but important advances appeared in the fourteenth century. Powerful bellows operated by water mills made it possible to heat the smelting furnaces to 1200°C and to pour the cast iron. The latter was then hammered and converted into iron proper thanks to hydraulic tilt hammers, which are heavy power hammers driven by a camshaft, itself operated by a mill. The network of Cistercian monasteries, a vast empire of agricultural enterprises and iron factories spread across Europe, played an important role in the diffusion of these new processes.[17]

Hood

Tongs

Bellows

Handle

Awl    Chisel

Forge

Counterweight

Sledgehammer

Hammer

Anvil

**THE BLACKSMITH**

Hacksaw

Hardsaw

Tenon saw

Side axe

Plane

Draw knife

Trying-plane

Adze

**THE CARTWRIGHT**

Water reservoir

Grain

Windmill with
adjustable sails

**THE MILLS**

Watermill
with vertical wheel

*Figure 7.4 Some Tools of the New Village Artisans and New Mills*

## The Mills

The use of water mills was also expanded to activities other than iron manufacturing. From the tenth to the fourteenth century, they became widespread in Europe, particularly in northwestern Europe. Operating winches, wheels, power hammers, saws, and bellows, the mills were used in oil factories, tanneries, cloth mills, sawmills, paper mills, breweries, and above all in flour mills. The peasantry was relieved of heavy manual tasks, but as the mills were dotted along streams and rivers, far from many villages, more substantial transportation was required, which became possible due to the development of heavy transport equipment. In flat countries lacking water power, windmills were used.

The first water mills, with either a horizontal or vertical wheel, precede the Middle Ages, going back to 100 B.C.E.[18] The Romans had already constructed some large hydraulic flour mills. However, at that time, lacking adequate land-based means of transportation, and perhaps also due to the abundance of slaves, water mills were not as widespread as in the Middle Ages.

Windmills date from the seventh century. They are originally Persian, but they were improved in the West in the twelfth century, with the development of adjustable sails. They spread through the cold regions where the rivers were frozen during several months each year, and in the southern areas where the flow of rivers or streams is insufficient or irregular. In the latter regions, because of inadequate means of heavy transportation and the persistence of transport by packsaddle, there were numerous windmills, spread across the land close to houses and thus of small size. This situation lasted until the nineteenth century and even after.

Cartwrights and blacksmiths, blast furnaces and mills: the agricultural revolution of the Middle Ages was thus inseparable from a veritable artisanal and industrial revolution.

## Commercial Growth

The increase in agricultural productivity and the growth in artisanal and industrial activities were also concomitant with a large development of commerce. Peasants sold their increased surplus, the nobility sold a large part of the products from their reserves and from the taxes in kind which they continued to receive, artisans sold the products of their work and newly cultivated regions exported their surplus. As exchanges took on importance, the number of merchants increased and markets and fairs spread into villages and cities. Consequently, the need for money grew to such a point that the availability of gold and silver became insufficient to maintain daily exchanges and coins were struck that weighed less and had smaller quantities of precious metals in them.

The enrichment of the lay and religious nobility, merchants and entrepreneurs, entailed a strong growth in demand for luxury products: wines, finely woven woolen cloth of unusual colors, but also spices, silk, and other Oriental products. These products, and even some more ordinary commodities (wheat, herring, wood, wool), became part of a maritime trade that grew around two large basins: the North and Baltic Seas and the Mediterranean. In the south, long-distance trade was in the hands of Italian merchants who had trading posts around the whole Mediterranean region. Genoa and Venice were the most active centers. These merchants imported from Asia spices, precious stones, ivory, perfumes, silks, jewels, and other handcrafted objects, as well as alum, a substance necessary for processing woolen cloths and hides. In return, they exported, above all, large quantities of high-quality woolen cloth. This trade resulted in a positive commercial balance for Europe and an influx of gold. It also brought in enormous profits for the Italian merchants who thus acquired the financial means to extend their control to a large part of European commerce.

Large-scale trade in the north, however, remained in the hands of the wealthy merchants of the Hanseatic cities (Cologne, Bremen, Hamburg, Lubeck, Danzig, Riga, Visby, Stockholm, etc.). Their boats transported wheat, furs, and honey from the German colonies in the east to ports in Norway, England, and the Netherlands. They also carried salt and wine from the Atlantic coast of France to England, the Netherlands, and Scandinavia. From there, they transported salted fish to the rest of Europe. Finally, they carried wool from English sheep to the cloth mills of Flanders and Artois, and they brought the cloth produced in these regions to the rest of Europe.[19] The fairs in Champagne (Provins, Troyes, Lagny, Bar-sur-Aube), which developed beginning in the twelfth century, were the meeting point between these two worlds: the Italian merchants exchanged Oriental products for the cloth provided by the Hanseatic merchants, which they then exported to the Orient.

Thus the two commercial centers of western Europe were not of the same nature. That of the north, with its cross-exchanges of grains, wine, salt, fish, wool, and cloth, was based on the agricultural, artisanal, and industrial revolution that took place in the regions around the Baltic Sea, the English Channel, and the North Sea. This trade integrated different sectors of activity as well as the already partly specialized regions that participated in this development. The southern center, which was more outward looking, played the role of an intermediary between Europe and the Orient.

As its trade with the Orient grew, Europe became wealthier in gold and the princes of the Occident had money struck from precious metal, which served as international money. The discovery of silver mines in central Europe made it possible to increase even more the stock of money in circulation. However, the increase in the volume of transactions was such that it turned out to be necessary to resort to diverse forms of bank money, developed by the merchants and bankers of the time, in order to facilitate them. The fairs in Champagne played an important role as financial markets in this regard.

## The Birth of Capitalism

The merchant trade was very lucrative, but also very risky. Convoys of merchandise were at the mercy of bandits along the main roads and pirates at sea, of accidents and bad weather, all of which caused numerous losses. To guard against these occurrences, merchants traveled in caravans and financed commercial expeditions with several people in order to share the risks. They also invested a part of their capital in less risky businesses: industrial workshops, mines, mills, property investments, loans against security, but also, as we have seen, in large land-clearing enterprises and agricultural and animal breeding estates. In devoting funds to these enterprises, merchants sought above all the profitability of their investments. The nobility, whether lay or ecclesiastic, did the same. They employed wage laborers who did not generally contribute to the financing of means of production. This was so for the mills in the Toulouse region, given as examples by J. Gimpel.[20] These mills gave rise to the formation of the first known joint stock companies in the twelfth century. In the following century, these stocks yielded an interest on capital ranging from 19 to 25 percent per year, and there was no longer a single miller among the shareholders. These were already true capitalist enterprises in which the search for profit motivated the investment of capital and where the wage laborers did not share in the capital.

## Urbanization

Beginning in the eleventh century, the population growth led to the consolidation of preexisting population centers, around villas, castles, monasteries, and ancient Roman cities. At the same time, new population centers multiplied in recently cleared territories. In France, the names of villages such as Villeneuve, Neuville, Neubourg, and Bourgneuf are often of medieval origin. At the end of the thirteenth century, the map of European villages was already almost the same as that of the nineteenth century.

Among these villages some were particularly favored by their location in the middle of a productive agricultural area, or at the crossroads of communication routes. Artisanal and commercial activities gradually supplanted agricultural activity and the villages were transformed into cities. However, at the time, the urban population did not exceed 10 percent of the total population and it was rare for cities to have more than 10,000 inhabitants. Paris was an artisanal, commercial, and cultural center whose influence was international. It already had more than 100,000 inhabitants at the end of the thirteenth century and was the largest city in Europe. It owed its exceptional situation to the fact that it was at the center of a highly productive grain-growing basin and was well served by the Seine and its tributaries.

## The Exemptions

The inhabitants of the market towns, employed in the new, increasingly more numerous "independent" artisanal and commercial occupations, remained subjected to the same taxes, corvées, military obligations, and tolls as those in the countryside. Beginning in the eleventh century, they formed associations in the form of corporatist guilds of artisans or merchants or even in the form of "communes," assembling all the family heads of a market town. These associations aimed to obtain from the local lord a charter of exemption[21] guaranteeing to the inhabitants their personal independence, freedom to trade and circulate merchandise, and reduced, regular, and clearly defined taxes. These charters were often obtained by negotiation, sometimes by revolts, but always at the price of turning over a large compensation. Just as peasant emancipation accompanied development in the countryside, *bourgeois* (in the original sense of the word) emancipation accompanied the artisanal and commercial development of the cities.

## Monasteries, Cathedrals, and Convents

After the year 1000, Europe became full of churches and other religious edifices. The new monastic orders contributed greatly to this vast activity of building these monumental structures. The order of Cluny, for example, built no less than 1,400 monasteries throughout Europe and the order of Cîteaux built close to 750 of them. According to J. Gimpel in *Les Bâtisseurs de cathédrales*, "In the space of three centuries, from 1050 to 1350, France quarried several million tons of rock to build 80 cathedrals, 500 large churches and tens of millions of parish churches. France carted more rocks in these three centuries than at any period of ancient Egypt's history."[22]

In this period, the Church played a central role in regulating the social system of the Christian West. It set the calendar as well as the days and hours of work and rest. It organized public meetings (masses, communions, and other assemblies). It controlled the Scriptures and their interpretation as it did all writing. It was the source of public and private morals, baptizing, catechizing, and confessing each and everyone. It crowned kings, blessed military expeditions, was in charge of the entire educational and hospital systems, and was the largest owner of estates and industries. In short, the Roman Church was indeed the primary economic and political power, and the true moral authority of the medieval West.[23] In a rapidly expanding European society, the Church possessed an enormous concentration of wealth. It gathered tithes, as well as the taxes on its own estates and the profits from its industries, without counting the numerous donations that it received from lay nobles and merchants. What is more, because of the clergy's celibacy, the Church was exempted from the responsibility of renewing its own population. That responsibility was incumbent on the rest of society (nobles, merchants, artisans, peasants) who supplied the Church with the "mature" men and women, from all ranks, that it needed. The Church's surplus was considerably increased, which allowed the clerics to concentrate mainly on indirectly productive tasks (study, prayer, education, preaching, care) and gave it an incomparable investment capability, both productive (agriculture, industry) and unproductive (churches, abbeys).

The Church benefited from a large part of society's surplus in this time period and used it to build lavish monuments, dedicated to the glory of God. St. Bernard was indignant about it: "Oh vanity of vanities, but even more madness than vanity! The Church glitters on all sides, but the poor are hungry! The walls of the Church are covered with gold, the children of the Church remain naked. ... The poor are left to shout famine and what could be used for relief is spent on useless luxuries." As religious buildings multiplied, their builders competed with each other to see who had the most technical prowess or could produce the greatest aesthetic masterpieces, demonstrating a veritable exuberance of artistic and architectural creation over nearly three centuries. The Roman style spread at the end of the eleventh century, soon replaced by the Gothic style. In the thirteenth century, new monastic orders rose up against this wealth of the Church and against its isolation from the rest of society. These mendicant orders set up their convents in the cities, preached poverty, and played a fundamental role in education.

## The Intellectual Renaissance and the Universities

The creation of universities and intellectual renewal closely followed economic and urban growth. In the eleventh century, places of learning were still found in the abbeys scattered across the countryside. But in the twelfth century, schools flourished in the cities and universities appeared in the following century. Masters and students rediscovered the great works of antiquity and the educational system became aware of other authors, notably Arab, and new disciplines such as mathematics, medicine, law, theology, and philosophy.

It is at this time that agronomy made its first appearance at the university. Walter de Henley, an English Benedictine experienced in the management of large agricultural estates, was invited by the University of Oxford to present lectures on this subject. He was the author of a celebrated book on agriculture, *Book of Husbandry*, in which he dealt with the good management of a farm, plowing, spreading manure, choosing seeds, managing livestock, and so on. In the second half of the thirteenth century, other treatises on agriculture were written in England. Those that were written in the vernacular language, aimed at landowners concerned about improving the management of their estates, were quite successful across Europe. But it should be emphasized that these manuals did not precede the agricultural revolution of the Middle Ages. Rather, they drew from the lessons of experience accumulated over the course of that revolution. Another treatise on agriculture, written by the Italian Pietro di Crescenzi, had a great effect in the fourteenth century. As opposed to the preceding works, this book was situated in the tradition of the Latin agronomists and relied on the agriculture practiced in southern Europe.

Thus, over three centuries, from the year 1000 to the year 1300, agricultural expansion contributed to a demographic, economic, urban, architectural, and cultural development that carried European society to the threshold of modern times. In the thirteenth century, the results of this "first Renaissance" prefigured in many respects the Renaissance of the sixteenth century.

But at the end of the thirteenth century, signs of decline were manifest. Agricultural growth slowed down, then stopped altogether. Intellectual production at the universities became ossified. The widespread activity of building religious monuments declined and some cathedrals under construction even remained unfinished (the spires of Notre-Dame of Paris, among others, were never built, though they were in the original plans). Artisanal and industrial activities regressed, trade collapsed, and the fairs of Champagne gradually faded away, while the population began to stagnate. Then in the fourteenth century, this cessation of growth led to an immense and multiform crisis, involving both agriculture and food, but also economic, social, and political factors.

## 5. THE CRISIS OF SYSTEMS BASED ON FALLOWING AND ANIMAL-DRAWN CULTIVATION USING THE PLOW AND ITS LATER REAPPEARANCE

In the fourteenth century agricultural production was on the decline and the population starved and started to diminish. Then the plague made its appearance and the population collapsed, leading to the fall of the rest of the economy. At the beginning of the fifteenth century, European society ended up with a population and level of activity close to that of the tenth century. It subsequently stagnated for more than a century because increasingly numerous and drawn-out wars were an obstacle to renewal. The crisis reappeared at the end of the sixteenth century as soon as the population was reconstituted, and lasted until a new, more productive agricultural system made its appearance. Hence this crisis was, above all, a crisis of agrarian systems based on fallowing and cultivation with the plow.

## Overpopulation, Overexploitation, and Collapse of the System

From the end of the twelfth century the signs of overpopulation were obvious in some regions of Europe. These indications spread to other areas and multiplied during the last decades of the thirteenth century. Shortages became increasingly frequent. Wood was in short supply, above all lumber, but also firewood, necessary for the inhabitants of both the country and city, as well as the iron industry. The clearings were pushed as far as possible, so far that, at the end of the thirteenth century, recently cleared and cultivated lands had to be abandoned because they turned out to be infertile.

Malthusian practices, which already stood out during earlier periods of overpopulation (late marriages, celibacy, abortions, infanticides) appeared again, but did not spread widely enough to block population growth, which continued to be stimulated by the customs and mentalities adopted over the three preceding centuries of rapid growth in production and population. There resulted a growing gap between the needs of an expanding population and the production capacity of agrarian systems based on cultivation with the plow, which could expand no further. Thus, just like in the tenth century, but with a population at least three times more numerous, western Europe was again, at the end of the thirteenth century, relatively overpopulated.

### Shortages and Famines

In the fourteenth century, shortages multiplied and worsened, to the point of becoming increasingly deadly famines. In 1305, an acute shortage affected the whole Parisian Basin. Another began in Germany in 1309–10, and then spread to several other areas of western Europe. In 1315–17, a terrible famine, both lengthy and extensive, decimated the population in numerous cities and country areas. Since existing societies lacked the potential to increase production and did not know how to limit the population increase, it was the famines that, initially, took on the harsh task of forcing the population and its needs into balance with the stagnant level of available food.

But as the famines multiplied, they became even more catastrophic. In Forez, for example, famine struck in 1321, 1322, 1332, 1334, 1341, and 1343.[24] When the

famines followed one another at less than five-year intervals, the decimated population did not have time to reconstitute itself for the next famine. That means,
then, that from one famine to another, the availability of food fell since, despite the
reduction in population, the famines continued to strike.

## Deterioration of the Cultivated Ecosystem

The fall in population observed at the beginning of the fourteenth century
resulted, then, in a lowering of production that can be explained by overexploitation and deterioration of the cultivated ecosystem. In effect, in order to
respond to a growing demand, which was expressed in higher prices, or in
order to avoid a shortage, every peasant, wealthy or poor, tried to increase production. To this end, they pushed the clearings too far, onto marginal lands
where, after several years of cultivation, the yields fell below an acceptable
threshold, because the stock of inherited organic matter on the moor or in the
forest began to be exhausted. These lands were then abandoned to the wild and
production declined.

Moreover, the extension of crops was often made to the detriment of pastures and hay meadows, which required a reduction in the size of livestock
herds. Consequently, the quantity of manure available to fertilize the arable
lands diminished at the same time that these lands were expanding into new
areas. Very quickly, cereal yields declined and the resulting fall in production
was much larger than the gains in production coming from the extension of the
sown land area. Even more seriously, the short supply of manure also led, in the
long term, to a decline in the organic matter content of the soil, which led to a
lasting deterioration in the fertility of arable lands and a new decline in production. Finally, if, in order to compensate for the decline of pasturage in relation to
arable lands, there was an attempt to expand the former into forested areas, then
the shortage of wood was aggravated.

Thus, when the expansion of systems based on fallowing and cultivation
with the plow attained their limits, all increase in the production of grain
obtained by an increase in the cereal-growing land area contributed to a shortage of fodder, a decline in the size of herds and availability of animal products, a
lack of manure, a lowering of fertility, and ultimately a fall in grain production
itself, not counting the shortage of wood.

Once more, it is clear that the maximum capacity of production for a cultivated ecosystem is attained when certain proportions, the optimal proportions, among its constitutive parts are achieved. It is possible to assume that, at
the end of the agricultural revolution of the Middle Ages, at the moment when
cereal production and the population reached their maximum everywhere, the

best proportions among cereal-growing areas, hay meadows, pastures, size of herds and forests were achieved nearly everywhere. But if the demographic growth in a particular cultivated ecosystem continues—the human population expands beyond the optimal proportions of the species that it eats (the cereals)—then it necessarily happens that there is a decline in fertility and production in that ecosystem. That is exactly what we mean by overexploitation and crisis of a cultivated ecosystem.

### Health and the Demographic and Economic Collapse

However, the agro-ecological crisis of the overexploited and weakened cultivated ecosystem also has harmful effects on the anatomy, physiology, and health of plants, animals, and humans. The living conditions of all species deteriorate. Deficient wheat wilts, the size of hungry livestock is reduced, and diseases of plants and animals proliferate.

Undernourishment and hardships of all kinds have grave effects on human populations. The vigor of the population, its capacity to work, its resistance to diseases, diminishes and, on this favorable terrain, increasingly deadly epidemics of plague, whooping cough, and smallpox multiply. During the Middle Ages, the plague, which, without ever having totally disappeared, had not massively struck the West since the 6th century (Justinian's plague) returned in force. A plague of Asiatic origin, carried by silk caravans and Italian ships trading with the Orient, spread to all of Europe from 1347 to 1351. This was the great "black plague," a major shock dealt out to a severely weakened population, which had already been dying for a long time from hunger, cold, and various diseases. This time, the population collapsed. Depending upon the place, death carried off between one-fifth and the whole of the population. Towns were wiped off the map. Cities and entire regions were devastated.

This demographic collapse led to a large-scale abandonment of farmland, the return of natural vegetation, and reforestation. It also led to an industrial, artisanal, and commercial disorganization and regression such that, even after massive human deaths, the shortages persisted. After several years of respite, the favorable conditions for disease were reconstituted and the plague returned once again. These epidemics of plague were not chance occurrences. They were linked to the crisis in the agrarian systems based on fallowing and cultivation with the plow, just as the plagues of the sixth century were linked to the crisis of the agrarian systems based on fallowing and cultivation with the ard. The plague, however, was not the last manifestation of the crisis. Social turmoil increased and wars multiplied. War was also "made part of the system" or, more exactly, of the crisis of the system.[25]

## Social and Political Crisis and War

### Rural and Urban Poverty

Shortages and famines affected the poor of the cities and countryside above all. In fact, because of the method of dividing the inheritance in land, many farms became too small to employ the family labor force completely and support its needs. In these conditions, many peasants had to buy a portion of their food on the market. As soon as a harvest was bad, those who had the means rushed to buy and stockpile large quantities of grain, whether to satisfy their own needs or to resell it later at a higher price. Conversely, just after the harvest, indebted peasants were forced to sell a portion of their grain at low prices, even if their harvest were poor. As the months went by, prices rose until reaching the maximum on the eve of the following harvest, sometimes rising to more than ten times the average prices for normal years. Bread became inaccessible to peasants and poor city people, who are always the first hit by hunger.

In this context, where free lands were uncommon and labor was overabundant, the property taxes demanded by the nobility continually increased, while wages tended to fall. In order to survive through difficult years, the rural poor were indebted to their lords or to merchants and, when they were no longer able to repay the debt, the only options remaining to them were to sell their tenure, submit themselves to debt serfdom or run away in order to escape the debt. Serfdom thus tended to develop again, although unequally from one region to another, and a relatively active property market developed from the end of the thirteenth century. Finally, as there were no more lands to clear to accommodate the poor, they became vagabonds who attempted to survive by begging and theft. Organized hordes of pillagers were formed, who killed and were killed in turn.

### Turmoil and Revolts

As production per inhabitant fell, the division of the fruits of labor became more difficult and conflictive everywhere. In the cities, the opposition between rich and poor, wage and antitax demands, and rebellions against speculators and usurers were often violent. All who were "foreign" were attacked: Jewish, Flemish, Italian, Hanseatic merchants, and the like. In the impoverished countryside, resistance was organized against exactions and pillaging. Riots broke out and spread, uniting large numbers into regional uprisings, such as the insurrection in maritime Flanders (1323–1328), the jacquerie of the Ile-de-France, the revolts of the armed bands of Wat Tyler in England, the Tuchins in Languedoc, etc.

Local powers, the nobility as well as city authorities, did not have the means to prevent unrest of this magnitude and had to appeal to those more powerful.

Thus a vast political reorganization took shape, which saw power concentrated in the hands of some great lords, dukes, princes, or kings.

## War

But this political reorganization itself was cause for conflict. For more than a century, Europe was torn apart by wars: wars of princes against rebellious lords; wars of princes among themselves over a new division of territory (the Hundred Years' War between the kings of France and England, which began in 1337, did not end until 1453); wars of pure and simple pillage. These wars were the product of the crisis of the system and contributed toward aggravating and prolonging that crisis. In order to finance war, the rulers levied taxes, direct ones like tolls or indirect ones like the salt tax, which fell on populations already exhausted by all sorts of scourges.

## Reconstruction

In a little more than a century, famines, plagues, banditry, pillaging, revolts, repressions, and wars led to a dramatic drop in population and production. By 1470, in most European villages half of the family households existing at the beginning of the fourteenth century had disappeared.[26] Now, as the number of persons in each home had also decreased, it is possible to assume that, at its lowest ebb in the first quarter of the fifteenth century, the population had decreased by more than one-half, falling to about the same level that it had been in the year 1000.

However, in the fifteenth century, economic conditions were very different than they were in the tenth century. The resources from the system of cultivation using the plow and from handicrafts and industry, although seriously weakened, were nevertheless present everywhere, while they were as yet almost nonexistent five centuries earlier. Furthermore, the money supply was also much larger, from a clearly inflationary conjuncture. Wages in particular were greatly increased, because of the labor shortage, as well as prices of labor-intensive products, such as wine, for example. In these conditions, the tremendous demographic and economic expansion in Europe between 1450 and 1550 was not an exact reproduction of the expansion during the central Middle Ages. In the fifteenth century, the revival began in the most fertile regions, i.e., the great silt-laden plains and the rich alluvial valleys. Beginning in those areas, the surviving population began cultivating the best lands again using preexisting means of production. Then peasants who had survived in marginal regions (mid-elevation mountains, high plateaus and valleys, chalky plateaus with thin soils, sandy lands, etc.) and were attracted by the unused good lands in the wealthy regions joined in the revival of production. Thus in the fifteenth century a migratory movement took shape that concentrated the remaining population on good lands. This movement was opposite to that which took place with the large clearings of the central Middle Ages.

Furthermore, because of the labor shortage, many nobles were led to institute tenant farming or sharecropping on their estates, with rather extensive plots. Thus a relatively well-off stratum of large tenant farmers and sharecroppers was formed. In several decades, plowing and hay meadows had taken over the uncultivated lands of the most favored areas, which had restored their human and animal populations. During this first phase of expansion, land was abundant and ground rent (the price of renting the land) remained low, while the farmed area per worker and yield were high. Consequently, the prices of grains and meat remained relatively stable, despite the abundance of money and high wages. Since the population and the demand for grain continued to increase, grain prices also began to rise, attaining, at the beginning of the sixteenth century, a high enough level to make the restoration of cultivation on marginal lands advantageous. Then the reconquest and restoration of the less favored areas, devastated and abandoned two centuries previously, began. Lands were cleared for the second time, villages were rebuilt and these areas were also repopulated with people and animals.

Thus in a little more than one century, Europe restored its agriculture and reconstituted its population, both of which had been devastated by the crisis. This agricultural restoration, of course, encouraged an artisanal, industrial, commercial, urban, and cultural renaissance whose achievements went well beyond those of the thirteenth century. This renewal also stimulated the formation of modern states, which developed at the expense of the lay and religious nobility.

## The Reappearance of the Crisis

From the end of the sixteenth century, however, the signs of overpopulation reappeared. Farms were too small, clearings were pushed too far, productivity and yields declined, prices rose, there were shortages, famines, and epidemics. The crisis reappeared and all the signs pointed to developments similar to those of the fourteenth century. However, this time, events took another turn. First, the plague, which returned to wreak havoc upon some Mediterranean cities in the sixteenth century (Rome and Naples in 1525, Venice from 1575 to 1577, Marseille in 1581), did not spread to the rest of Europe. The crisis, even if it resulted in some devastation, did not lead to another collapse of the population and production. On the contrary, despite shortages and famines that lasted until the beginning of the nineteenth century, the European population did not stop growing for more than two centuries. Certainly, it only grew slowly and unequally depending upon the area, but it did grow.

This slow population growth is explained in part by further progress of the agrarian systems based on fallowing and cultivation with the plow. Triennial rotation, which was not widespread in the Middle Ages, spread greatly in the seventeenth and eighteenth centuries. Under the impetus of the new states, interior marshes and coastal polders continued to be developed. Finally, progress in navigation and the digging of canals opened up some regions that could then export their surpluses more easily in good years or, conversely, be supplied in case of shortages. The best-served areas began to specialize, which made it possible for them to increase output markedly for products in which they had an advantage. All in all, production from all regions and the global availability of food increased.

But the latest improvements in the agrarian system could not go very far. In fact, beginning at this time, the slow growth in production and population resulted in part from the emergence of new agrarian systems that did not use fallowing and were twice as productive. They had begun to develop in the sixteenth century in the Netherlands and in the seventeenth and eighteenth centuries in England and in several other areas of Europe. Thanks to this new agricultural revolution, which was just beginning, the availability of food in Europe increased enough to allow a minor growth in the population. However, this availability did not grow quickly enough to avoid shortages and famines completely, which did not truly disappear until the nineteenth century, when this new agricultural revolution reached other areas of Europe. At that time, the population could then double again.

## 6. CONCLUSION

The study of the genesis and crisis of the agrarian systems based on fallowing and cultivation using the plow in the medieval West invites us to draw some conclusions regarding the conditions of emergence and the limits of development of an agrarian system.

There is no doubt that without the demographic pressure which took place from the tenth to the thirteenth centuries, the agricultural revolution of the Middle Ages could not have developed so rapidly and attained such a large scale. In fact, as long as the older systems based on cultivation using the ard did not reach the maximum number of people and animals they could support, the necessity to invest in the new equipment and to take on the great expenses involved in large-scale clearings was not imperative.

But it is also clear that demographic pressure would not be sufficient in and of itself to lead to such a development. Without the material means to change the system, which were slowly developed at the end of antiquity and during the early Middle Ages, the relative overpopulation of the year 1000 would have led, as in the sixth century, to a crisis of the old system, a crisis which, by reducing the resources available per inhabitant, would certainly not have favored development.

But, as we have also seen, the growth in population and the technical means to meet its needs would not have been sufficient alone to enable the rapid development of cultivation based on the plow. In order for that to happen, the social forces that had the means to invest in such development had to have the interest to do so. And it is likely that if wars to capture slaves had remained profitable, in any case more profitable than the new investments, the nobility would not have committed its resources to the clearings and to agricultural or industrial equipment and it would not have found as many partners to participate in these enterprises.

In the eleventh and twelfth centuries, once the agrarian revolution had started, production increased faster than the population. A large marketable agricultural surplus appeared, which influenced the development of cities and nonagricultural activities, and improved the diet as well. However, it is necessary to emphasize that this surplus represented undoubtedly less than half of the average production and it was quite variable. Overabundant during the good years in relation to the needs of the nonagricultural population, which was still not very numerous, it led to lower prices, which was discouraging for the producers. Conversely, during the bad years, the surplus was reduced a little, which slowed down general development and sometimes even caused shortages. Although already very substantial, the level of surplus derived from the agricultural revolution of the Middle Ages was still inadequate to guarantee support for nonagricultural activities during bad years.

Moreover, all indications are that from the end of the thirteenth century, agrarian systems based on fallowing and cultivation using the plow had reached their maximum extension and maximum human population. The strong demographic growth of the three preceding centuries continued to the beginning of the fourteenth century, causing not only shortages but also excessive clearings and the deterioration of the ecosystem. The resulting ecological, health, social, and political crisis led, in the fourteenth and fifteenth centuries, to a drastic fall in population. Then, in the sixteenth and seventeenth centuries, after the population was restored, the crisis reappeared and death was again responsible, during the dark times, for adjusting the size of the population to the volume of the available subsistence.

It appears certain, however, that death from hunger, cold, disease, or war was not the only regulator of the population size. Birth control was also practiced, the same as in all periods of overpopulation. As E. Le Roy Ladurie explains so well:

It would be absurd to explain everything by death. Even animal sociology has long refuted the "Malthusian" idea (in fact pseudo-Malthusian) according to which the numbers of animals, in the wild, are regulated only by the volume of available subsistence. According to this idea, once the latter is consumed, the result is that poverty, famines and epidemics are automatically triggered *ad hoc*, which limit, through individual misfortune and in the general interest, the number of individuals taking part in the great banquet of life. In fact, animal species, from the penguin to the centipede, the elephant and the whale, have a principle or at least an enforcement of order, intelligent though instinctual, that regulates the size of the population. The group is allowed to evolve numerically to a demographic *optimum* and not to a *maximum* or a *pessimum*. The same is true *a fortiori* for our French peasants in the seventeenth century, except that this enforcement is not purely biological or unconscious, but culturally determined.[27]

Finally, it appears that the relation between agricultural growth and demographic growth is not a simple and univocal relation, but, on the contrary, a contradictory relation, changing according to the conditions of agricultural development. When the technical, economic, and social conditions of development of a new agrarian system that is more productive than the previous one are brought together, there is no doubt that demographic pressure, even if it can cause momentary difficulties, pushes the development of this new system forward, as was the case in Europe in the tenth and beginning of the eleventh centuries. But that does not mean, for all that, that population growth is the principal motive force of agricultural change, as some have supposed.[28] When a necessary condition of development of a new agrarian system is not fulfilled, as was the case in Europe in the fourteenth century, the tendency for the population to increase becomes a cause of overpopulation, ecological disequilibrium, famine, disease, and death.

But one should not conclude from that, as Thomas Malthus did in *An Essay on the Principle of Population*, that without voluntary limitation of births, population necessarily increases faster than production. When the totality of conditions necessary for the rapid development of a more productive new system are brought together, agricultural production can well increase faster than population, which means that the productivity of agricultural labor increases and an agricultural surplus appears, which not only makes it possible for the population to increase but also improves its diet and spurs the development of nonagricultural activities and cities. Such was the case in the eleventh and twelfth centuries in northwestern Europe with the development of a cultivation system using the plow. And such will be the case, as we will see in the next chapter, in the eighteenth and nineteenth centuries, with the development of systems that do not involve fallowing.

# 8

# Agrarian Systems without Fallowing in the Temperate Regions: The First Agricultural Revolution of Modern Times

Without fertilizer, no harvests, without livestock, no fertilizer, which has such an immediate impact; without seeded pastures, no livestock; finally, without the elimination of fallowing, no or very few seeded pastures; all is linked in agriculture; its system must be total.

—Instruction to the National Convention, 1794

By the end of the Middle Ages, Europe had already experienced three agricultural revolutions: the Neolithic, the ancient, and the medieval. Those revolutions had given birth to three major types of agriculture: systems of temporary, slash-and burn-cultivation; systems based on fallowing and cultivation using the ard; and systems based on fallowing and cultivation using the plow. From the sixteenth to the nineteenth century, most areas of Europe were the scene of a new agricultural revolution, the first agricultural revolution of modern times. We call it that because it developed in close relationship with the first industrial revolution.

For reasons that will be explained later, we refer below to the the first modern agricultural revolution as the first agricultural revolution. The first agricultural revolution gave birth to so-called systems without fallowing, which originate in the agrarian systems based on fallowing from the preceding period. Through this transformation, the fallow lands, which still occupied a large place in the old triennial and biennial rotations, were replaced either by pastures seeded with grass, such as ryegrass, or fodder legumes, such as clover or sainfoin, or fodder roots and tubers, such as turnips.

In the new rotations, fodder crops alternated with cereals with almost no break, such that arable lands henceforth produce as much fodder as the pastures and meadows did together in the old system. The development of these rotations went hand in hand with the development of raising herbivores, which supplied more animal products, draft power, and manure. This increase in animal manuring led, in turn, to a strong growth in cereal yields and made it possible to introduce other crops, which have higher fertility requirements, into the rotations. As they developed, the new rotations were expanded to include row crops for food, such as the turnip, the cabbage, the potato and corn or industrial crops, such as flax, hemp, sugar beets, etc. Moreover, the improvement in animal diet and in crop fertilization made it possible to begin selecting more demanding and productive animal breeds and plant varieties capable of taking advantage of these improvements.

All in all, as a result of this vast transformation, with the number of livestock and the volume of manure almost doubled, the new systems produced at the very least twice as much as the older systems and made it possible to feed a much larger total population better than in the past. On the other hand, since these gains in production were generally obtained with little additional investments or work, they were characterized by a strong increase in the productivity of labor and in marketable agricultural surplus. From the end of the nineteenth century, more than half of the active population in the industrialized countries could devote themselves to rapidly developing nonagricultural activities, such as mining, industry, and services.

These gains in production and productivity put an end to the crisis of the systems based on fallowing that erupted in the fourteenth century and lasted until the eighteenth century. The new systems that did not use fallowing began to develop in Flanders in the fifteenth century. One can then ask why these systems took so long to spread in a Europe where it would still be possible to die from hunger, cold, and disease for several centuries. The slowness of this development is not explained by technical reasons. The true obstacles to the development of this new agricultural revolution lay elsewhere. Indeed, as long as juridical obstacles, such as the right of common grazing on fallow lands and obligatory plot allotment, were not discarded in favor of the establishment of the exclusive right to property and the right of using cultivated lands freely, cultivating fallow lands was hardly possible. As long as what was left of feudal charges, obligations, and taxes, was not abolished, a peasantry crushed by various obligations did not have the possibility of embarking on such a development. Finally, this new agricultural revolution could develop only insofar as industrial, commercial, and urban development made it possible to absorb the large marketable agricultural surplus that it produced. Indirectly, then, the development of the new agriculture was also conditioned by the suppression of obstacles to the development of industry, such as the feudal and corporatist monopolies, and by the suppression of obstacles to the development of commerce, such as provincial customs and local tolls.

The combined development of the agricultural, industrial, and commercial revolutions could only have occurred in each country as a result of reforms establishing the free disposition of land, freedom of enterprise and trade, and free circulation of people and goods. Enacted by enlightened or constitutional monarchies or by revolutionary assemblies, these reforms were made, depending on the country, under varying pressure from the social groups directly concerned, that is, the bourgeoisie, the landed property owners, and the peasantry. But the ground for the reforms was equally prepared by enlightened minds during the period of the Enlightenment. Agronomists and economists (the physiocrats), witnesses of the success of an agriculture in Flanders and England that did not use fallowing, made themselves the theoreticians and propagandists of this new agriculture and the necessary reforms for its progress. Counselors to princes, organizers of learned societies and specialized governmental commissions, they informed and influenced a small stratum of large property owners and farmers, as well as other, equally limited intellectual milieus and circles of power. Sometimes, in a number of countries such as France, Prussia, and Denmark, the ideas of the agronomists and physiocrats helped to create the necessary political awareness and greatly influenced the long-awaited reforms.

As a result of these reforms, each European country inherited a particular agrarian social structure. Almost everywhere there were large and small property owners, farms worked by the owners, by tenant farmers or sometimes sharecroppers, farms with wage laborers and family farms. But, from one region to another, the proportions among these different categories of farms varied enormously. There was a world of difference between countries such as Prussia of the *Junkers* or Great Britain of the landlords with their large farms worked by wage laborers and countries such as Denmark, the Netherlands, and the largest part of France and western Germany, where peasant farms using family labor predominated. In all these countries, agricultural entrepreneurs and peasants took up the new agriculture as soon as there were strong reforms, growing industrial and urban outlets, steady prices and tolerable taxes.

But in the southern and eastern regions of Europe ( the south of Portugal, Spain and Italy, Slovakia, Hungary, eastern Prussia, Russia), far from the great centers of industrialization, where large latifundia-type estates maintained agricultural labor in a state of quasi-servitude, the first agricultural revolution did not take place. These regions were plunged into underdevelopment and crisis.

What is the origin of the new systems that did not use fallowing? What are their structures, modes of functioning, results, and limits? What were the juridical, economic, political, cultural conditions, and the consequences of their development? Such are the questions studied in this chapter.

## 1. THE BIRTH OF THE NEW AGRICULTURE

## The Limits of Systems Based on Fallowing

As we have seen, in both the thirteenth and sixteenth centuries, the agrarian systems based on fallowing and cultivation with the plow reached their limits. At the end of the thirteenth century, once the large clearings were completed and the best proportion among arable lands, meadows, pastures, and forests was achieved, the production of grain reached a maximum. Then shortages, famines, plague and war exterminated more than half of the European population. In the sixteenth century, after the restoration of the economy and the reconstitution of the population, shortages and famines again made their appearance and were rife throughout the seventeenth and eighteenth centuries.

During the latter period, attempts to increase grain production occurred, but most failed. The extension of cereal-growing lands at the expense of pasture lands certainly made it possible to obtain additional grain momentarily. But it also just as surely led, in the end, to a reduction in the number of livestock and a decline in manure production, resulting in a lowering of cereal yields and production. In the same way, the replacement of the long fallow period by a cereal crop made it possible to obtain an additional harvest immediately. But this suppression of fallowing disrupted the regular use of animal excrement to transfer fertility to cereal-growing lands. Also, by reducing the number of plowings and harrowings that were carried out, it favored the proliferation of weeds on cereal-growing lands. Again, the result was a lowering of cereal yields and production.

All these difficulties reinforced the old "agronomic" myth according to which fallowing allowed the soil to "rest" in order to restore its "strength." However, the fallowing that periodically occupied arable lands, which were some of the best lands of the cultivated ecosystem, formed the only significant margin possible for increasing production. But this new agricultural "frontier" could only be conquered on condition that new crops were able to contribute to the renewal of fertility and to the battle against weeds more effectively than fallowing. That was precisely the case with fodder row crops such as turnips, which simultaneously make it possible to feed more livestock, produce more manure, and thoroughly clear the lands due to the frequent required hoeings. Such was also the case with sown pastures of grasses and legumes, the rapid growth and early mowing of which limit the proliferation of weeds. Such was also the case with corn from America, which can be cultivated in hot and humid southern regions and which, while supplying an additional grain harvest, presents the double advantage of producing fodder from its leaves and male panicles and of requiring frequent hoeing, which results in clearing the soil.

## Principles of Agrarian Systems That Do Not Use Fallowing

In principle, the first agricultural revolution of modern times consists precisely in replacing fallowing with fodder row crops and seeded pastures, thereby encouraging the development of animal breeding and the production of manure.

By replacing the long fallowing of 15 months in the old triennial rotation with a seeded pasture and the short fallowing of 9 months with a fodder crop at the end of summer and in autumn, a new rotation without fallowing of the following type is obtained:

| Old Triennial Rotation with Fallowing | | | |
|---|---|---|---|
| **YEAR 1** | **YEAR 2** | | **YEAR 3** |
| August........October | November........July | August........March | April........July |
| *long fallowing* | *winter cereal* | *small fallowing* | *spring cereal* |
| <<< 15 months >>> | <<< 9 months >>> | <<< 8 months >>> | <<< 4 months >>> |
| **New Triennial Rotation "without Fallowing"** | | | |
| *seeded pasture* | *winter cereal* | *fodder catch crop grown for a few months in summer and in autumn* | *spring cereal* |

The fodder crop planted between the winter and spring grains at the end of summer and in autumn is a crop with a short cycle (turnips, for example), which, when planted after the harvest, can produce a harvest before the winter. This crop, which takes up only a portion of the time previously given over to the short fallowing between the two principal crops, i.e., the winter and spring cereals, is called a "catch crop."

The real advantage and success of the new rotation, which includes both fodder and cereal crops, comes from the fact that it produces practically as much fodder as the pastures and meadows together. The replacement of fallowing by fodder crops makes it possible to roughly double the number of livestock, the production of manure, the animal draft power, as well as all other products that come from animal raising (wool, hides, meat, milk, etc.). Finally, in systems without fallowing, cereal yields can be increased as a result of the doubled manuring.

## An Old Agronomic Tradition

Taking these advantages into account and that most of the fodder plants used in the new rotations were known for a long time, it is surprising that, in a Europe where people had frequently died from hunger from the thirteenth century on, the new system spread very slowly, between the sixteenth and nineteenth centuries.

Moreover, the benefits from rotations that alternate cereal and fodder crops were known since early antiquity. In Egypt, where there were no natural pastures, clover was cultivated every other year, alternating with wheat or barley. This crop, which by itself improved the fertility of the soil, also made it possible to feed live-stock and produce manure, primarily intended for the irrigated crops (see chapter 4). This tradition, which was maintained and developed in Egypt in the Hellenistic, Roman, Byzantine, and Arab periods, was passed on to Europe. The ancient Greeks were not unaware of it (Theophrastus), the Latin agronomists (Columelle) advocated alternating a cereal crop with a legume crop on the best lands, and the Andalusian agronomists of Arab origin also praised the merits of such a practice.[1] This tradition was not unknown to Western agronomists of the Renaissance, such as Torello the Venetian or Olivier de Serres, the Frenchman who at the end of the sixteenth century in turn advocated this practice. Finally, the agronomists of the eighteenth century, English, French, and others who were enthusiasts of the new agriculture, also fell within this tradition.

This ancient tradition was, on the other hand, probably unknown in regions (Artois, Normandy, England) where beginning in the thirteenth century a portion of the spring cereals, or even winter cereals, was replaced by food legumes.[2] At the same time, it was also probably unknown on some English country estates where broad beans and peas began to be planted as a replacement for fallowing.[3] This practice, advantageous both for human diet and the fertility of the soil, undoubtedly produced better results than a simple increase in cultivation of cereals. However, in that context, it was only one attempt among others of expanding the cultivation of cereals intended for human consumption, in order to face immediately the growing food shortages of the time. It is indeed wrong to choose to see in that practice some beginning of the first agricultural revolution.

## Increasing the Production of Fodder in Order to Increase Grain Production

The first agricultural revolution does not consist of searching for an *immediate* increase in food production by *directly* replacing fallowing with cultivation of grains intended for human consumption, even if the crop be a hoed legume. It consists, and this is completely different, of indirectly pursuing an increase in cereal yields by replacing fallowing with fodder crops that, in turn, make possible the development of animal breeding and the production of manure. In a certain way, this new agricultural revolution continues that of the Middle Ages, which thanks to the use of hay had already experienced an increase in cereal production by means of an increase in livestock and manure. By developing the cultivation of plants entirely or partially intended for animals, the first agricultural revolution went one step further in the direction of an increasingly tighter integration between cultivation and animal breeding.

Undoubtedly, cereal-fodder rotations were occasionally carried out locally, beginning in antiquity and all through the early Middle Ages, even if history does not say so. But this practice became noticeably widespread and persistent only from the end of the Middle Ages. In the fourteenth century, the peasants of Flanders and the Netherlands began, by empirical means, to reduce the place of fallowing in the rotation. The long fallowing, which alternated every two out of three years with cereals, was carried out no more than every four, five, or six years and finally disappeared. It was replaced by peas and vetch, then in the sixteenth century by clover, feed turnips (the latter could also be planted between the spring and winter cereals in place of the short fallowing), and by various industrial crops.[4] In the sixteenth century, the cultivation of corn was extended into the valleys of the Po, Ebro, and Garonne. In the seventeenth century rotations without fallowing, alternating grains and fodder crops, became widespread in England and in the Rhine Valley. In the eighteenth and nineteenth centuries, these rotations spread to the rest of Europe. Subsequently, various rotations also developed, in which fallowing was, in part, replaced by non-fodder crops for either food or industry.

Systems that exclude fallowing are more productive in fodder, livestock, manure, and in cereals and other food products than systems that use fallowing. Ultimately, such systems without fallowing are quite diverse. How are they organized, how do they function, what are their results, and how are they explained? We will now take up these questions.

## 2. ORGANIZATION AND FUNCTIONING OF SYSTEMS THAT EXCLUDE FALLOWING

### A More Effective Mode of Renewing Fertility than the Former One

In order to explain the gains in yields and production that are obtained by replacing fallowing with fodder crops, several factors can be pointed out. The increase in the quantity of manure, more intensive exploitation of the soil by the new crops, and, if need be, the contribution of nitrogen by legumes when they are part of the new rotations are usually cited, and rightly so. The relative effectiveness of seeded pastures and row crops in the struggle against weeds is also cited. But all these good reasons do not explain, at bottom, where the additional quantities of fertilizing minerals incorporated into the additional plant and animal production come from, minerals which, in the end, are exported outside of the cultivated environment. To say that these additional quantities of exported minerals come from the manure and thus from the new fodder crops, says nothing about their origin. These fodder crops do not produce these minerals, they take them from the soil solution. In order for the soil solution to be able to supply the additional quantities of regularly exported minerals over the long term, it is necessary that it either receive a new contribution of fertilizing minerals through increased solubilization of the parent rock and fixation of atmospheric nitrogen or that it experience smaller losses through leaching and denitrification. What happens exactly?

## Reduction in Leaching

On land that is subjected to fallowing for fifteen months, plowed three or four times, frequently grazed and trampled on by livestock, the natural groundcover cannot take root thickly and deeply and therefore is unable to produce a large biomass. That is even truer for a short fallowing of eight to nine months. The quantity of fertilizing minerals that this meager natural ground cover absorbs and fixes is thus relatively low. As a result, a large portion of the minerals in the soil solution is not absorbed and fixed by this vegetation and is subjected to severe leaching during the rains of autumn, winter, and spring. In addition, repeated plowing of fallowed lands leads to a sharp reduction in soil organic matter.

Conversely, in the new rotations, the seeded pastures and the fodder row plants that replace the fallowing develop quickly on land well prepared for this purpose. Their roots expand widely and deeply, they exploit the soil solution intensely and absorb great quantities of fertilizing minerals, which thus avoid both drainage and denitrification. It is precisely these minerals, shielded from losses resulting from drainage and denitrification, that are incorporated into the biomass from the new fodder crops, consumed in the stables by an increased number of livestock, and found, for the most part, in the additional manure produced. Gathered with care, preserved well, and applied properly and at the appropriate time, this manure decomposes slowly during the summer. The minerals are continually available in small amounts throughout the growing season, subject to few losses and absorbed by the crops as they grow.

## Green Manure

It is not essential that the additional biomass produced by the new crops be consumed by the livestock in order to improve the fertility of the soil. This biomass can also be directly incorporated in the soil, where it forms what is called green manure. So long as several precautions are taken in order to facilitate its decomposition (preliminary crushing and drying, burying twice, the first time shallow, then deeper), green manure is no less effective than manure. The use of green manure even makes it possible to avoid the export of minerals that occurs when animal products are sold. These exports may not be very large, but they are still real. It also makes it possible to avoid losses of minerals caused by the transport and preservation of fodder and manure. But when animal products sell well, fodder crops, transformed by the livestock, are more advantageous for farmers than green manure.

### Enrichment of the Soil with Organic Matter

Whether it comes from green manure or manure, the additional quantity of organic matter buried each year leads, in the long run, to a significant increase in the organic matter content of the soil. In ten or twenty years, this content can double or triple. As a result, the storage capacity for soil nutrients increases, drainage and leaching are reduced, the structure of the soil is improved, its porosity and water storage capacity increase, microorganisms abound, and the solubilization of parent rocks and the fixation of atmospheric nitrogen are favored.

The basic reasons why the mode of renewing fertility in agrarian systems that do not use fallowing is more effective than in the older ones include intensified occupation of the soil; reduced drainage and leaching; much larger biomass produced and recycled; increased humus content in the soil; and, ultimately, a strong increase in the availability of exportable minerals through the harvests.

### The Case of Legumes

It is important to note that when fodder legumes are part of the new rotations, which is frequently the case, they improve the fertility of the cultivated lands noticeably. As is well-known, legumes host nitrogen-fixing bacteria in the nodules of their roots (the *rhizobiums*), which absorb atmospheric nitrogen in order to synthesize the nitrogen elements upon which the plant directly feeds. As a result, legumes do not suffer from a lack of nitrogen in the same way that other crops do and can grow more vigorously by absorbing larger quantities of all kinds of fertilizing minerals. The biomass produced is much larger and the availability of exportable minerals through harvests is also increased.

Moreover, when a fodder legume is sown under cover of a spring cereal already in place (clover under cover of barley, for example), this cereal can, to a certain extent, feed on nitrogen through contact with the roots of the legumes. Finally, when the roots and nodules of a legume decompose, they enrich the

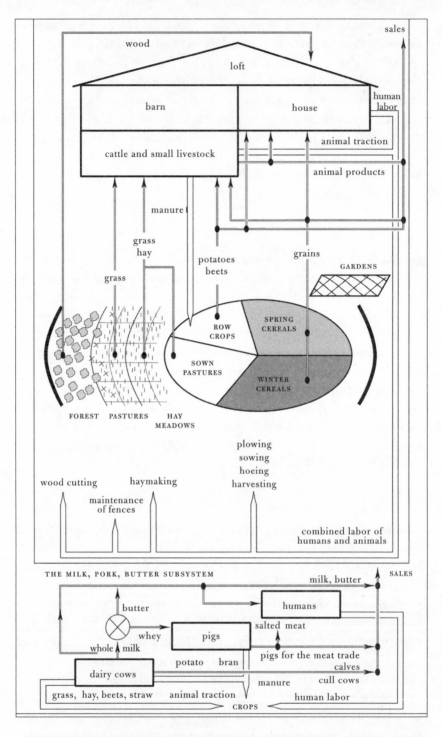

*Figure 8.1 Diagram of the Structure and Functioning of the Cultivated Ecosystem in Agrarian Systems without Fallowing*

soil with nitrogen, so much so that the winter cereal crop that follows immediately after the legume benefits from this additional contribution of nitrogen.

## The New Rotations Without Fallowing

### *Triennial Rotation and Derived Rotations*

Most of the fodder crops rotated with the grain crops are so enriching for the cultivated soil that, on good soils, it is not always necessary to extend these fodder crops to all of the fallow lands to obtain good yields. In this case, the cultivation of demanding food or industrial crops—that is, strong exporters of nutrients—can replace a portion of the fallow lands. Varied rotations can develop combining in diverse ways fodder crops, whether legumes or cereals, and industrial crops.

In many areas of middle Europe, the new triennial rotations were obtained by replacing the long fallowing by a sown pasture of legumes such as red clover (on acid soils), sainfoin, birdsfoot trefoil (on chalky soils) or vetch, or by a sown pasture of fodder grass such as ryegrass or even by a sown pasture of grass and legumes mixed together. Moreover, in many regions, the short fallowing was replaced at the end of summer and beginning of autumn by a hoed crop of turnips (transplanted in September), rapeseed, or fodder cabbage. Thus a new triennial rotation of the following type was obtained:

| Triennial Rotation without Fallowing | | | | |
|---|---|---|---|---|
| YEAR 1 | YEAR 2 | | YEAR 3 | |
| April......October | November.....July | Aug....Nov.    Dec.....March | April.....July | |
| clover | wheat | catch turnips    short fallowing | barley and clover (under cover of barley) | |
| <<< 15 months >>> | <<< 9 months >>> | < 4 months >    < 4 months > | < 4 months > | |

While in the former triennial rotation with fallowing the crops occupied the soil only 13 months out of 36, or an occupation rate of 36 percent, in the new rotation, the soil was occupied 32 months (15 + 9 + 4 + 4) out of 36, or an occupation rate of 89 percent.

On the other hand, as we pointed out above, on very fertile soils of loess and alluvium, it sufficed to replace half of the fallowings with fodder crops in order to obtain a large tonnage of manure and cereal yields of more than 15 quintals per hectare. The other half of the fallowings could then be devoted to hoed plants intended for human consumption, such as turnips, cabbage, potatoes, or to industrial plants such as sugar beets, flax, hemp, or rape oil-seed.

When the long fallowing was replaced half by a seeded pasture and half by a hoed crop, a rotation of six years was carried out, in which the enriching crops and the demanding crops alternated regularly. Such a sextennial rotation could be formed in the following way:

| Sextennial Rotation Without Fallowing | | | | | |
|---|---|---|---|---|---|
| YEAR 1 | YEAR 2 | YEAR 3 | YEAR 4 | YEAR 5 | YEAR 6 |
| clover | wheat followed by turnips | spring barley | potatoes | wheat followed by turnips | spring barley |

In an analogous manner, but on less fertile soils, the long fallowing could be replaced two-thirds or three-fourths by seeded pastures, the remaining one-third or one-fourth could then be devoted to hoed plants. This led to the practice, at least in principle, of long rotations in a ternary rhythm of nine or twelve years.

In high-elevation areas and in northern areas, the early cold prevented the practice of planting catch crops in autumn. New rotations with a ternary rhythm, of the type clover-wheat-barley followed by potato-wheat-barley, were both simpler and less productive in such a context than the previous ones.

### The Norfolk Rotation

When the autumn crop of feed turnips following a grain crop was difficult, it was possible nevertheless to plant feed turnips for a complete season. They were thus planted between the winter grain and the spring grain.[5] But in order to balance this succession of three years of demanding crops, it was necessary also to replace the long fallowing by a crop of enriching fodder legumes, clover, for example. In this way a quadrennial rotation of the following type was obtained:

| Quadrennial Rotation | | | |
|---|---|---|---|
| YEAR 1 | YEAR 2 | YEAR 3 | YEAR 4 |
| clover | winter cereal | full season of feed turnips | spring cereal |

This type of quadrennial rotation first developed in the seventeenth century in the county of Norfolk, England, in connection with the raising of sheep and the production of wool for the rapidly expanding textile industry. Then it spread into many other areas of northern Europe. But in addition to this well-known quadrennial rotation, other rotations, combining in diverse ways one year of turnips with two or three years of grain and one or two years of fodder legumes, sometimes combined with a cereal, were also practiced in English counties at this time. There were quinquennial rotations of the type wheat-turnips-barley-(two years of clover and ryegrass) and sextennial rotations of the type wheat-(barley or rye)-turnips-barley-(two years of clover and rye-grass), etc.[6]

## Biennial Rotation and Derived Rotations

Otherwise, at the time when the first agricultural revolution developed, the old biennial rotation and the ard were still predominant in most of the southern regions. There the replacement of fallowing by seeded pastures, of vetch oats or Egyptian clover, for example, led to the formation of biennial rotations without fallowing and the introduction of the plow. In certain cases, fallowing was replaced half by a seeded pasture and half by a hoed plant, resulting in a quad-rennial rotation of the following type:

| Quadrennial Rotation | | | |
| --- | --- | --- | --- |
| YEAR 1 | YEAR 2 | YEAR 3 | YEAR 4 |
| *vetch-oats* | *winter wheat* | *early potatoes* | *winter wheat* |

The old biennial rotation with fallowing could then serve as the basis for development of rotations without fallowing of four, six, or eight years.

Lastly, in the new rotations, the use of fodder legumes such as alfalfa, a perennial crop that can last two, three, or four years, white clover, a perennial crop sometimes grown in association with English ryegrass, and red clover, a biannual crop sometimes grown in association with Italian ryegrass, led to the practice of various rotations, breaking with the old biennial or triennial sequences. Alfalfa, a crop that is particularly good for improving the soil, demands deep, non-acid, well-drained soils, while red clover can endure more acid soils.

## The Expansion of Arable Lands

As we have seen, the new rotations, as varied as they were, always included fodder crops that made it possible to expand animal breeding and the production of manure. Sometimes such a large quantity of fodder was produced that it was possible to eliminate some of the natural pastures and convert them into arable lands. Certainly, this expansion of arable lands was not possible everywhere. Pastures located on lands that were too hilly or too rocky and meadows located on difficult-to-drain wetlands lent themselves poorly to cultivation.

But on the plains of plateaus covered with a thick layer of loess or alluvium, the entire former *saltus* was cultivable and, insofar as the new cereal-fodder rotations were so productive, the arable lands could occupy the whole of the land area. On the loess lands of the Alsace plain, for example, at the end of the nineteenth and beginning of the twentieth century, seeded pastures, grains, and hoed food, industrial and fodder plants mixed with and succeeded one another so closely that they were talked about in gardening terms, which would have been inappropriate in another context. In this system, practically without pasturage, the animals remained in the stable year-round, where they were fed directly by humans morning and evening everyday of the year. Thus not one blade of grass and not one cow patty were lost.

In other cases, the former *saltus*, not very fertile but nevertheless plowable, lent itself advantageously to the cultivation of perennial, also called temporary, seeded pastures that were turned over and renewed every three to six years. Many poor heaths used for sheep, on porous and leached sandy soil on the oceanic side of northern Europe, from Brittany to eastern Prussia, became relatively prosperous areas for animal raising in this way. These soils, copiously manured and enriched with organic matter, sometimes even became fertile enough so that one or two years of cereals, such as oats or rye, or a hoed plant, such as the potato, could be intercalated between two crops of temporary pastures.

## 3. THE CONSEQUENCES OF THE FIRST AGRICULTURAL REVOLUTION

Overall, the first agricultural revolution led to a doubling of production and in the productivity of agricultural labor, which resulted in both a large increase in the availability of food and in a marketable agricultural surplus. These gains would contribute to a demographic expansion, an improvement in the diet, and an unprecedented industrial and urban development. But the development of systems that did not use fallowing and the resulting increase in crops and herds required, first of all, much additional labor.

## An Overcrowded Agricultural Calendar

In systems based on fallowing and cultivation using the plow, the calendar of agricultural labor was already quite full. With the new system, it was necessary to add to the calendar one to three harvests of hay, as well as sowing, hoeing, and harvesting of row crops, whether they were crown between two principal crops or planted for a full season. What is more, it was necessary to lead and care for a herd that was twice as numerous, cart and spread more manure, and cut, bundle, transport, and thresh harvests twice as large.

Henceforth, the principal nondeferrable seasonal tasks imposed on farmers practicing the new systems included plowing, harrowing, sowing the "wheats" of autumn; harvesting the catch crops; grooming the animals all winter; sowing the spring "wheats"; planting potatoes and beets, then hoeing them; harvesting the hay; harvesting, sowing, and weeding the "catch" crops; threshing the harvest; and cutting the second crop of hay. But it was also necessary, in the meantime, to maintain the forests, prune and weed vineyards and orchards, harvest grapes, gather, garden, etc. As one can see, very little time remained to carry out the multiple tasks that could possibly be postponed but were no less necessary, such as maintenance of equipment and buildings, repairing fences, cleaning out ditches, making tools, spinning, weaving, and all the household tasks. Consequently the work calendar for a peasant family tended to be overloaded. As always in agriculture, it is the busy periods with the heaviest and most restrictive work that limit the development of new systems. In the new system without fallowing, it was a question principally of harvests, hay making, hoeing, plowing, other work for preparing the soil, and sowing, all tasks that henceforth had to be carried out in increasingly narrow time frames and sometimes ended up accumulating and even overlapping one another.

In general, with the labor force and teams of draft animals previously available, a farm could, up to a certain point, expand the cultivation of fodder crops and animal-raising activities by filling in the gaps in the old agricultural calendar. But as the new crops and herds increased, time was increasingly lacking to effect even the most urgent tasks, and the need for new, more effective equipment, which would make it possible to save time during busy periods of work, made itself felt. That is why, from the beginning of the nineteenth century, a whole gamut of new mechanical equipment for animal traction (swivel plow, mower) and new machines for processing harvests (threshers, sorters, grinders, creamers) were developed. The industrial manufacture of this mechanical equipment and its diffusion into the recently industrialized countries of Europe and overseas became much greater at the end of the nineteenth and beginning of the twentieth century (see chapter 10).

## The Performance of Systems that Exclude Fallowing

Depending upon the regions and the farms, the gains in production and productivity resulting from the first agricultural revolution were certainly diverse. It is difficult to know how to give an account of this diversity, but one can try to explain why the new rotations without fallowing generally made it possible to obtain at least a doubling of production and productivity.

To this end, let's consider a small, basic unit of production of 5 hectares, in the cold temperate region, composed of 3 hectares of arable lands in rotation, 1 hectare of meadow, and 1 hectare of pasturage. In the old system, with the triennial rotation fallow-wheat-barley, such a unit could have fed, rather poorly, a pair of draft cows, which supplied fifteen tons of manure. This would have made it possible to produce 10 quintals of cereals (6 quintals of wheat and 4 quintals of barley), or hardly enough to support the needs of a family of five (see chapter 7).

In the new rotation without fallowing (clover-wheat followed by a catch crop of turnips and then clover under cover of barley), the clover crop sown under cover of barley provided a first cut of hay in the autumn and it provided two or three the following year, which made it possible to feed more than one additional head of large livestock. Beyond that, the autumn-grown crop of fodder turnips that followed wheat made it possible to feed more than one half-head of additional livestock. Hence, the number of livestock and thus the animal products and the production of manure could actually double. With 30 tons of manure (in place of 15) for two hectares of cereal, an average yield of 10 quintals of grain per hectare was obtained (12 quintals per hectare in wheat, 8 in barley), or double that formerly obtained. Moreover, it is known that the yield of a cereal following immediately after a legume increases again by about 2 quintals. It is then no exaggeration to estimate that the new system made it possible to double the output of both plant and animal products.

Thus in the former system, with an area of 5 hectares, a family of five was hardly self-sufficient in cereals, and it had neither a pair of draft cows nor even a calf to sell each year. In the new system and on the same area of land, the same family could, without additional equipment, more than double its previous output and sell half of its production, while feeding itself better.

This doubling of production required, as we have seen, additional labor. This additional labor had different origins and costs depending upon the category of farm in question: midsize peasant farm, large farm with wage labor, or very small family farm. The most favorable case was that of the midsize peasant farms of 5 to 10 hectares, already owning a team of draft animals and a complete set of necessary equipment, and employing only family labor. The new system could then be adopted by employing the preexisting equipment and family labor force more intensively and, if necessary, some additional animals. In any case, it was not necessary to resort to wage laborers. In these conditions, the doubling of production entailed few new responsibilities and resulted in a quasi-doubling of productivity per worker.

However, in a large farm having recourse mainly to seasonal wage labor, the increase in the volume of work resulting from the adoption of the new system increased the volume of wage expenses and the profit of the farm was reduced by as much. The return on the capital invested to carry out this agricultural revolution (fixed assets in livestock and buildings, tools, seeds, etc.) was clearly better in the midsize family farms, at the price of an obviously overworked family, than in the large farms with wage laborers. As for the very small farms whose income was less than the needs of the family, they had, consequently, the greatest interest in increasing their herd, but they did not always have the means to do so.

On the eve of the agricultural revolution, there were different types of farms almost everywhere in Europe, combined in varying sizes from one region to another. There existed a strong contrast between areas where small, medium-size, or large family farms predominated (northwestern Europe) and those where large or very large farms using wage labor occupied most of the land side by side with a multitude of mini-farms that supplied them with the day-to-day wage laborers they needed (eastern and southern Europe). The proportions among the different types of farms, the more or less unequal division of the land and other means of production, these "agrarian structures," as they are called, played a large role in determining the forms of the agricultural revolution as well as the rapidity with which those forms were adopted.

## Population Growth and the Improvement in Diet

Contrary to earlier agricultural revolutions, the results of which can only be roughly estimated, the development of the first agricultural revolution can be followed using rather reliable records pertaining to the evolution of the surface areas and yields of crops to the growth in size of herds and their output, to the increase in rural and urban populations.

In France (considered within its current boundaries), toward the middle of the eighteenth century, on the eve of the agricultural revolution, fallow lands occupied 10 million hectares out of 24 million hectares of arable lands and were divided between around 4 million hectares out of 12 million in a triennial rotation for the northern half of the country and around 6 million hectares out of 12 million in a biennial rotation in the southern half. The cultivation of these fallow lands began

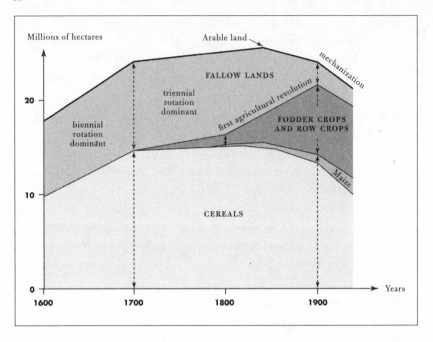

*Figure 8.2 Development of Arable Land and the Cultivation of Fallow Lands
in France (current boundaries) from 1600 to 1940*

at this time, but by 1800 had spread only to some territories in Flanders, Alsace, and the Garonne Valley. One century later, at the end of the nineteenth century, 75 percent of these fallow lands had been cultivated. There remained no more than 2.5 million hectares of fallow lands in 1900, which gradually disappeared in the twentieth century.

Moreover, from 1800 to 1900, grain production in France went from 80 to around 170 million quintals, or 2.1 times greater, while the production of meat was increased by a factor of 3 and the production of milk more than doubled.[7] At the same time, the population grew from 27 to 39 million inhabitants, or 1.4 times greater, while famines disappeared and the average dietary ration rose from some 2,000 to 3,000 calories per person per day, or a 1.5 increase. In total, in one century, consumption, like production, increased by more than two times (1.4 x 1.5 = 2.1).

In 1900, the use of mineral fertilizers was still quite limited and the net foreign trade balance did not surpass 10 percent of production. From that, it can be concluded that the growth in production and consumption in the nineteenth century was indeed due, essentially, to the development of the first agricultural revolution. Moreover, since, 25 percent of the fallow lands had not yet been brought under cultivation by 1900, it can be estimated that the hypothesis that we put forward, of a doubling of production because of the first agricultural revolution, is in fact a modest one.

Finally, since this increase in production was obtained with an active agricultural population that did not increase, it can also be concluded that, from 1800 to 1900, the gross productivity of agricultural labor more than doubled.[8] It was this doubling of agricultural productivity that made it possible for the nonagricultural population in France at this time to grow from less than 10 million to more than 20 million.

An analogous evolution of production and population occurred in European countries affected by the agricultural revolution, beginning with England in the eighteenth century, followed by France, northern Italy, Germany, and the Scandinavian countries in the nineteenth century. From 1750 to 1900, the population of central and western Europe went from around 110 million to 300 million people.[9]

## Industrial and Urban Growth

As a result of the first agricultural revolution, a form of agriculture appeared that was capable of extracting a marketable agricultural surplus on a long-term basis, representing more than one-half of the total production. For the first time, agriculture in the West could thus support the needs of a nonagricultural population that was more numerous than the agricultural population itself. Mining, industrial, and commercial activities could develop to the point of occupying more than half of the total active population. In the earlier agricultural system, the surplus did not exceed, on average, 25 percent of production and it was at the mercy of a bad harvest.[10] In bad years, the surplus was in fact nonexistent, characterized by both shortages and a standstill in nonagricultural activities. In other words, so long as the average agricultural surplus remained low and uncertain, as in the Middle Ages, industrial development inevitably remained not only limited but also fragile.

The increase in agricultural productivity and the formation of a large surplus (on the order of 50 percent of production), without risk of falling below a certain level because of the least accident, was an indispensable prerequisite for a large and lasting development of industrial and commercial activities. In other words, the first agricultural revolution indeed conditioned the growth of the first industrial revolution. Thanks to its high productivity, the new agriculture could reliably supply to nascent industry raw materials, labor, provisions, as well as capital, in sufficient quantities and at low cost. In return, this more productive agriculture, which was a large consumer of iron, tools, etc., became an increasingly important outlet for industrial products.

## 4. THE CONDITIONS OF DEVELOPMENT OF THE
### FIRST AGRICULTURAL REVOLUTION

Though it was a child of the old agricultural system, the new agriculture nonetheless differed in essential respects. Characterized by the absence of fallowing, it also included an enriched cultivated ecosystem, new ways of maintaining fertility and fighting against natural vegetation, and much higher performance. The movement from one to the other is typically one of the large changes that we call agricultural revolution.

This agricultural revolution was far from being the first in Europe, since it followed from the Neolithic, ancient, and medieval agricultural revolutions, even though it is often not recognized as such. But it was the first of modern times, and since it coincided with the unprecedented industrial development that is commonly called "first industrial revolution," we call it the "first agricultural revolution."

From the sixteenth to the nineteenth century, this first agricultural revolution spread to the Netherlands, England, France, Germany, Switzerland, Austria, Bohemia, northern Italy, northern Spain, and northern Portugal. In all these areas, its development was influenced by important reforms in the "anciens régimes," and it was intimately linked to the growth of industry, commerce, and cities. But in southern Europe (in the Alentejo, Andalousia, and the Mezzogiorno) and in the East (Hungary, Slovakia, many areas of Russia), wherever archaic social conditions survived, the use of fallowing lasted until the beginning of the twentieth century and the industrial revolution did not take place. Thus a pronounced contrast formed between the center of a developed Europe, on the agricultural as well as industrial levels, and its underdeveloped southern and eastern periphery.

The first agricultural revolution was a change that went well beyond the simple modifications in cultivation practices involved in the "suppression" of fallowing or its "replacement" by a crop, to which it is often reduced. Rather, it was a complex agricultural development, inseparable from the development of other sectors of activity, and whose conditions and consequences were ecological, economic, social, political, cultural, and juridical, much more than technical. Indeed, just like the means and practices of the agricultural revolution of the Middle Ages were known long before they were generalized, rotations without fallowing were known in Europe several centuries before the development of the first agricultural revolution. It was not because of technical obstacles that this agricultural revolution took so long to develop. Certainly, time was necessary to develop all sorts of new rotations and new procedures and to make them known. But it would be absurd to think that it was due to "technical" reasons that entire countries remained separate from this movement for centuries. Moreover, Europe did not take long to export its own crops, such as wheat and barley, and domestic animals, such as cows, sheep, and horses, to other continents and to develop sugarcane and cotton crops in the Antilles and in America. Other, more significant obstacles must have impeded the development of the first agricultural revolution.

## Juridical Conditions

### *The Right to Cultivate Fallow Land*

The largest and most widespread of these obstacles was the right of common grazing on the fallow lands, which was nothing other than the right of each and every person to pasture his animals on all of the fallow lands in the immediate area. But, in return for this right, each farmer was obliged to open his own fallow lands to the herds of others. This right of "common" use marked the limit to the right of "private" use of cereal growing lands. Each farmer indeed had the right to plow, sow, and harvest his grain on "his" fields, but once the harvest was completed, these fields returned to the common domain, and everyone else could then glean and pasture their herds there. The right of use for the owner or the tenant was then far from being a right of absolute use, i.e., a private and exclusive right to use and abuse.

As long as these arrangements prevailed, no one could cultivate "his" fallow lands for fear of seeing the fruits of his labor trampled and devoured by the livestock of others. The only means of escaping that consequence was to protect his lands against common use, in order to reserve their exclusive use for himself and, consequently, the possibility of cultivating them as he wished. This exclusion from grazing could be imposed forcibly by a powerful person, who thus dispossessed neighbors of their rights of use, but it could also be "consented to" by eligible parties in the neighborhood, sometimes against payment of a fee.

In regions with open fields, the exclusion from grazing was possible for large farms that had extensive and consolidated fields. This was the case for some seigneurial estates and some farms that had been patiently enlarged and consolidated by wealthy farmers. Sometimes, the exclusion from grazing the fallow lands of these large farms encountered resistance from those having rights to common use of the land. Even after it was imposed, decreed, or consented to, this exclusion from grazing long remained poorly accepted, so much so that it was necessary to enclose the lands with quickset hedges, rock walls, or ditches in order to enforce respect for the exclusion.

When the non-enclosed parcels were small, dispersed, and mixed together, however, the exclusion from grazing was difficult to carry out. It required either a preliminary consolidation of the properties to be enclosed or the outright abolition, pure and simple, of everyone's right to common grazing on the fallow lands. Naturally, in regions where the fields were previously enclosed, as in the *bocages* of western Europe and in the hilly areas of central and Mediterranean Europe, the abolition of common grazing rights was carried out much more easily.

The collective decision to abolish common grazing rights was not just advantageous for the large farms. It was advantageous for all farmers who were sufficiently well provided with equipment, livestock, and land to undertake the new agriculture. On the other hand, this decision often encountered the opposition of small farmers who owned little livestock and had to count on the animals of others to manure their lands, large-scale stockbreeders, who had little or no land and thus were large users of common pasturage, and landless peasants, who had small herds.

## The Abolition of Other Collective Obligations

The obligation to give everyone access to the fallow lands was not the only collective obligation prohibiting the full and free private use of the land. Common grazing right was extended also to fruit plantations (olives, almonds, etc.) and sometimes to vineyards. It was also often extended to meadows after the first cut of hay, which prevented the possessors of these meadows from obtaining a second cut of hay. What is more, in regions with regulated plot allotments, the obligation to plant grains one year out of two in a biennial rotation or two years out of three in a triennial rotation, also restrained the free choice of crops. This obligation was sometimes maintained even after the fallow lands were cultivated, particularly in regions with fields laid out in strips where, because of the narrowness of the parcels, which often had only a few furrows, fieldwork had to be synchronized and coordinated. That explains, at least in part, the persistence of binary or ternary rhythms in the new rotations.

From the end of the Middle Ages, however, a large movement was formed in several regions of Europe against common grazing rights in all its forms and, more broadly, against all collective obligations opposed to the free use of cultivated lands and the right to enclose them. This movement in favor of the right to use and abuse one's lands, and to exclude all other users, was at bottom nothing other than one particularly strong moment in a broad, centuries-long movement that saw private property in land emerge, develop, and finally triumph over the old undivided "common property," that is, the absence of property.

## The Decline of Joint Possession and the Development of Private Property

The movement of land appropriation began in the Neolithic epoch, with the construction of the first permanent dwellings and the enclosure of the first private garden plots. In slash-and-burn systems of temporary cultivation, the right to cultivate a wooded parcel was only a temporary right of use. All the idled lands and forests located around a community formed its common property. Initially, the same type of land arrangements characterized systems based on fallowing. The *silva* and the *saltus* were still types of permanent natural idled lands open to all, where each person could herd animals, gather, cut wood, and hunt. Fallow land, properly speaking natural fallow land of short duration subjected to gleaning and common grazing after the harvest, returned to the common domain in the same way as wooded idled lands of long duration in the old systems of slash-and-burn cultivation.

However, in early antiquity in the Mediterranean, beginning with the formation of city-states, neighboring communities were sometimes dispossessed of all or part of their joint rights.[11] Institutionalized private property in land was then extended by means of conquest to a good part of Europe and North Africa. But as immense as this first development of land ownership was, it was far from affecting the whole West. Many Celtic, Germanic, Scandinavian, and Slavic communities remained separate from this development, even if a process of private appropriation of the old common rights appeared among them also. Moreover, even in the interior of the Roman Empire, particularly in poor regions and in those that had been occupied only for a short time, communal rights remained very much alive. Subsequently, the great invasions from the north and east effaced the Roman right of property and superimposed various forms of communal right, even in the countries of southern Europe and North Africa.

Consequently, during the Middle Ages, regions in which common law retained vestiges of Roman law were quite uncommon and even in these regions, the serfs and free peasants benefited from rights of pasturage and wood collecting on common lands. Private property, excluding all collective obligations, was far from occupying all the land in these areas. In some Slavic and Germanic communities, the original joint possession of cereals-growing lands even lasted until the beginning of the twentieth century. These communities carried out periodic redistributions of arable lands among families based on the size of those families, so that the right of use accorded to each family was only temporary.

What emerges from this long history is that, beginning in the Neolithic, "property" in land was gradually extended to different categories of land *insofar as that land was altered in some way by human labor*. This included developed sites first of all, then gardens and enclosures enriched and cultivated each year, cleared lands bearing a harvest, developed meadows, lands cultivated between two periods of idling or plowed fallowing, continually cultivated lands, improved grazing lands, and planned and maintained forests. The old rights of common use (hunting, gathering, gleaning, common grazing, firewood collecting) still existed in areas where the forest, natural pastures, wild grasses on fallow lands and game continued to develop spontaneously, *without any particular labor*.

*Property and System of Land Tenure.* Viewed from this perspective, the private possession of land appears first of all as a means to collect the fruits of labor invested there. For the peasant, access to property is a sure means to be assured the benefits of his own labor. But monopolization of the land by some is also a means to claim a part of the fruits of others' labor, because the affirmation of the right of property also justifies the right of owners of large and medium-size properties to lease their lands either to tenant farmers or sharecroppers.

In general, the tenant farmer owns the equipment, livestock, and all the necessary working capital. The tenant farmer rents from the property owner the land and buildings, for which a fixed land rent is paid, most often in money but sometimes in kind. This land rent varies depending on the quality of the land and the income that can be drawn from it. The sharecropper possesses only a small portion of the working capital. The property owner supplies the sharecropper not only with land and buildings but also farm implements and livestock, as well as some of the everyday expenses of the farm. The rent paid by the sharecropper includes, beyond the land rent, payment for the right to use the capital furnished by the property owner, interest included. The rent, which is both for the land and the capital, is generally paid in kind and proportional to the harvest.

Land rent in the strict sense followed from the right to land property, and it was consequently of a different nature than feudal taxes, which were tributes in kind (part of the harvest) or in labor (corvées) that the lord imposed on the peasant serf because of a political and military relation of force. Land rent, on the other hand, resulted from a relation of property between lessor and lessee. It was essentially a market relation, though undoubtedly an asymmetric and unequal one. Land rent was not the heir, as some think, of the so-called feudal rent, even if in certain areas (England and Prussia, for example), after the dissolution of the feudal regime, landowners were far more often descended from nobility than from serfs.

Naturally, a landowner can also directly exploit his own lands by using the labor of his own family or of wage laborers, permanent or temporary, drawn from the family labor force of those peasants who are the most lacking in land and working capital. But whether these lands were exploited directly by the owners, by tenant farmers, or by sharecroppers, the important thing for the development of the new agriculture was that these lands, henceforth private property, were freed from all obstacles to their use and the farmer was able to derive the benefit from these new possibilities. Such is basically the reason why the progress of private property influenced the development of the first agricultural revolution so powerfully, just like the preceding advances in the right to use the land, had, beginning in the Neolithic, influenced the development of prior agricultural revolutions.

*Individualism and Cooperation.* The only way to get around the necessity of private property as a condition for pursuing the new agrarian practices would have been to exploit all the jointly possessed and underutilized lands of the villages in a cooperative manner (i.e., shared investments, work, and profits). That could have been possible if there had existed in the countryside of the West a true cooperative tradition. Contrary to a naively upheld myth (the famous "primitive communism" of peasant communities), the farming of joint lands by the villagers was not collective, it was essentially individual and, moreover, unequal. Each peasant used the pastures and the fallow lands subjected to common grazing in proportion to his wealth in livestock, and the profit was not shared. But at the same time this well-rooted agrarian individualism was not the same as a "fight of each against all." It excluded neither a certain solidarity (right to glean, mutual aid) nor certain common undertakings (threshings, road maintenance, cheese dairies, shared shepherding), when they were useful and not at variance with the well-understood interest of each person.

By the end of the Middle Ages, in most areas of Europe, the ancient rights of use and joint possession of common lands constituted obstacles to the development of the first agricultural revolution. The first attacks against these rights generally began in this time period and were pursued over the following centuries. Depending on the dates and the various forms in each country, these attacks generally led to the abolition of the right to common grazing lands and other collective obligations and to the institution of the right to enclose and cultivate freely (or have cultivated) one's own lands, that is, the introduction of a true right of private property in agricultural lands. In addition, a large portion of the jointly possessed forests and pastures were also divided. All things considered, this entire movement took the form of a large increase in private property in land.

## The Economic Conditions of the First Agricultural Revolution

But if the double movement of decline in collective obligations and growth in the development of private property and right of use was a necessary condition for the development of the new agriculture, it was far from being a sufficient condition for that development. As Marc Bloch shows in *Les Caractères originaux de l'histoire rurale française*, in certain southern regions like Provence, where Roman law had left some traces, this movement of appropriation had begun very early in the Middle Ages.[12] However, agriculture was deeply transformed only in the nineteenth century.

Conversely, beginning in the sixteenth century, in Normandy and some English counties where the textile industry was fast expanding, successful attempts to consolidate and enclose the lands increased, thereby removing the fallow lands from common grazing, replacing them with fodder crops, and demolishing the old rights of use still in force. These undertakings were the accomplishment of some lords and a stratum of well-off farmers, both grain growers and sheep raisers, well placed to take advantage of the new agriculture. They profited both from the demand for wool from factories and the demand for bread by the nascent working population. In these areas, it was indeed under the pressure of agricultural and industrial development that the old customs declined. It is possible then to conclude that when market conditions favorable to the agricultural revolution were brought together, its development began, and this despite the juridical obstacles whose fall that development precipitated.

Thus the increase in demand for agricultural products due to industrialization and urbanization appeared as a driving force in the development of the first agricultural revolution, while the juridical conditions, as necessary as they were, were basically nothing other than the suppression of institutional and customary obstacles to this development.

## First Agricultural Revolution and First Industrial Revolution

The first agricultural revolution was a vast evolutionary development that led to a doubling of agricultural production and productivity. Even if the improvement in the peasant diet absorbed a part of these gains, it remains the case that around half of the total agricultural production could henceforth constitute a marketable surplus. The agricultural revolution could develop fully only on condition that this surplus actually met an adequate solvent demand, coming from a nonagricultural population as large as the agricultural population itself.

For the first time in the history of the West, a society in which workers, artisans, merchants, employees, and persons of independent means made up more than half of the population became not only possible but necessary in order to absorb the surplus of production coming from the new agriculture. That is why, in the sixteenth and seventeenth centuries, the agricultural revolution developed first around textile centers in Flanders and England. In the eighteenth century, it continued to expand in England as the first industrial revolution reached the mining and iron manufacturing regions. It began to spread in France, Germany, and the Scandinavian countries. Finally, in the nineteenth century, it developed completely in all the industrialized regions of northwest Europe. The first agricultural revolution and first industrial revolution advanced together. They proceeded at the same pace because they existed hand in glove with one another.

Some processing industries used raw materials from the agricultural sector. Thus the development of sheep raising, based on the new fodder rotations, supplied the growing quantities of wool necessary for the expansion of the textile industry in Flanders and England. In the same way, the sugar beet crop was at the origin of an important rural industry, which developed in the plains of central Europe. These agricultural and industrial products were the origin of the wealth and large investment potential of these regions.

Many other industrial crops played a similar role, though less important: flax and hemp for the manufacture of canvas in northern France and Germany; hops and brewing barley in all of northern Europe; potatoes for the manufacture of alcohol and starch in Prussia; plants used for dyeing, such as woad and madder, the production of which expanded proportionately with the textile industries. In many regions, the development of an agricultural and industrial production and processing network of this type played a decisive role in the development of the agricultural revolution.

There would have been no revolution without the possibility of selling at satisfactory prices the surpluses of vegetable and animals products whose production was made possible by it. But there would have been no agricultural revolution either if farmers-stockbreeders did not have the investment potential to double the number of livestock, construct new buildings, build enclosures, and, if necessary, buy equipment and pay additional laborers.

## The Social Conditions of the Agricultural Revolution

### *The Possibility of Investing*

In most of the countries of northwestern Europe (Netherlands, western Germany, France), the agricultural revolution was mainly the accomplishment of the middle peasantry, using essentially family labor and little wage labor, if any at all. But it was also the accomplishment of some of the large European farmers: large farmers working with their families and employing some wage laborers, country gentlemen, the English gentry, the Prussian Junkers, etc.

This revolution was beyond the reach of small farmers who were poorly endowed with equipment, land, and livestock, and too poor to invest. They were often excluded from the process and forced to become wage laborers or leave the land. Such was the case particularly for "minifundist" peasants of eastern and southern Europe, marginalized by the large "latifundist" estates. At the same time, in these same regions, the absentee owners of these large estates, who had the possibility of investing more profitably outside of agriculture, had no incentive to get involved in the new agricultural revolution, either. Let's see more precisely in which particular social conditions the first agricultural revolution unfolded in England, France, and other European countries.

## The Case of England

*Enclosures and Large Estates.* From the sixteenth century in England, the landed nobility began to enclose the range lands, until then open to herds from the vicinity, thereby reserving exclusive use in order to extract a profit from the growing demand for wool from the textile industry. They encountered resistance from the villagers, and the confrontations and negotiations that followed generally led to a division, advantageous for the lord, between pastures henceforth enclosed for the estate and pastures remaining common.

But the lords also undertook to transform a portion of their arable lands into sheep pastures and to replace their fallow lands with fodder crops of clover or turnips, again depriving the villagers of their right to common grazing on these fallow lands. In order to enforce respect for this unilateral decision, it was necessary to consolidate their own parcels, frequently interspersed among those of small tenant farmers, and enclose them. Quite often, the lords profited from the occasion by appropriating the best lands, and many small tenants were marginalized and ultimately evicted. The enclosures did not stop at the boundaries of the estate. Many peasants were deprived of their lands in all sorts of ways: nonrenewal of limited leases, seizing lands upon deaths and transfers, and abusive evictions.[13] This enclosure movement continued even more in the eighteenth century, at the height of the agricultural and industrial revolution, this time with the support of Parliament, the majority of whose members were landowners. From 1700 to 1845, no less than 4,000 acts of enclosure authorizing the lords to divide the commons, consolidate their lands, and enclose them were enacted by Parliament. Representatives were named who were responsible for dividing the lands and who often allocated to small peasants only the most infertile lands.

Thus the majority of the English peasantry disappeared (the *yeomen*), were forced to become agricultural wage laborers, beg, migrate toward the cities, become industrial wage laborers, or emigrate to settler colonies. In the middle of the nineteenth century, as a result of this large movement of land appropriation and concentration, a large part of the lands was in the hands of a small number of large landowners (the *landlords*). Two thousand among them owned quite vast properties ranging from 100,000 to 400,000 hectares, which covered in total one-third of the country, where there were some 200,000 manor houses.

*The Corn Laws.* The landlords had the laws on cereals modified to their benefit. The famous Corn Laws, conceived at the beginning of the Middle Ages, limited the exports of cereals in order to avoid shortages and speculations during bull markets. Beginning in 1660, the Corn Laws were used for protectionist ends.

Imports were subjected to customs duties and, beginning in 1815, imports of cereals were prohibited when prices fell below a certain level. That made it possible to maintain the internal price of cereals at a high level and increase, consequently, agricultural profits and land rents received by the landowners.[14] Some landlords lived directly on their estates. They employed agricultural wage laborers to work on these estates under the direction of foremen and superintendents, who were also wage earners. These landlords often divided their lives between the country house and the city, where they invested a part of their income in real estate, mining, industry, commerce, and banking. However, most of the large property owners leased land at high prices to tenant farmers. In the middle of the nineteenth century, nearly two-thirds of these farmers were small owners, who employed an average of four wage workers. The remaining third were made up of family farmers employing no wage laborers. The social structure of English agriculture cannot be reduced, as is often done, to large landowner, capitalist entrepreneur, and wage laborers.

Thus, in England, the dissolution of the agricultural old order, with its seignorial reserves, peasant tenants, and rights of common use, and the advent of private property and enclosure rights led to the predominance of large landowners, tenant farmers, and wage laborers, and the elimination of the majority of peasants. However, family farms remained numerous. As for farms employing wage laborers, they were most often of relatively modest size and had no resemblance to the large agricultural estates employing hundreds of wage laborers or quasi-serfs, such as those found in eastern and southern Europe.

## The Case of France

*The Predominance of the Peasantry.* In France, unlike England, the dissolution of the agrarian old order led, in many areas, to the predominance of small and medium-size peasant property and to the decline of large seignorial property.

More than any other, the French nobility was shattered by the wars at the end of the Middle Ages, the Hundred Years War in particular. Many nobles abandoned their estates to join the court of some prince. Numerous noble estates changed hands and bourgeois property developed around the cities. Only the ecclesiastical nobility resisted these upheavals. In order to repopulate their fiefs, which had been deserted following famines, plagues, and wars, the nobles granted to their tenants increasingly liberal conditions and, lacking labor power, leased all or part of their own estates. The bonds of dependence between lords and tenants were weakened, serfdom and corvées almost completely disappeared. But above all, the tenants increasingly behaved as quasi-owners of their tenures. They were not only inheritable, but the tenants could also freely sell, lease, and mortgage them. At the time of these sales, the lord collected only modest transfer taxes and his preemptive right was hardly more than a menace to dissuade the sellers from fraudulently declaring a selling price lower than the real price.

The status of French peasants in the eighteenth century appears as one of the most favorable in Europe, even if many of them were compelled to pay fees, such as quitrents and tithes, and even if some still had to perform corvées. Everything that survived as privileges for some and subjection and taxes for others was abolished by the Revolutionary Assemblies. Duties in person were suppressed from the night of August 4, 1789. However, the permanent real duties owed by the tenant peasants to the local lord (quitrents and other feudal taxes) were maintained, declared redeemable in twenty annual payments. This measure was strongly contested by the peasantry that had for generations considered itself as owner of its tenures and only saw in these services and taxes traces of subjection and servitude. Moreover, if these taxes became negligible in some places, they could also, in other places, reach one-third of the rental value of the lands. The taxes were finally abolished without redemption by the Convention in July 1793. By this revolutionary act, the Convention turned subjugated tenants into a free peasantry who were quasi-owners of their lands, except for the collective obligations.

On the other hand, by nationalizing the goods of the Church and expatriate nobles, and by selling them to the bourgeoisie and peasantry, the French Revolution also dismantled the huge estates of the lay and ecclesiastical nobility and strengthened medium-size property. Even if these measures conformed to other political motives, they led to a transfer of property large enough to be considered a sort of agrarian reform.

Moreover, throughout the Revolution, the opposition of the peasantry to large property continued to be expressed. In the work that he published in 1789, *De la religion nationale*, the abbé Claude Fauchet expressed:

One of the strangest errors made by economists is to believe that small properties are generally less useful, and less productive than large ones, because of scarcity of manure and lack of ability at farming. What an unbelievable illusion! These economists have written volumes which have convinced no one, because the principles of common sense and the evidence of facts contradict them. One cow is adequate to manure one small field and the neighbor's oxen plow it for a small payment. Don't worry that there might remain one bush, one pothole, one corner without value. Look at the domain of a small agricultural landowner. His buildings are well maintained and at no great expense, because he makes repairs when the need is slightest. Look how his herds prosper, with what ingenuity he prepares his dairy products and makes them marketable and with what care his small fields are cleared, manured, sown, weeded and made ready for all the subsequent produce that one can expect! On the contrary, take a look at the vast lands of the wealthy. Most of the buildings of each sharecropper's holding are dilapidated. The herds are neglected. Large parts of the lands are uncultivated. The cultivated parts of the land have gaps of scrub and the plowings are poorly done. Losses of all types are incalculable and repairs are made only when everything falls into ruin and at enormous expense.[15]

However, the land redistribution conducted so well by the revolutionary assemblies benefited the bourgeois and wealthy peasants more than the small and landless peasants. It was then far from the much more democratic and radical, egalitarian even, agrarian reform advocated by the Babouvists (from the name of their inspirer, Gracchus Babeuf). The "agrarian law" proposed by them was rejected by the Convention and Gracchus Babeuf was even assassinated, as were his illustrious namesakes and Roman precursors, the Gracchus brothers (see chapter 6).

*The Right to Enclose and Common Grazing Rights.* Beyond that, the revolutionary assemblies generally accepted the ideas of progress in agriculture and the economy. They advocated the new agriculture, encouraging the cultivation of fodder crops in order to develop stockbreeding and the use of manure. They encouraged the development of hoed food crops such as the potato in order to meet the immediate needs of the population. The Committee for Public Safety even sent into the countryside commissioners responsible for making propaganda, organizing demonstrations and distributing seeds and plants for the new crops, organizing, undoubtedly for the first time in France, a true program of agricultural development and popularization.

The revolutionary assemblies also denounced common grazing rights and obligatory fallowing as "barbaric" customs, "tyrannical" laws, and "feudal" servitude. The Constituent Assembly also proclaimed the absolute right to enclose land and enlarged that right to all of French territory. Consequently, the Assembly instituted for each and all the right to cultivate fallow land as one pleases, on one's own account, protected by one's fences.

However, although every successive government was concerned about it, no general law could ever abolish the right to common grazing on the grass regrowth and the fallow lands. Certainly, common grazing was prohibited, in fact and in law, everywhere where individual property could establish itself through enclosures. But everywhere the decision to abolish common grazing remained within the jurisdiction of each community. It was not until the Third

Republic that common grazing was suppressed in principle, each community retaining nevertheless the right to demand its continuation. Thus, throughout the nineteenth century, common grazing declined, but with difficulty, and in some villages it persisted until the beginning of the twentieth century, which quite certainly slowed down the spread of the new agrarian systems.

*The Difficult Division of the Common Land.* As convinced as they were of the virtues of the new agriculture and private property, the revolutionary assemblies did not make a complete clean sweep of the survivals of common property, that is, of common grazing and common land. On the contrary, since nobles and villagers often contested "common" forests and pastures, the Constituent Assembly initially consolidated common property by clearly establishing that these forests and pastures indeed belonged to the village and not to the lord.

However, the question of dividing the common land was on the agenda and strongly contested. In effect, it was a question of knowing if these common lands would be distributed to all who were eligible for them, either in equal parts or in proportion to the size of their property and their herd or even if they would be sold in plots as national goods. In 1792, a law making the division of common pastures obligatory was passed. But in the following year, a less restrictive law, in which the division was optional, replaced it. It could be decided by only a third of the population.

In 1803, under the Empire, the distributions were stopped. The privatization of the communal lands occurred subsequently, not under the form of a division among eligible parties, but by sale to private individuals. Notwithstanding these arrangements, some common lands lasted until the present. This is still the case with some pastures close to settlements, and above all with distant pastures in the mountains or low valleys, which are still used for summer or winter transhumances. Undivided forests remained the property of the villages, but were removed from common use and subjected to the Forest Regime imposed by the state. This restriction of rights of use of the forest, which extended to pastures in the process of reforestation, provoked actual peasant guerilla war in some regions at the end of the nineteenth century (Pastouraux in the Cévennes, War of the Demoiselles in the Ariège).

Thus in France, unlike what happened in England, the Revolution seriously caused the decline of large ecclesiastic and secular seigniorial property, leaving a place for small and medium-size peasant property. Nevertheless, in some regions, large property retained a significant presence. In either case, the small and landless peasants benefited little from the reorganization of landed property. In France as in England, the agricultural revolution profited medium-size and large family farms and those farms that employed same wage laborers, although in France, and this is the second great difference from England, family farms that did not employ wage laborers were predominant.

## Other Countries of Europe

In central Europe (Germany, Austria, Bohemia, Switzerland), Scandinavia, northern Italy, and northern Spain, the first agricultural revolution developed beginning at the end of the eighteenth century with the emancipation of the peasantry, the resolution of landed property issues, and industrial development. In most of these countries, the agricultural revolution was the accomplishment, as in France, of middle or wealthy peasants. However, there were also areas where farmers using same wage laborers played an important role. Even in Prussia, the *Junkers*, who were both large landowners and entrepreneurs, took advantage of, on the one hand, the expanding western European market and, on the other, the underpaid, quasi-servile labor force of eastern Europe. From the end of the nineteenth century, some had made their estates into veritable agro-industrial complexes manufacturing beet sugar, or potato-derived alcohol or starch, for example.

Throughout industrialized northwestern Europe, farms employing same wage laborers and family farms—whether middle income or wealthy, owned or leased—demonstrated their strong ability to develop the new agricultural system. The peasantry even succeeded in spreading the new system into remote and difficult mid-elevation mountain regions, as well as in the Alps, Apennines, and Pyrenees, by developing quality products that could be preserved and transported, such as long-lasting cheeses, dried and smoked meats, and alcoholic beverages.

*Latifundism and Underdevelopment.* In areas distant from the large centers of industrialization, in eastern and southern Europe (Hungary, Slovakia, Russia, the Alentejo, Andalousia, and the Mezzogiorno), large, underequipped estates, employing a labor force that was paid very little or not at all, were still, in the middle of the twentieth century, carrying out rotations with fallowing. This inability to carry out the agricultural revolution has often been attributed to the archaic technical and social heritage of these areas, which had little or no experience of the agricultural and industrial revolutions of the Middle Ages, and where more or less attenuated forms of serfdom persisted. The absenteeism of the latifundia owners, more inclined to squander their incomes in the large metropolises, casinos, and fashionable seaside cities than to invest, was also denounced as a defect impeding all progress.

But the reasons why the peripheral areas dominated by latifundia did not carry the first agricultural revolution are socioeconomic before being socio-psychological. These reasons stem first of all from the fact that these regions, which were distant from the large consumer centers resulting from the first industrial revolution, had to bear the high costs of marketing their products, which reduced the profits the latifundia owners could extract from their agricultural investments by a like amount. Also, these latifundia owners, who were socially close to the bourgeoisie of the central regions, had the possibility of investing in all types of industrial, commercial, banking, or colonial businesses that were much more profitable for them than cultivating fallow lands. Thus a large latifundia-dominated periphery was formed in eastern and southern Europe, devoid of an internal market and underindustrialized. In brief, this was an underdeveloped periphery whose profits were based on the underpayment of a subjugated labor force, firmly kept away from the industry of the central regions and its relatively higher wages.

## Political and Cultural Conditions

Everywhere the first agricultural revolution developed, it was closely linked with the industrial revolution. Next to the conditions directly necessary for its own development (abolition of collective obligations, development of private property, expansion of the market, and possibilities of profitable agricultural investments), it is also necessary to take into consideration all of the political and cultural conditions that made possible the development of the industrial revolution itself, and the immense economic and social transformations with which it was linked.

English industry, for example, could not have developed and triumphed over the craft industry if it had to respect the old craft regulations that obligated employers to use fully qualified professional workers, confirming that they had first had a long and complete apprenticeship. In England these regulations had fallen into disuse well before their legal abolition, at the beginning of the nineteenth century. In France, the system of guilds also imposed a long apprenticeship, limited access to each guild, restricted innovations, commerce, and competition and consequently was a heavy obstacle to industrial development. In 1776, Turgot attempted to abolish this quite contested system, but royal power was weak and backed away from the reforms, parted from its reforming minister, and reestablished the old system. As for many other reforms, it was necessary to wait for the Revolution in order to establish firmly the freedom to invest, work, and trade.

Finally, what was at stake behind the juridical questions affecting the right to work, conduct business, and own property was the freedom to undertake some activity or other, not only for peasants and agricultural employers but also for industrial and commercial employers. The institution of this freedom of action formed the most essential rupture with the constraints and powerlessness of the old regimes, which had generally remained prisoners of both the conservative social forces on which they relied and a form of government that had become ineffective. Only powers of a new kind, enlightened despotisms, constitutional monarchies, or republics, strong in their alliance with the forces of progress, proved capable of imposing this revolutionary rupture.

These political disruptions, which swept Europe from the seventeenth to the nineteenth centuries, had been prepared by the new ideas that had developed and spread since the Renaissance in every area: art, philosophy, religion, politics, economics, science, and technology. Among these new ideas, some gave birth to doctrines, to movements of opinion and powerful social movements, which became true political forces. It is among these doctrines that the "new agriculture" and "physiocracy" must be placed, which undoubtedly had little direct influence on the way farmers carried out cultivation but greatly inspired the legislative work and agricultural policy of reformist or revolutionary governments.

## Supporters of the "New Agriculture"

Agriculture that excludes fallowing had been practiced since the fifteenth century in Flanders, Brabant, and Artois, without being the creation of any agronomists. Some English ambassadors responsible for economic espionage in the Netherlands, England's principal competitor at the time, wrote some reports on this subject. Certainly, Olivier de Serres in 1600 in *Théâtre d'agricultre et Mesnage des champs* had given an account of his own attempts to cultivate fodder legumes on his estate of Pradel, but he seems to have been completely unaware of the immense agricultural revolution being born in the Netherlands. It was not until the eighteenth century that English and French agronomists began to formulate the principles of the new agriculture and advertise them. Thus, in 1731, while English rotations that combined clover, wheat, barley, and turnip crops in various ways had already been carried out for several decades, Jethro Tull, lawyer turned gentleman farmer and proud of his knowledge of agricultural progress, went so far as to proclaim the uselessness of manure and of crop rotation.[16] However, Tull foresaw the use of a sowing machine, sowing in a line, the good management of seeds, and the increase in plowings and hoeings using animal power. He thought that plants fed themselves through direct contact of their roots with soil particles, and that it was sufficient, consequently, to pulverize and loosen the soil constantly in order to increase the surface area in contact with the roots and facilitate the roots' penetration. In this way, Tull obtained several successive years of good yields, which could be explained, as he thought, by the looseness and cleanness of the lands, though also by the reduction in capillary rising and evaporation of water, by the accelerated mineralization of the organic matter, and by a more intense exploitation of the soil solution.

But after several years of practicing this type of cultivation, yields fell and Tull suffered some setbacks. He was not aware that increasing the working of the soil, very effective for exploiting its organic, hence mineral fertility, necessarily leads to exhausting that fertility if this exploitation is not offset by contributions of additional manure, as the later experiments of his compatriots Home and Dickson demonstrated. The new rotations with fodder crops clearly made it possible to obtain the additional manure. In the end, the new agriculture that triumphed in England in the second half of the eighteenth century combined both the methods of more intense exploitation of the soil's fertility advocated by Tull and the enhanced renewal of the fertility advocated by Home and Dickson.

In France, Duhamel du Monceau was the first to formulate the principles of this new agriculture: "Repeated plowings increase yields and make it possible to cultivate turnips and other hoed plants. But it is necessary to compensate: in order to suppress fallowings, fertilizers are needed, hence livestock are needed. Finally, in order to feed the livestock, more pastures are necessary and since natural pastures are lacking, sown ones are necessary."[17] Landowner and farmer, translator of and commentator on Tull, promoter of English methods of agriculture, Duhamel du Monceau was undoubtedly the most influential agronomist of his time. Member of the Academy of Sciences, of which he was also the director, his activities extended to numerous areas: diseases and growth of plants, botany, arboriculture, meteorology, chemistry, wood destined for naval construction, and naval construction itself.

## The Physiocratic Doctrine

The physiocrats, who were partisans of the new agriculture, added an economic analysis and proposed a development policy. As M. Augé-Laribé in *La Révolution agricole* emphasizes, physiocratic ideas in matters of political economy were born "from a reaction against the industrialism and mercantilism of Colbert."[18] They shared in the fascination with nature, with rural and pastoral life, which were popular among the French elites in the eighteenth century. They also shared in the strong revival of interest in agriculture that had been forgotten by the rulers since Henry IV and Sully.

One century after Sully's proclamation of his celebrated formula, "Plowings and pastures are the two teats which feed France, the true treasure mines of Peru," de Boisguillebert was undoubtedly the first to reaffirm that "the source of people's income is the sale of excess food stuffs, which leads to all the incomes of industry, which rise and fall in proportion to this sale." Another precursor,

Vauban, military engineer, chief overseer of the kingdom's fortification work, was the originator of the first investigations and statistics on agricultural production and incomes. He denounced the excessive and disorganized taxation as the cause of the agricultural crisis, and he proposed a unified system of taxes on incomes, which was rejected by Parliament.[19] Many other authors contributed to outlining, developing and spreading physiocratic ideas (Melon, Cantilon, de Vivens, the Marquis of Mirabeau), but François Quesnay is recognized as having formulated the physiocratic doctrine in the most complete form. Master surgeon, the king's primary regular doctor, Quesnay came late to the study of agriculture, taxation and economics. He was the author of the articles "Farmer" and "Grain" in the *Encyclopedia* and of the celebrated *Tableau économique* (1758), the first attempt at national accounts and a theory of taxation. He was also, on the eve of the Revolution, leader of the group of "Economists," an influential group in political circles in France and later elsewhere.

In order to illustrate Quesnay's thinking, we cite some extracts from his *Maximes générales du gouvernement économique d'un royaume agricole:*

Land is the unique source of wealth, and it is agriculture which increases that wealth; property is the essential foundation of the economic order of society .... A nation ... which has a large territory to cultivate and the possibility to have a large trade in local foodstuffs, should not increase the use of money and men in manufacturing and luxury trade too greatly, to the detriment of agricultural work and expenses; because preferably to everything, the kingdom should be well populated with rich farmers .... Each person should be free to cultivate in his own fields such products as his interest, his abilities, and the nature of the land suggest to him as a way to extract the greatest possible product .... The increase in livestock should be favored ... because this is what supplies the lands with the fertilizer which leads to abundant harvests .... Lands used for grain cultivation should be consolidated as much as possible into large farms exploited by wealthy farmers. Foreign trade of local foodstuffs should not be prevented; because as go the sales of goods, so go their reproduction ... Do not believe that a good market is profitable to humble people ... because the low prices of foodstuffs cause the wages of common people to fall .... Complete freedom of trade should be supported ... because the surest, most rigorous, most profitable policing of internal and foreign trade, for the nation and the state, consists in complete freedom of competition.

The physiocrats broke with mercantilism in that they considered that the true wealth of a nation rests on the products extracted from its soil and on their redistribution within the social body, not on the accumulation of metallic money through an unequal exchange with other nations and particularly with colonies. They were, in a certain way, the precursors of classical political economy.

The physiocrats saw in high agricultural prices the source of the wealth of property owners and tenant farmers, the condition for investments and sustained agricultural development, the basis for a profitable tax system, but also the source of high wages and of a purchasing power proportional to these prices. Without falling into their agrarian fundamentalism, which denies to nonagricultural activities all possibility of producing wealth (and this on the eve of the industrial revolution!), it is possible to maintain that all activity, whether it be agricultural or not, and all leisure time can exist in a society only on the basic condition of being sustained by products of the land. The development of nonagricultural activities is only possible in proportion to the "surplus" produced by the farmers beyond the satisfaction of their own needs. The physiocrats also saw correctly that property, the free disposition of productive goods, and free internal and external circulation of commodities as stimulants to production could present advantages.

However, these enthusiasts of the "English way" were also defenders of large property and of large farms employing wage labor. They underrated and scorned the possibilities for development of the peasant family farm, which was going to prevail in France and in northwestern Europe, as well as in the European settler colonies of North America, Australia, and New Zealand. This error explains why they lost much of their influence in these countries as they continued to be listened to by enlightened princes yearning for reforms in countries with large estates in eastern and southern Europe.

Without any doubt, agronomists and economists contributed to the diffusion of the new ideas and they inspired laws that greatly facilitated the development of the agricultural revolution. But it is necessary to emphasize that their conceptions did not precede social practice. In many places, farmers had enclosed their fields, cultivated their fallow lands, and encroached upon the undivided village lands, as local powers had abolished common grazing and other collective obligations, and all this occurred decades, even centuries, before the new doctrines were formulated.

The genius of the new agronomists was not then to produce *a priori* normative theories (from which previously established science would they have been able to do so?). Rather, both the experiences and needs of the society at that time were expressed in their analyses and in their proposals. By doing this, they outlined a scientific, agronomic, economic, and social analysis of agriculture, its transformations and its place in the economy, and an analysis of the policies and other means that made it possible to influence agricultural development, thus laying the basis of a true political economy of agriculture. These agronomists, who participated in the large intellectual movement of the Enlightenment, thus contributed to preparing the way for reformist and revolutionary policies that subsequently facilitated the blossoming of the market economy and capitalism.

But as necessary as the technical, juridical, economic, political, and ideological conditions of this revolution were to the development of the agricultural revolution, they were not the true causes. At bottom, the agricultural revolution was nothing other than the most effective means, in that time period and in that part of the world, to pursue the development of animal raising and cultivation begun centuries earlier, nothing other than a particular moment of this vast economic movement of capital accumulation and increase in production, trade, and population, whose "cause," if one insists on using this term, was necessarily of an economic nature itself. This "cause," or rather this "driving force" of the agricultural revolution, resides basically in the particular economic and social dynamics of the human species, a species that works, which develops its means and methods of production without respite, which multiplies the domesticated plants and animals it increasingly exploits in order to multiply itself and continually improve the conditions of its existence. Everything that we have considered as "conditions" of development for the agricultural revolution were not the "moving causes" of this development, but the material, organizational, or ideal "means" that millions of people gave themselves, at one moment in their history, consciously or not, to pursue this immense adventure.

The miracle is that this ensemble of new things, which expressed in a confused and sometimes contradictory manner the aspirations of Western society at that time, ultimately found a political outlet that made possible an economic transformation of this magnitude.

## 5. CONCLUSION

In the Neolithic and at the beginning of the age of metals, deforestation of a portion of the temperate forests of the Mediterranean region and Europe, because of too frequent use of slash-and-burn techniques, had reduced the cultivated ecosystems of these areas to a state of extreme degradation.

The agrarian systems based on fallowing and cultivation with the ard in antiquity inherited these degraded ecosystems, composed of a mosaic of fields, some cultivated, some fallow, of grazed meadows and heaths, and of residual forests, the total biomass of which certainly did not exceed 10 to 20 percent of the original biomass. However, lacking strong tools, farmers using the ard for cultivation left aside vast forested expanses, located in areas that were too cold or on soils that were too heavy, too wet or not fertile enough, as well as marshes and other lands subject to inundation.

In the Middle Ages, with cultivation using the plow, these quasi-virgin ecosystems were each cleared in turn, and new cultivated ecosystems developed in the northern half of Europe. With their hay meadows, livestock, and increased harvests, these new ecosystems were wealthier than those of antiquity. In the same time period, the cultivated ecosystems of the southern regions were enriched, thanks to arboriculture, terracing of slopes, and irrigation. But despite these advances in the cultivated biomass, it remains the case that with the large clearings of the Middle Ages, the total biomass of western Europe had once again diminished.

From early antiquity to the beginning of modern times, all the advances in agricultural production and the increase in European population were characterized by a comprehensive drop in the total biomass. During this whole period, the biomass temporarily increased only during periods of crisis and population collapse.

With the first modern agricultural revolution, on the other hand, the population and biomass together increased for the first time in the agrarian history of western Europe. Enriched by new crops and larger harvests, the biomass of the cultivated ecosystem doubled at the very least. Certainly, this biomass was much smaller than that of the original forest, but the annual production of plant biomass in the new ecosystems was nevertheless high. Moreover, it was entirely useful. A large part (fodder and by-products) was consumed by livestock and recycled through manuring and, as a result, the other part, directly consumed by people, was greatly increased. The enlarged possibilities for nutrient exports via the harvests were ultimately accounted for by a higher rate of occupying the soil and by a greater recycling of organic matter, which effectively counterbalanced the losses of minerals through drainage and denitrification.

In most of the industrializing temperate countries, the gains in production obtained through the first agricultural revolution were more rapid than population growth. These gains were characterized first of all by the disappearance of shortages and famines, then by a lasting improvement in diet. Finally, they led to the formation of a growing marketable surplus, capable of supplying the rapidly expanding nonagricultural and urban populations, now able to make up more than one-half of the total population.

The agricultural revolution indeed influenced the growth of the industrial revolution, but there is no doubt that without large-scale industrialization and urbanization, the first agricultural revolution could not have developed so completely. Finally, neither one of these revolutions could have appeared without the profound juridical, social, political, and cultural transformations which put an end to the ancien régime.

Thus was born a new economic and social system whose most striking originality was industrial, agricultural, commercial, and banking capitalism.

However, strictly capitalist enterprises, using wage labor, were far from occupying the whole terrain. Indeed to the contrary, in most industrialized countries, the peasant family economy remained clearly predominant. Even in England and Prussia, it had not disappeared. Moreover, in all the other areas of the economy—craft industries, trade, transport—the non-wage labor family enterprise continued to occupy a major place.

# 9

# The Mechanization
of Animal-Drawn Cultivation
and the Transportation
Revolution: The First World Crisis
of Agricultural Overproduction

No longer do natural forces traverse his field, but economic forces, social forces,
human forces.... From harvest to harvest, his labor remains the same, the price of
his wheat drops almost constantly.... For a half century, in the great plains of
India, Russia, the American West, other men work, at lower cost, and all of this
production, quickly brought closer by the speed of great ships, weighs constantly on him. Thus there are distant peoples and continents appearing suddenly
now from the mist, as hard and massive realities, and it is perhaps the quantity of
wheat sown by a farmer in the American West, the wage distributed to the poor
day workers in India, and even the tariff, tax and money laws promulgated in
every part of the world upon which the price of his wheat, the price of his labor,
his liberty perhaps and his prosperity will depend.

—JEAN JAURÈS, speech to the Chamber of Deputies, 1897

From the sixteenth to the nineteenth century, the development of agrarian systems that do not use fallowing doubled production and doubled the productivity of agricultural labor in most of the temperate regions of Europe and overseas. This new agricultural revolution, the first of modern times, made possible
a large increase in population, a significant improvement in diet, an unprecedented development in industrial, mining and commercial activities, and large-
scale urbanization, all at the same time.

However, as successful as these new systems were, their productivity was limited by the nature of the equipment and means of transport inherited from the

Middle Ages. Certainly, the equipment associated with cultivation based on the plow (scythe, carts, plows) was sufficient to allow, up to a certain point, the development of new agrarian systems. Using this old, and when all is said and done, not very effective equipment, the agricultural calendar was quickly saturated, which, in turn, limited the maximum surface area cultivable by each worker and therefore the labor productivity of the new systems.

Carts, wagons, fodder, and manure made it possible to take complete advantage of the local possibilities for renewing the fertility of the new cultivated ecosystems. But the weakness and high cost of land-based transportation by carts and wagons and maritime transportation by sailboats severely restricted the use of amendments and fertilizers from distant places. At this point in the development of agriculture in the temperate countries, amendments and fertilizers became the most direct means to bring the fertility of cultivated lands to a still higher level. Finally, this weakness in means of transportation also greatly limited the possibilities of long-distance sales of the growing marketable surpluses resulting from the agricultural revolution.

Up to the end of the eighteenth century, industry primarily provided consumer goods. However, at that time it also began to produce new machines and, with the use of the steam engine, industrial mechanization took on great importance. In the nineteenth century, a rapidly growing iron and steel industry produced all sorts of new machines, for industry first, but also for agriculture and transportation. Thus from the first half of the nineteenth century, industry began producing a whole range of new equipment for animal traction such as metallic plows, Brabant plows, and harrows, sowers, reapers, harvesters, threshing machines, as well as all sorts of small farm equipment such as winnowing machines, sorters, chaff-cutters, root-cutters, grinders, churns, creamers, threshing machines with a crank, etc. The use of these machines, which were more effective than the older ones, saved precious time, in particular during the heaviest periods of work in the agricultural calendar. Gradually, they formed a new, comprehensive system of equipment that made it possible to double the surface area per worker and the productivity of labor in systems that exclude fallowing. In the second half of the nineteenth century and at the beginning of the twentieth, this equipment was made in large quantities and widely distributed, in the United States first, then in other colonies of European origin in the temperate regions (Canada, Argentina, Australia, New Zealand, South Africa, etc.) and in Europe.

At the same time, industry revolutionized transcontinental transportation with the development of railroads and transoceanic transportation with the development of steamships. More extensive new territories were opened to European agricultural colonies, and European markets were brought within reach of those colonies' exports. At the same time, the agricultural regions of Europe were opened up by the new means of transportation, thereby making it possible for them to obtain amendments and fertilizers and sell their produce in more distant markets.

In this chapter, we retrace the genesis of this new system of mechanized animal traction. Also, we investigate the reasons it developed as comprehensively as it did. In addition, we explore how the mechanization of terrestrial and maritime transportation, by adding its effects to those of the mechanization of animal traction, led to an enormous growth of marketable surplus and to the first world crisis of agricultural overproduction, beginning at the end of the nineteenth century. Finally, we will indicate how some European countries (United Kingdom, Denmark, France, Germany) reacted to this fierce competition and how they were led to adopt very different commercial policies, as a function of their geographic conditions, their colonial empires and their level of development.

## 1. THE MECHANIZATION OF ANIMAL-DRAWN CULTIVATION AND HARVEST PROCESSING

As we have seen (chapter 8), agrarian systems that exclude the use of fallowing first developed in regions that were already well provided with the equipment associated with cultivation using the plow, without the need for new equipment. In regions where cultivation with the ard predominated, the progress of the new agrarian system often necessitated the adoption of the cart for harvesting hay and the plow for turning over the soil in seeded pastures properly and at the required time.

But beyond these first developments, the calendar of agricultural work became full and there was extra work at peak periods, mainly because the increased development and use of seeded pastures, hoed fodder, food, or industrial plants, and stockbreeding were all so labor intensive. Some advances in the design of agricultural equipment made it possible to ease these constraints a bit. Equipment was reinforced with iron or steel. Animal-drawn hoeing machines and ridgers for row crops were manufactured, as were more powerful plows to plow the soil more deeply. Up to this point, agricultural equipment continued to be manufactured by village cartwrights and blacksmiths, who could fashion "made to order" tools inexpensively, adapted to the draft animals and lands of each farmer, by using, in part, wood and old iron supplied by the clients themselves.

On large and medium-size farms, however, there was never enough time to perform the important work: plowing and sowing, hay making, harvesting, threshing and sorting grain, preparing feed for the livestock, etc. That is why, beginning at the end of the eighteenth century and throughout the nineteenth century, farmers, artisans, and agronomists competed to perfect existing

equipment and manufacture new equipment that was more effective. Among
the new inventions that abounded at that time, only those that made it possible
to go beyond the system's limits were successful and hence could be
profitable. History has remembered only the latter, among the many other
inventions that were abandoned.

## New Mechanical Equipment

As long as factories only copied the agricultural equipment made by artisans and
added innovations that the latter could easily imitate, they could only seize a limited part of the market. In order to expand its share of the market, industry had to
design and develop new equipment that either led to enough saving in labor or
increase in production to justify the replacement of artisanal equipment by this
more costly, factory-made equipment. This was accomplished with the appearance of a whole range of new animal-drawn machines throughout the nineteenth
century: metallic plows, Brabant plows, mechanical sowers, reapers, tedders,
windrowers, side delivery reapers, grain binders, hoeing machines, ridgers, grain
threshers, and all types of manual machines for processing the harvests.

### Equipment for Tilling the Soil and Sowing

*Metallic Plows and Brabant Plows.* The wooden plow inherited from the Middle
Ages was only superficially outfitted with iron. It was one of the first instruments
to be improved. In France, the Dombasle plow, part wood, part iron, equipped
with precise adjustment mechanisms, experienced some success from the beginning of the nineteenth century. In the same time period, the entirely metal plow,
perfected by John Deere and manufactured industrially in the United States, was
sold by the hundreds of thousands. But the most remarkable of the new instruments for working the soil, the one that had the greatest impact in Europe, was
undoubtedly the reversible Brabant plow, a machine that met all the plowing
needs of the new agrarian systems.

The reversible Brabant plow is a machine made entirely of iron, cast iron, and
steel. It is composed of two complete plows (colter, plowshare, and moldboard),
which are symmetrical in relation to a horizontal plane and supported by a front
axle assembly. These two plows turn around an axle formed by a beam in such a
way that one plow turns over the earth on the left on the outward run and the
other turns it over on the right upon returning. This arrangement is particularly
useful on sloping lands, because it makes it possible to plow by turning over the
earth toward the downhill side on the outward run as well as on the return. With a
simple plow, which turns over the earth only on one side, one can plow on the

outward run only and it is necessary to return "empty," so to speak, because the earth cannot be turned over in an uphill direction. Moreover, as the Brabant plow is outfitted with precise and stable mechanisms for adjusting the depth, width, and incline for the plowing, there is no need to guide it by hand, contrary to the ordinary plow. As a result, a single person, guiding the team, can plow with a Brabant plow, while it is generally necessary for two to guide a regular plow. On sloping lands it is possible to virtually reduce the working time of animals by a factor of two and the working time of humans by a factor of four. On flat land, the advantage of the reversible Brabant plow is less clear, because the simple plow can work on the outward run as well as on the return if one carries out a conventional plowing by turning around the first furrow. But the Brabant plow, guided by one person, makes it possible to economize on one worker. Lastly, the Brabant plow, which is quite short and easy to handle, turns better at the end of the field than the older plows. It was able to adapt to slightly longer fields in wooded and hilly regions, where the large plow had barely been used in the past.

*Harrows, Rollers, and Sowers.* Other metallic implements were also made, which added to the time savings in soil preparation or sowing. Several types of articulated harrows and cultivators, which were equipped with adjustments to control how deeply the soil could be worked, replaced the old harrow. Smooth rollers in cast iron and cambridge rollers, of corrugated or ribbed metal, replaced the wooden rollers ringed with iron. Mechanical sowers replaced broadcast sowing, symbol of the old agriculture.

*Hoeing Machines, and Ridgers.* Summer or autumn row crops, which require much labor to maintain the soil, could not have developed fully without the aid of animal-drawn hoeing machines and ridgers, which replaced regular hoes and hand hoes. The mechanical hoeing machine is an instrument fitted with small blades or teeth, mounted on a diamond or triangular-shaped frame of variable width. Pulled by one or two animals, it pulls up or cuts the roots of weeds, loosens and aerates the soil, and slows down the capillary rising and evaporation of water from the soil.

Moreover, some row crops such as the potato need to be ridged. The earth must be raised up around the base of the plants in order to encourage the formation of new roots and prevent roots and tubers turning green. The towed ridger, which makes it possible to carry out this work much faster than with manual hoeing, slightly resembles a small plow, but it consists of a large plowshare and two symmetrical moldboards that throw the earth to both sides of the furrow opened up at an equal distance from two rows of crops.

## Equipment for Hay Making, Harvesting, and Threshing

*Reapers.* Another symbol of the mechanization of animal-drawn cultivation was the reaper equipped with a lateral cutter bar, which revolutionized hay making and harvesting. It replaced the scythe for cutting grass and the scythe fitted with a rake or sickle for cutting grains. The simple reaper yields ten to twenty times more than manual tools. It is composed of a lateral cutter bar more than one meter long that is mounted on a two-wheeled vehicle and can be tipped up. This cutter bar, formed from a blade equipped with sharpened and replaceable teeth, is a sort of animated saw that moves back and forth in a transverse fashion. This movement is produced by a connecting rod driven by the wheels as the machine moves forward and angle transmission gears transmit that movement. The cutter bar is supported and guided by a cutter head equipped with fingers that form a sort of comb and hold grasses and stalks in an upright position thereby facilitating their cutting. The driver of the team can lower or raise this cutter bar from the seat.

Today it is difficult to perceive all the inventiveness, trials, errors, and corrections that were necessary in order to develop this fine piece of engineering. Reapers with circular blades and ones with a frontal cutter bar, which were pushed (like the Gallic harvester) and not drawn by animals, were manufactured. Also manufactured were reapers in which the cutter bar was located behind the animals, which trampled the harvest as a result, and reapers having the blade and teeth in one piece, so that it was necessary to replace the whole thing at the slightest damage, as well as reapers lacking the combs.

But while the reaper made it possible to reduce the time it takes to cut the grass by a factor of ten, the workload peak of hay making was far from disappearing, because tossing and stacking with a pitchfork and rake still took a great deal of time. Thus there was a demand for equipment able to carry out these activities quickly, so manufacturers also developed tedders, rake-stackers, and rake-tossers that supplemented the work of the reapers.

*Loose-Sheaves Grain Reapers.* For cutting grains, the reaper was equipped with a supplementary mechanism allowing the stalks and ears to be gathered and arranged into small bundles called loose sheaves. For this operation, a wooden or canvas apron is located just behind the cutter bar. A second worker, sitting on a second seat, holds a long rake with which he pulls the cut stalks and ears onto this apron as the machine moves forward. When the apron is loaded with a large enough pile of ears to form a loose sheaf, this worker deposits it onto the ground by lowering the apron. The loose sheaves are then assembled by hand, bound into sheaves, put together into small stacks, carted and put into the barn in good weather, then threshed in the weeks following the harvest. Some sophisticated loose sheaves grain reapers were equipped with a rotating winch-operated reel, which accomplishes the same task as the second worker.

*Grain Reapers Binders.* Just as the reaper did not sufficiently cut the peak work-load of hay making, the loose sheaves grain reaper did not reduce the peak work-load of the grain harvest. The manual creation of sheaves still took much time. A harvester capable of mechanically preparing the sheaves was eagerly awaited. This extraordinary machine, clearly more complex than the simple side-delivery reaper from which it derived, was the grain reaper binder. Like the loose sheaves grain reaper, the grain reaper binder is mounted on two wheels that operate all the mechanisms while it moves forward. In addition to the cutter bar, it includes a reel, which mechanically gathers the cut stalks and ears onto the apron, and a mechanism for making the sheaves. This mechanism is composed of a canvas rolling apron and a knotter that ties the string holding the sheaves together.

## Threshers

The thresher joined the loose sheaves grain reapers and grain binders in radically changing the conditions of harvesting grain. Neither threshing in the barn with a flail, as in the southern regions, nor threshing in the open air by animals, which trampled the ears or pulled a roller or sled over the threshing area, as in the northern regions, was especially effective. It took days and days to fill the granaries. On farms that had been able to reduce their labor force or grow larger thanks to the mechanical harvester, it became impossible to devote as much time to the threshing because of the extra labor at the end of the summer and beginning of autumn. The mechanical thresher, which actually offered some relief for this period of heavy work, is a large, complex machine, composed of a thresher with flails or a drum, and various other mechanisms for sorting, winnowing, and discharging the grain, straw, chaff, and fine grains.

The first threshers, brought into service at the end of the nineteenth century, were operated by a hand crank. These threshers were subsequently replaced by larger ones operated by draft animals. Some threshers were even operated by steam engines. Few large farms had the means to acquire, fully employ, and secure a return on this heavy machinery. But the steam-powered thresher benefited small farms thanks to threshing companies that moved their machine from farm to farm, for several days or hours depending on the size of the farm.

Finally, the diffusion of small machines rotated by a hand crank, intended for processing grain harvests (winnowing machine, sorter), preparing feed for live-stock (chaff-cutter, root-cutter, grinding mill) and undertaking the primary

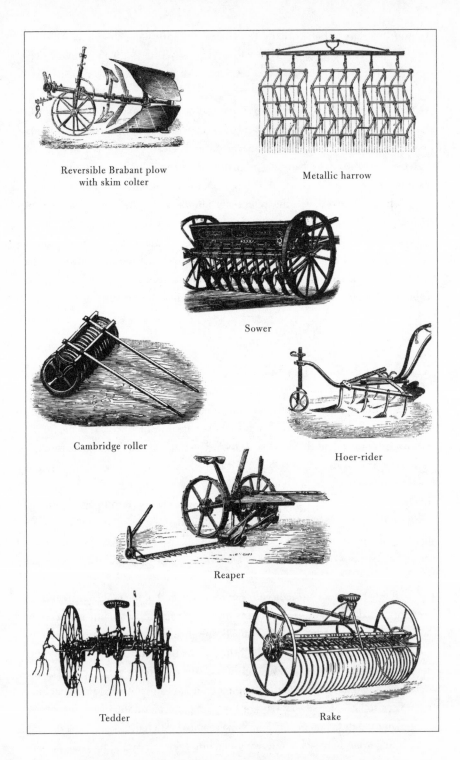

Reversible Brabant plow
with skim colter

Metallic harrow

Sower

Cambridge roller

Hoer-rider

Reaper

Tedder

Rake

*Figure 9.1 Animal-Drawn, Steam-Powered, and Manually-*
*Operated Mechanical Equipment*

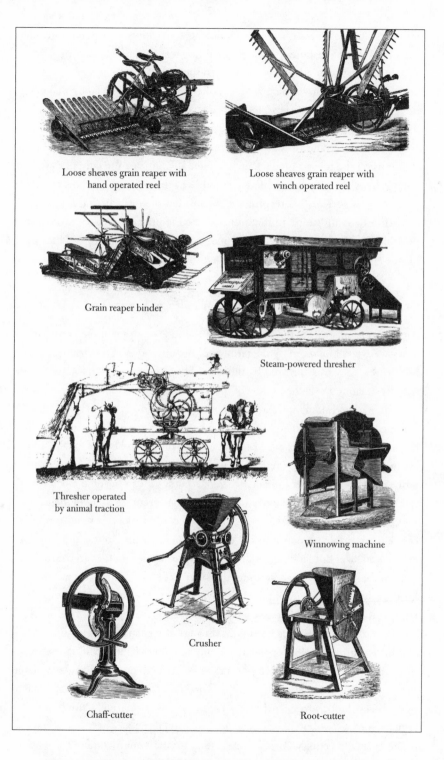

Loose sheaves grain reaper with hand operated reel

Loose sheaves grain reaper with winch operated reel

Grain reaper binder

Steam-powered thresher

Thresher operated by animal traction

Winnowing machine

Crusher

Chaff-cutter

Root-cutter

*Figure 9.1 (continued) Animal-Drawn, Steam-Powered, and Manually-Operated Mechanical Equipment*

transformation of animal products (centrifugal separator, churn, meat grinder) must be seen as a response to the proliferating tasks resulting from the increase and diversification of plant and animal products.

With Brabant plows, harrows, metallic rollers, sowers, hoeing machines, ridgers, reapers, tedders, rakes, loose sheaves grain reapers, grain reaper binders, threshers—in brief, with the mechanization of tilling, sowing, hay making, harvesting, threshing, and various other farm tasks—a new technical system of cultivation using the plow and mechanized manual labor saw the light of day. This comprehensive system, completely suitable to the agrarian systems resulting from the first agricultural revolution, constituted the ultimate improvement of animal-drawn cultivation in the industrialized temperate countries. It made it possible to reduce by half the labor force necessary in agriculture and thus double the cultivated area per worker and the productivity of labor.

## Diffusion of the New Agricultural Equipment

American farmers were the first to adopt the new equipment on a large scale. They were soon followed by the farmers of the other new countries because, in both cases, they were favored by the size of their farms and stimulated by the scarcity of labor. From the middle of the nineteenth century, the large American, Canadian, Australian, and Argentine farms had adopted this equipment. In Europe, notably England and Prussia, the large farms employing wage labor also began to equip themselves in a like manner from the second half of the nineteenth century, and consequently reduced the number of their wage laborers. Peasant farms of more than 10 or 15 hectares were equally able to become mechanized, reducing the seasonal labor force to which they had recourse up until then.

On the other hand, midsize peasant farms could make this new equipment cost effective only on condition of either reducing their family labor force or growing in size. But lacking employment opportunities on the large farms that were in the process of mechanizing, the surplus family labor force had to leave for the city and change occupations. Having no new lands to clear, the midsize farms could grow only on condition that other farms disappeared. This development could only take place gradually, through the enlargement of some farms and the exodus of no-longer-needed family laborers. That is why, in the European countries dominated by a peasant economy with no virgin lands available to clear, the process of acquiring mechanical equipment, the correlative freeing of the peasant labor force, and industrialization was relatively slow. In many regions, the mechanization of animal traction had taken place only in the first half of the twentieth century. Also, it is necessary to point out that these transformations were also slowed down by the First World War and the crisis of the 1930s.

In France, for example, the reversible Brabant plow appeared toward 1850, but there were still only 200,000 for more than 5 million farms in 1900. It was not until the years 1950–55 that the maximum number of 1,450,000 was attained. The reapers and the side-delivery reapers also appeared around 1850, but there were still only 50,000 in 1900, less than one reaper per 100 farms. In most regions, reapers had neither been seen nor heard of and it was only in 1955, with 1,450,000 reapers for 2,200,000 farms, that this equipment attained its maximum distribution.

The grain binders, presented in 1855 by MacCormick at the Universal Exposition in Paris, spread at the same time as the reapers, but remained less numerous. There were 30,000 at the beginning of the century, only 530,000 in 1950–55, or three times less than the reapers. In fact, with a price clearly much higher than the side-delivery reaper and requiring a much more powerful team of draft animals, the grain binder was not suited to farms of fewer than 10 hectares. Small farms remained without them, as well as farms in areas that were too hilly or that already specialized in non-grain products.

Mechanical threshers became widespread from the middle of the nineteenth century. There were some 100,000 hand-crank threshers or threshers operated by draft animals moving in a circular path in 1850 and close to 250,000 at the beginning of the twentieth century, some of which operated with steam engines. Steam-engine threshers reached a maximum of around 220,000 units in 1950–55, after which the combine-harvester gradually replaced them, but there still remained 60,000 of them active in 1970.

In a country such as France, sufficiently representative of continental Europe, the process of acquiring new equipment was not completed in 1950. Brabant plows and reapers were present in only two out of three farms and there were grain binders and mechanical sowers in only one farm out of four. Cartwrights and blacksmiths had become sellers and repairers of this new equipment, but they still continued to manufacture carts, wagons, plows, and other part-wood, part-iron equipment for small farms. In the interwar period, however, and above all after 1945, animal traction began to disappear in favor of motorization. The mechanization of animal-drawn cultivation was not then completely generalized. Nevertheless, this mechanization went far enough to demonstrate that a peasantry thus equipped, representing around one-third of the population, was able to feed a whole nation properly. The nation could then devote the largest part of its resources to non-agricultural activities. Thus, by liberating close to one-half of the labor force previously employed in agriculture during the first half of the twentieth century, the mechanization of animal traction and of some farm work provided the workers necessary to the initial development of the second industrial revolution.

## 2. THE STEAM ENGINE AND THE TRANSPORTATION REVOLUTION

From the end of the nineteenth century, the steam engine began to replace animal power in some agricultural work. On some large estates of the plains, these powerful machines, installed at the end of a field, used a cable to pull heavy plows with multiple plowshares and, in the first half of the twentieth century, operated most of the grain threshers. But these steam engines were indeed too large to be self-powered. They had to be pulled by oxen or horses from one threshing site to another. That is why use of the steam engine in agriculture remained quite limited.

However, by revolutionizing terrestrial and maritime transportation, the steam engine profoundly modified the possibilities of providing agriculture with soil enriching amendments and fertilizers from great distances, as well as the possibilities of selling basic agricultural products in distant markets. These radical changes prompted, simultaneously, the expansion of production in the new countries, some increase in yields, the expansion of competition, and, finally, the first world crisis of agricultural overproduction.

### The Transport of Amendments and Fertilizers

In the old agriculture, the nutrients (nitrogen, phosphorus, potassium, as well as calcium, magnesium, sulfur, and trace elements) necessary for the growth of plants essentially came from the cultivated environment, through solubilization of parent rocks, fixation of atmospheric nitrogen, and mineralization of the soil organic matter. For a long time, it was known how to concentrate the fertility of the ecosystem on cultivated lands by bringing to them mineral or organic matter taken from uncultivated lands (ashes, leaves, heath earth, marine algae, animal manure, etc.). It was also known, since early antiquity, how to improve the texture and structure of the cultivated soil and correct its acidity with amendments (marl, sand, lime, urban sludge). It was also known how to fertilize the cultivated soil by bringing to it fertilizers taken from outside the cultivated ecosystem (the *tells* of the Nile Valley, guano from the Peruvian coast, various quarries).

The use of carts and wagons had long made it possible to carry to some areas a range of amendments and fertilizers from outside sources. But as long as these areas were not served by the railroad, those located very far from quarries and waterways were not able to obtain them. Thus farms at the center of Europe's large crystalline massifs could be supplied with calcareous amendments once there was a wide enough railroad network. Also, the use of nitrates from Chile and guano from Peru, which began in the first half of the nineteenth century and remained limited to the neighboring areas of ports and waterways, grew at the end of the nineteenth century due to steamships, which made it possible to import increased quantities at reduced costs, and the railroads, which made it possible to distribute to most areas of Europe.

The systematic exploitation of phosphate materials (bones from butchery, fish bones, phosphate nodules from some sedimentary sands, phosphates from decalcification of chalk, fossil phosphorites from natural cavities of the Causse plateau) and their use as fertilizer also began at this time. Since these phosphates were not soluble, it was necessary either to grind them finely in order to facilitate their solubilization in the soil or treat them with sulfuric acid (a procedure proposed by Liebig and exploited by Lawes) in order to obtain soluble superphosphates. The first superphosphate factories began to operate in 1843 in England, in 1855 in Germany, in 1865 in the United States, and in 1870 in France, where the first workshop for grinding phosphates had been established in 1856.

Beginning in 1870, the potassium mines of Germany began to be exploited. The fact that potassium is the last great mineral fertilizer to be exploited is not the result of chance. The primary factor limiting agricultural yields in the eighteenth century was, in most soils, nitrogen. The quite noticeable beneficial effect of legumes in the new rotations demonstrates that quite well, as, moreover, does the success of nitrogen fertilizers, which were the first to be marketed. But the increase of yields obtained due to nitrogen fertilizers then encountered a second limiting factor, which was generally a lack of phosphorus. In order to increase yields even more, it was necessary to supply phosphates. Above a certain level of yields, external sources of potassium fertilizer became necessary, at least on some soils.

Thus, at the end of the nineteenth century, mineral fertilizers had arrived on the scene, though their use still remained limited. No more than 25 percent of the agricultural areas and farms in the industrialized countries used them in quantities nowhere near what they are today. All things considered, it is possible to conclude that in 1900 only 10 to 15 percent of the nutrients exported through the harvests came from mineral fertilizers. All the rest still came from the cultivated ecosystem itself. Mineral fertilizers were far from having revolutionized agriculture at this point in time.

## The Opening Up of Different Regions and Specialization

From its beginnings, the agricultural revolution could only fully develop in areas situated close to industries using agricultural raw materials and centers of urban consumption or in areas well served by waterways. These areas had long ago begun to specialize, at least partially, by developing commercial products that were the most advantageous for their needs. Thus the large plains of northern Europe sold their grains. Coastal areas and some large valleys sold their wines and alcoholic beverages (sherry, port, Bordeaux, cognac, wines from the valleys of the Loire, Rhine, Moselle, Saone, and Rhone). Denmark and the Netherlands exported products from dairy and pig-raising operations, while other countries bordering the North Sea sold wool, flax, and hemp. Outlying suburbs produced perishable commodities: fruits, vegetables, and dairy products. Mountainous areas exported their livestock on the hoof, wool, long-lasting cheeses, or, lacking any salable product, young men and women who went down into the lowlands to work as chimney sweeps, builders, peddlers, servants, or nannies.

Even so, diversified production intended principally for supplying the local market was the rule. Beginning in 1850, the railroad opened up the most poorly served European regions one after the other and gave them access, at lower cost, to broader outlets for their products and to all kinds of supplies from other areas. Consequently these areas could develop the practices associated with the agricultural revolution much further, increase their marketable surplus, and specialize in a more advantageous manner. But the culmination of the agricultural revolution and strengthening of specialization also threw onto the market increasingly larger quantities of agricultural commodities.

## The Conquest of New Countries

At the same time, the railroad opened up immense territories in the United States, Canada, Australia, New Zealand, South Africa, North Africa, Argentina, and southern Brazil to European agricultural colonization. In these temperate areas, recently immigrated farmers did not suffer from lack of space, survivals of serfdom, or the heavy burdens associated with land ownership to which the peasants of old Europe were subjected. From the middle of the nineteenth century, they were better equipped, more productive, and generally had lower costs of production than European farmers. Moreover, as their active agricultural population was continually expanded with new arrivals who had few family responsibilities, the new countries had large marketable surpluses in grains, wool, meats, butter, vegetable oil, and so forth. The markets in these countries were unable to absorb these surpluses, a large part of which had to be exported. As soon as steamships with screw propellers made possible a significant reduction in the costs of transoceanic transportation (between 1870 and 1900, the price of transporting American wheat to Europe was reduced by a factor of three), basic agricultural products from these countries arrived in Europe in quantity and at prices lower than the costs of production in many European regions and farms.

## 3. COMPETITION, OVERPRODUCTION, AND CRISIS

Overseas agricultural products, the prices of which continually fell, invaded European markets. Between 1850 and 1900, wheat exports from the United States to Europe increased nearly forty times, going from some 5 million bushels to close to 200, while, at the same time, the price of the imported wheat fell by more than half. The imports of wool from Australia, South Africa, and South America tripled and the prices plummeted, just like grain prices. From 1875, refrigeration techniques made it possible to import frozen meat in growing quantities from America, Australia, and Argentina. The fall in the prices of meat and other perishable animal products nevertheless occurred much later and was not as large as with grains and wool.

These massive imports of basic agricultural commodities caused a huge drop in prices of production in Europe, which led to a fall in agricultural income and land rent, the cessation of investments, the ruin of the most fragile farms in the least productive areas, a decline in production, and an increased exodus from rural areas. In brief, they put whole areas of European agriculture into crisis. Confronted with this new configuration of the international agricultural and food economy, European countries reacted in different ways.

### The Case of the United Kingdom

Some countries chose to favor their industry by importing agricultural raw materials and food products at low prices, which allowed them to maintain low wages and thus low costs. But, by doing this, they chose at the same time to sacrifice a part of their agriculture. Such was the particularly revealing case of the United Kingdom. From the middle of the nineteenth century, English industry began to be threatened by competition from European and American industries. In order to maintain its competitiveness, Parliament decided, in 1846, to abolish the Corn Laws and gradually suppress most of the import taxes on cereals and other agricultural products. The long and difficult political battle that led to the abolition of the Corn Laws contributed greatly to consolidating support for the free trade doctrine in a good portion of public opinion and among the English political class.

From then on, low-priced agricultural imports contributed to a strong increase in industrial activity and employment. On the other hand, although English agriculture was at the time one of the most advanced in the world, these same imports caused a crisis and significant decline in agricultural production. There was a fall of more than one-half in production of wheat, barley, and wool. More than 5 million hectares of plowed lands became natural pastures and heath lands, i.e., *saltus*, which returned in force. There was a drop by more than one-third in agricultural income and land rents, a reduction of nearly 30 percent in the active agricultural population, a drop in wages, and a rural exodus. It was the height of irony and a just reward, some will say, that the landlords and agricultural employers now had to bow down before the peasants, most of whom were descended from those British peasants that the landlords' and employers' ancestors had chased from the lands at the time of the enclosures. Output of potatoes, vegetables, and animal products other than wool was generally better maintained since it was less threatened with competition. However, since the country had lost a large portion of its peasantry, the output of labor-intensive products could not keep up with the increase in population and they were thus imported from continental Europe, notably the Netherlands and Denmark.

Although it was the great pioneer of the agricultural revolution, English agriculture had to evolve toward forms that used more space and fewer workers, just like agriculture in the new countries, whose example it was constrained to follow. But since its territory was much more limited in relation to its population, the United Kingdom sank into a lasting food dependence that, even today, weighs on its balance of payments. Finally, contrary to commonly accepted opinions on this subject, low-priced agricultural imports hardly seem to have benefited British consumers. At the beginning of the twentieth century, a working-class British family had, it appears, a smaller and less diversified consumption of food than a French family with the same income.[1]

## The Case of Denmark

Among the North Sea countries, Denmark had developed the new agriculture in a remarkable manner. Indeed, the reforms at the end of the eighteenth century went much further in this country than in the other countries of Europe. These reforms not only had abolished serfdom and the corvées, instituted the right of enclosure and consolidated peasant property, but a true agrarian reform had given a few hectares to landless peasants. The duration of the leases was for fifty years or life. State credit was introduced and legislation particularly favored the preservation of peasant farms of medium size, by preventing both the concentration and the splitting up of farms.

The early institution, in 1814, of mandatory primary education and the development, from the middle of the century, of secondary and higher agricultural training schools, as well as the organization of cooperatives for credit, sales, and processing contributed to strengthening those farms that combined the output of both vegetable and animal products. Seeded pastures, row crops, and a portion of the cereals were effectively used in the raising of dairy cows, poultry, or pigs, the latter, moreover, increasing the value of the by-products from manufacturing butter and cheese. From the middle of the nineteenth century, Denmark exported cereals, butter, cheese, pork, and other animal products to the United Kingdom and Germany.

When, in the last decades of the nineteenth century, European markets were invaded by low-priced agricultural products, Denmark's problem, itself an agricultural exporter, was not to protect itself but to succeed in maintaining and developing, if possible, its exports. There was no chance that protection measures would resolve this problem. The Danes imported large quantities of cereals at a low price to use more intensively in animal feed, and they replaced a portion of their cereal production by fodder production while still improving the quality of their products. Therefore, since the prices of perishable animal products did not fall by much, Danish agriculture succeeded in increasing its production and exports of butter, pork, eggs, etc., by a considerable amount. Even though this agriculture experienced some real difficulties in the last years of the nineteenth century, the development of animal products and row crops greatly compensated for the decline in wheat and mutton, to such a point that the population living from agriculture actually increased by more than 10 percent between 1880 and 1900. For all these reasons, Danish agriculture was justly considered for decades an exemplary case of the technical and social success of peasant agriculture.

## France and Germany

Most of the European countries would have been incapable of enduring low-cost agricultural imports over a long period of time, as did England, or of benefiting from them, as did Denmark. France and Germany did not have the maritime, colonial, and industrial power that would have made it possible for them to ensure their food security through imports. Moreover, the agricultural population of France and Germany was, at around 50 percent of the total population, high enough that its ruin would have definitely caused massive unemployment and, in the circumstances of the time, rebellions that could have taken a revolutionary turn. On the other hand, since the European market was limited, these countries could not, like Denmark or the Netherlands, take advantage of the new conjuncture by becoming massive exporters of animal products, vegetables, or flowers. It was then economically and socially inevitable that France and Germany took steps, sooner or later, to provide more or less significant protection for their agriculture.

In these two countries, agricultural protectionism appeared rather late. It was inspired by the English Corn Laws as well as by the industrial protectionism implemented by Napoleon, who had strongly limited and taxed the imports of English manufactured products into the continent. In the nineteenth century, and more particularly during the crisis at the end of the century, these two countries took important measures to protect their cereal and animal products. However, at the demand of industrial circles, this protection was not extended to agricultural products serving as raw materials for industry (wool, flax, hemp, oil-producing plants, etc.).

In France and in Germany, like the United Kingdom, unprotected production collapsed. The sheep herd, for example, diminished by half in France and by a factor of five in Germany between 1870 and 1914, because of massive imports of wool. However, protected production of meats, milk products, and cereals were not only maintained but continued to increase, sometimes considerably.

Agricultural protectionism thus allowed France and Germany to limit imports and the fall in prices of basic agricultural commodities, a fall which, in these countries where the agricultural revolution was less advanced than in the United Kingdom, would have led to an agricultural exodus much larger than industry and settler colonies would have been able to absorb. Moreover, contrary to current opinion, the partial protection of French and German agricultures did not prevent them from developing. On the contrary, due to growing outlets because of the population increase and thanks to the preservation of sufficiently remunerative agricultural prices, the first agricultural revolution continued to progress and the mechanization of animal traction expanded rapidly.

This agricultural protectionism was less massive than commonly considered. In Germany, during the last decade of the nineteenth century, it was on several occasions partially challenged under the pressure of industry, which imposed a series of bilateral accords liberalizing trade with various European countries. The accords favored both the exports of manufactured products and the import of low-priced agricultural products. In France, in the middle of the crisis at the end of the nineteenth century, as the world prices of wheat collapsed by nearly half in fifteen years, falling well below the costs of production of most farmers in Europe and even in America, custom duties on imports prevented neither the exertion of pressure from external competition nor a significant fall in internal prices. Between 1880 and 1895, the price of wheat in France fell by more than 20 percent.

It is possible nevertheless to conclude that the progress of French agriculture and the French economy in the first half of the twentieth century was slowed down by protectionism. But that resulted undoubtedly as much from industrial protectionism as from agricultural protectionism. In fact, duties on industrial products were very high (two to three times higher than duties on agricultural imports), which made the industrial products bought by farmers much more expensive and slowed down their investments.

In the last analysis, whether it be by using to their advantage the imports of low-priced cereals, as Denmark and the Netherlands did, or by partially protecting themselves against the collapse of prices, as France and Germany did, the industrialized countries of the European continent were able to limit the damage from an agricultural crisis which was caused mainly by excess production due to the expansion of the railroads and the mechanization of agriculture, as well as the progress of maritime transport.

## The Eastern and Southern Regions of Europe

At the end of the nineteenth century, some relatively underdeveloped peripheral areas of southern and eastern Europe, where latifundia estates dominated, had still not carried out the first agricultural revolution. The old rotations, which included fallowing, were still practiced in those areas, which suffered the full effects of the drop in prices and the reduction in outlets for their products. The economic crisis of peripheral latifundism thus became particularly violent and was transformed into a social and political crisis: hardening of the conditions of labor, lowering of wages, strikes, land occupations, calls for agrarian reform, repression, and rebellion.

In the first decades of the twentieth century, there was a political radicalization of workers' and peasants' movements as well as of the most retrograde landowning and employers' oligarchies in all these countries. This confrontation led to the establishment of "fascist" (Italy, Hungary, Germany, Portugal, Spain) or "Bolshevik" (Russia) totalitarian regimes. Significantly, dictatorial tendencies of this type have continued to appear in the latifundist countries of Latin America, while in countries characterized by medium-size peasant farms or farms employing wage laborers, totalitarian regimes, even if imposed from the outside, have never established themselves. These countries generally remained democratic. The democratic distribution of the land appears then to have been, as in Athens of the sixth century B.C.E., a condition of political democracy. That is so true that, in the aftermath of World War Two, agrarian reforms backed by the Allies in the defeated countries (Japan, Germany, Italy, Hungary, Romania) were notably aimed at reducing the influence of landed oligarchies that had supported the fallen regimes and were indeed conceived as an indispensable first step to the establishment of democracy in these countries.

### 4. CONCLUSION

Beginning in the middle of the nineteenth century, industry in the developed countries mass-produced more effective new agricultural machines as well as new means of transportation that were able to supply fertilizers and amendments to agriculture and ship large quantities of heavy and bulky commodities cheaply. Gradually conquered by transcontinental railroads and linked to Europe by transoceanic steamships, the large white-settler colonies in the temperate regions of the Americas, Australia, New Zealand, and South Africa began agricultural production. Having large amounts of space for very few people, these agricultural colonies rapidly adopted the new mechanical equipment and their low-priced surpluses then began to invade the one great solvent market of the time, the European market, where they squeezed out the marketable surpluses resulting from the first agricultural revolution. The excess supply and the fall in prices that resulted, particularly for products that are easy to preserve, such as cereals, wool, oils, and fats, plunged entire sectors of European agriculture into crisis.

Despite the modernity of its agriculture, the United Kingdom, which observed the principles of free trade, experienced a significant decline in its cereal and wool productions and a new rural exodus. It then sank into long-term food dependence. However, small countries such as Denmark and the Netherlands, having a large and experienced peasantry, took advantage both of the low prices of cereals and of the relatively good performance of prices for rapidly perishable products by specializing in animal products or in the production of vegetables and flowers. Under the shelter of both selective and limited protections, countries such as France and Germany succeeded, to a certain extent, in escaping the crisis, completing the first agricultural revolution and adopting the mechanization of animal-drawn cultivation.

Thus at the end of the nineteenth century, for the first time, industry produced powerful enough means of transportation to open up the Old and New Worlds to competition with each other, made it possible to use amendments on a large scale, and begin to use mineral fertilizers obtained from distant locations. Also for the first time, industry began to produce machines capable of significantly increasing the cultivated area per worker, which, in the old agricultural countries of Europe, would lead to a large reduction in agricultural labor and the disappearance of many small farms.

At the beginning of the twentieth century, the scene was set and the actors were in place for a new agricultural revolution to take off—the second agricultural revolution of modern times.

# 10

# The Second Agricultural Revolution of Modern Times: Motorization, Mechanization, Synthetic Fertilizers, Seed Selection, and Specialization

Scientific precision is first attainable in the most superficial phenomena, when it is a question of counting, of calculating, of feeling, of seeing, when there are directly verifiable quantities. ... These are processes of schematization and abbreviation, a way of seizing multiplicity thanks to an artifice of language—not to "understand," but to name in order to arrive at an agreement....What would one have grasped of music, once one had calculated all that is calculable in it and all that can be abbreviated in formulas?

—Friedrich Nietzsche, *The Will to Power*

In a little more than three centuries, from the sixteenth to the nineteenth, the first agricultural revolution, which was based on the replacement of fallow lands by seeded pastures, row crops, and an increased number of livestock, doubled agricultural productivity in the temperate countries and accompanied the rapid development of the first industrial revolution. Then, at the end of the nineteenth century, industry produced new means of transportation (railroads, steamships) and new animal-drawn mechanical equipment (metal plows, Brabant plows, sowing machines, reapers, grain binders), which led agriculture in these countries to the first "world" crisis of agricultural overproduction in the 1890s.

The second agricultural revolution continued this first phase of mechanization into the twentieth century, but it rested on the development of new means of agricultural production stemming from the second industrial revolution: motorization (internal combustion or electric motors and increasingly powerful motorized tractors and engines); large mechanization (increasingly complex and effective machines); and chemicalization (purified and synthetic fertilizers). It also rested on the selection of plant varieties and domestic animal breeds, both adapted to these new means of industrial production and capable of making a profit on them. In the same way, motorization of truck, rail, boat, and air transportation opened up farms and agricultural regions, making it easier to supply them with fertilizers from distant locations and for them to ship their products to more distant markets.

Freed from the necessity of obtaining supplies of various consumer goods and essential production goods (tractive force, fodder, manure, seeds, reproductive animals, tools), farms began to specialize. Diversified plant and animal production was abandoned in order to produce just a few products intended for sale. Taking into account the physical and economic conditions of each region, as well as the particular means and conditions of production belonging to each farm, these products would be the most advantageous. Thus a vast multiregional agrarian system was formed, composed of complementary specialized regional subsystems (large crop-growing regions, regions with pasturage and stockbreeding for dairy or meat products, grape-growing regions, vegetable- and fruit-growing regions).

The new system resulting from this second agricultural revolution is situated between a set of extractive, mechanical, and chemical industries located upstream from agricultural production which supplies it with its means of production and a set of downstream industries and activities which stock, process, and sell its products. The horizontal division (interregional) and the vertical division (between agricultural production and upstream and downstream activities) of labor specific to this system is coupled with an extensive separation of the tasks of conceptualizing, developing, distributing, and using the new means of production. This distinction between the material and intellectual tasks of production is also reflected in the systems of agricultural education and information, themselves specialized and hierarchical.

The productivity gains resulting from this immense mutation cannot be compared with those of earlier agricultural revolutions. As far as grain production is concerned, for example, with a tenfold increase in yields due to fertilizers and selection and more than a tenfold increase in cultivated area per worker due to motomechanization, the raw productivity of agricultural labor has increased by more than 100 times. Thus, in our day, a working agricultural population reduced to less than 5 percent of the total working population is sufficient to feed the whole population in the industrialized countries better than ever before.

Starting in the first half of the twentieth century, the second agricultural revolution spread to all of the developed countries and some limited sectors of the developing countries in only a few decades after World War Two. It was much more rapid than earlier agricultural revolutions, which had taken several centuries to develop. For all that, however, it did not happen suddenly. On the contrary, it advanced in stages, in step with industry's production of increasingly powerful motomechanical equipment, with the increase in the potential for industrial production of fertilizers, chemical treatments, and livestock feed, with the selection of plant varieties and animal breeds capable of valorizing of the growing quantities of fertilizers and livestock feed, and, finally, with the development of farms capable of acquiring all these new means and of making a profit on them.

It is easy to understand that large capitalist farms employing wage labor would have the necessary capital to acquire these new means of production, and that they could easily enough part with their labor force in order to replace them with machines. On the other hand, it is not as easy to understand how family farms, which had only a few hectares at the beginning of the century, could traverse all the steps of the second agricultural revolution in order to transform themselves into highly capitalized farms, much larger and several dozens of times more productive. Until the recent past, numerous agronomists and economists thought that the new agriculture could only develop in large capitalist or collective units of production. However, it is indeed in the developed countries where peasant family farms predominated, and still predominate, that motomechanized agriculture has triumphed. This vigorous progress of the second agricultural revolution has not been a generally harmonious process of development, however. On the contrary, in a peasant economy, development has been essentially unequal and contradictory. Among the multitude of farms that existed at the beginning of the century in the developed countries, only a tiny minority succeeded in getting through all the steps of this development. At the same time, the large majority of the farms existing at the beginning of the century ultimately ended up in difficulty and disappeared.

The study of the developmental mechanisms of the second agricultural revolution in a peasant economy shows that, at each stage of this development, the only farms that can continue to invest and advance are those that are already sufficiently equipped and are large enough and productive enough to attain an income per worker greater than the market price of unskilled labor. This income level forms the *threshold of capitalization* or *threshold of renewal*. In general, developing farms invest and advance in proportion to how high their income is above this threshold. The development of these farms is thus unequal.

Studying these mechanisms also shows that underequipped and relatively unproductive small farms, whose income per worker is less than this threshold of renewal, can neither invest nor renew their equipment nor remunerate their labor force at market price. In fact, farms that do not renew themselves completely regress. They are in crisis, even if, at the price of heavy sacrifices, they most often happen to survive until the retirement of the head of the farm. After that, having no family or outside successors, these farms are broken up and their lands and other still usable producers' goods are taken over by developing farms.

From the beginning of the twentieth century, step-by-step, the dividing line between the unequal development of some and the crisis and elimination of others has been displaced by increasingly greater levels of capitalization, size, and productivity. The gains in productivity achieved by developing farms have led to a secular tendency of lower agricultural prices, in real terms, and lower incomes for farms that have not adequately invested and developed. In the same way, the threshold of renewal for farms has continually increased, because of productivity gains in industry and resulting higher real wages. Set back by the fall in agricultural prices and the higher threshold of renewal, most farms have been gradually thrown into crisis and have disappeared.

Study of the mechanisms of development of the second agricultural revolution also shows that there exists, in each region, one specialized system of production that is more successful than all the others. This system, which depends on the physical and economic conditions of the region, is precisely the one that most of the developing farms of the region tend to adopt, which leads to a noticeable regional specialization. But regions also exist in which no specialization is economically viable. These regions are characterized by the abandonment of farms and the return of the land to its natural state.

At the end of several decades of agricultural revolution, it is necessary to recognize that government, at the expense of a multitude of farms dispersed among quite different regions, has led the agricultural economy of the developed countries to an accumulation of capital and to an efficient distribution of the means of production, of the activities of cultivation and stockbreeding and even of people. But it is also necessary to recognize the enormous disadvantages to this type of development: large inequalities in earned income between farms and regions; elimination of the majority of farms through impoverishment; large inequalities of agricultural and rural population densities, with an excessive concentration of activities in some regions and the abandonment of other regions entirely; pollution; disequilibria of supply, and demand and vast fluctuations in the prices of agricultural products. This is why, after having applied policies aimed at encouraging the development of the second agricultural revolution, most developed countries also ended up carrying out policies intended to correct some of these disadvantages.

What are the stages and the economic mechanisms of the development of motomechanization, fertilizer and treatment product use, selection, and specialization? How could millions of dispersed peasant farms, guided essentially by prices, achieve such a huge accumulation of capital and such an efficient distribution of means, products, and people? What are the negative aspects of these developments and what policies are likely to remedy them? These are the principal questions to which we try to respond in this chapter.

## 1. MAIN STAGES OF THE DEVELOPMENT OF THE SECOND AGRICULTURAL REVOLUTION

Not long ago, in many European villages one could still meet some old peasant who have worked all their lives under the conditions of the old agricultural system. Every morning, in order to warm up, they often began by drinking a drop of plum gin, or cherry, apple or pear brandy, of their own making, chosen from among those that they still had the good fortune of being able to distill each year. Then came the morning meal, composed of a few slices of homemade bread made from their own grain, baked in a wood-burning oven, which they dipped into a soup of garden vegetables. This was later followed by a snack of the same bread, accompanied by a piece of salted meat or a piece of cheese made on the farm and a glass of local wine. Everything on this menu, except the salt, was the result of the close collaboration of a tiny piece of land, the rain and sunshine that showers on it, and the plants, animals, and people who lived there.

Today, there is a change in scenery. The up-to-date European farmer eats his or her breakfast in the English fashion, the same as his or her American colleague and the Singaporean businessperson: orange juice reconstituted from concentrate imported from California; long-term skimmed milk; all-purpose sandwich bread; Danish butter; graded eggs laid by recently selected "super chickens" fed by the thousands with cassava from Thailand, corn gluten and soycakes from Iowa, dehydrated alfalfa from Champagne and mineral and vitamin supplements, according to a diet planned day to day by a computer "up-to-date" on the prices of raw materials; and "Italian" coffee composed of a strong blend of Ivoirian robusta and Brazilian arabica. In short, the whole world on a plate! The minerals, the sun, the water, and the labor come from the four corners of the world, combined and recombined several times and this in absolutely innumerable proportions.

## The "Old" Agriculture

At the beginning of the twentieth century, in Europe and in the new countries of the temperate regions, systems of diversified crop and stockbreeding without the use of fallowing, which resulted from the first agricultural revolution, were predominant. Certainly, some of the farms had already adopted the new mechanical equipment for animal traction produced by industry (metallic plows, reapers, rakes, tedders, grain binders, threshers), but many farms still used the equipment associated with the earlier system of animal-drawn cultivation, of medieval origin, manufactured by artisans (scythes, wagons, plows). In the Mediterranean regions, the old system of cultivation with the ard was still the dominant mode of agriculture. In most areas, even manual cultivation (spades, hoes, sickles) had still not completely disappeared.

In the still relatively unspecialized systems that did not rely on fallowing, farms produced a large variety of products intended to satisfy directly, through self-consumption, most of the needs of the agricultural population itself. Each farm sought to produce its own grain, potatoes, vegetables, fruit, pigs, poultry, eggs, milk, wine, cider or beer, firewood, etc., and attempted to make its butter, cheese, and salted meats, bake its bread, press its oil, spin and weave its flax, hemp, or wool, and distill its brandy.

In these systems of plant and animal polyproduction, many products and by-products were also intended for self-supplying, that is, for supplying the farm with its own means of production. Each farm itself renewed most of its seeds and reproductive animals, and produced its silage, hay, roots, tubers, fodder grains, litter, manure, lumber, and some of its tools.

Despite the progress in water and rail transportation, most small localities were still protected from the competition of distant regions by the high cost of land transportation via carts and wagons. Through the sale of surpluses, diversified plant and animal production also provided a large supply for local markets.

For all that, even largely self-sufficient farms and regions poorly served by transportation did not live in autarky. They bought iron, salt, special tools, cloth, cheap trinkets, and a few luxury products. They paid taxes, ground rents, and interest on borrowed funds. In return, they contributed to the supplying of cities and other regions by selling their products. To this end, they developed a particular product that was especially advantageous for them, taking into account their physical conditions of production, the conditions for selling their commodities, their equipment, their know-how, and the reputation of their products. But this specialization (wine growing, grain growing, cheese making, etc.) of farms and regions was only partial. Diversified production continued to meet the demands of self-consumption and self-supplying almost everywhere.

The productivity of labor was quite obviously very different from one farm to another. For example, the marketable surplus of grains could vary from a few quintals per worker on farms using manual cultivation to more than 100 per worker on farms using mechanized animal-drawn cultivation.

This brief evocation of an agriculture that was still alive in many areas after World War Two makes it possible to evaluate the road taken over the last few decades.

## "Modern" Agriculture

Today, farms most often specialize in a reduced number of particularly profitable products. They are equipped with heavy tractors and large machines; they require huge amounts of synthetic fertilizers, treatment products, livestock feed, and specially selected plant varieties and animal breeds. These farms sell almost all of their products in multiregional and multinational markets and buy almost all of their means of production. Self-consumption and self-supplying are only a small part of their operation. In large-scale cereal growing, for example, the cultivated area per worker varies from 50 to 200 hectares and the yields range from 50 to 100 quintals per hectare. The gross output per worker oscillates between 2,500 and 20,000 quintals, which is between 25 and 200 times the maximum gross output per worker attainable at the beginning of the century (10 hectares x 10 quintals per hectare = 100 quintals).

How could small and medium-size farms of a few hectares, practicing diversified production using animal traction and largely self-sufficient, be converted in a little more than half a century to large-scale motorized, mechanized, and specialized commodity production? How could they increase their yield in such proportions? As rapid as it was, this immense agricultural revolution was in fact not at all a rapid metamorphosis. Looking at it more closely, it appears as a sequence of gradual transformations that developed one after another, one from the other, in step with successive advances in large mechanical and chemical industry, the selection of domestic plants and animals, and the expansion and specialization of farms. Let us look at the principal stages in these transformations.

## The Stages of Motomechanization

Agricultural motomechanization began to develop in the interwar period in large areas of the European settler colonies established in different temperate regions of the world (United States, Canada, Australia, Argentina) and, to a lesser extent, in the large agricultural areas of Europe. But it is necessary to emphasize that, in 1945, animal traction was still overwhelmingly predominant in most industrialized countries, and motomechanization was deployed in all of these countries only after World War Two.

When motorization and mechanization developed depended on the products in question. Grains and other major crops (rapeseed, sunflowers, seed legumes) were the first to use tractors and harvester-threshers. These crops always set the tone for this whole movement. They occupied a large part of the arable lands and thereby offered a large outlet for the agricultural machinery industry. Motomechanization then spread to the harvesting of row crops such as beets and potatoes, which, high in water and bulky, are less easy to handle. Subsequently, it spread to milking dairy livestock, harvesting fodder, providing fodder to stabled livestock and the removal of their manure, wine growing, and vegetable and fruit crops.

With grain-cropping, for example, it is possible to distinguish five stages in the process of motomechanization, stages that are conditioned by the increasing power of tractors. The first stage, which we will call *motomechanization I,* is characterized by the replacement of draft animals and a few rare steam tractors with tractors driven by low-power internal combustion engines (10 to 30 horse-power). These tractors were generally attached to preexisting animal-drawn mechanical equipment (Brabant plows, reapers, grain binders) and to old arti-san-made transport equipment (carts, wagons, and other tipcarts), but some-times also to new equipment better adapted to motorized traction. This first wave of motomechanization, which began before the Second World War, spread rapidly at the end of the 1940s and beginning of the 1950s to farms of more than 15 hectares which were capable of buying and making a profit with the use of a tractor. Although not very powerful, these tractors were much faster than ani-mals and, above all, indefatigable. These tractors indeed made it possible to increase the maximum area per worker from ten hectares, using mechanized animal traction, to 20 or 30 hectares, typical of large-scale farming at that time.

The second stage, which we will call motomechanization II, is characterized by the use of medium power tractors (30 to 50 horsepower), generally fitted with lifting mechanisms making it possible to carry some tools, such as the plow, instead of simply pulling them, and with power takeoff, capable of operat-ing some other machines. This new generation of tractors made it possible to use equipment with a working capacity two to three times higher: plows with two plowshares, harrows, sowers, rollers, spreaders, and rake-tedders 3 to 4 meters long, and lateral cutter bars of 2 meters, for example. New equipment, some of which combined several operations, could also be attached to these tractors: harvester-threshers, low-density pickup balers for hay and straw, beet harvester-tippers, potato harvesters, corn pickers, silo loaders. But since these heavy machines were tractor-drawn, their working pieces were often arranged laterally in relation to the tractor and, as a result, their potential remained limit-ed. In Europe, farms motorized in the preceding stage generally adopted moto-mechanization II at the end of the 1950s and in the 1960s. Compared to moto-mechanization I, motomechanization II made it possible to double the farming area per worker, reaching 50 hectares in large-scale farming.

The third stage, which we will call *motomechanization III,* rests on the use of tractors of 50 to 70 horsepower, able to carry three-furrow plows and to pull implements 5 to 6 meters long. It also rests on the use of large, self-propelled combined machines such as harvester-threshers, whose working pieces are arranged frontally and whose width of cut can greatly exceed that of tractor-drawn machines. This third wave of motomechanization developed at the end of the 1960s and in the 1970s. It made possible an increase in the area per worker in large-scale farming to some 70 to 80 hectares.

The fourth stage, or *motomechanization IV*, rests on the use of tractors of 80 to 120 horsepower. Such tractors make it possible both to use four-furrow plows and pull several machines at once, whether those machines are doing identical work or carrying out several complementary operations. They also make possible the use of harvester-threshers whose width of cut attains 5 to 6 meters. Motomechanization IV spread in Europe in the 1970s and 1980s and led to an increase in the number of hectares per worker to more than 100.

The fifth stage, or *motomechanization V*, is characterized by the use of four-wheel drive tractors of more than 120 horsepower and the use of associated equipment that makes it possible, for example, to carry out, in only one run, all the operations involved in preparing the soil and sowing grains. It developed first in the United States and in the other "new" countries, as well as in the large state-owned or cooperative units of production in the USSR and other countries in the East. In the past few years, it has spread to western Europe. This type of motomechanization makes it possible to increase the area per worker in large-scale farming to more than 200 hectares.

From manual cultivation of grains to cultivation with the ard, the plow and the mechanized plow through motomechanizations I, II, III, and IV to moto-mechanization V, the fixed capital per worker increases from around $50 to almost a half-million dollars, and the surface area per worker increases from 1 to more than 200 hectares. In the same way, the average yield of grains, which is on the order of 10 quintals per hectare in a manual agriculture without the use of fertilizers, exceeds 50 quintals per hectare in a mechanized agriculture that uses chemicals. In order to measure the enormous gap in the productivity of labor between these two types of agriculture and take into account the stages that it is necessary to go through to move from one to the other, gross productivities (measured in quintals of grain produced per worker) and the farmed areas per worker corresponding to each of the levels of equipment we have distinguished can be represented on the same graph (*Figure 10.1*).

In the other main types of specialized production, motomechanization advanced through analogous stages. We will take only one other example, that of milking dairy cows, which also went through a succession of increasingly more powerful machines, making possible just as many significant increases in labor productivity. While a farmer can milk by hand a dozen cows twice a day, he can milk double that number with a portable milking can, and he can milk

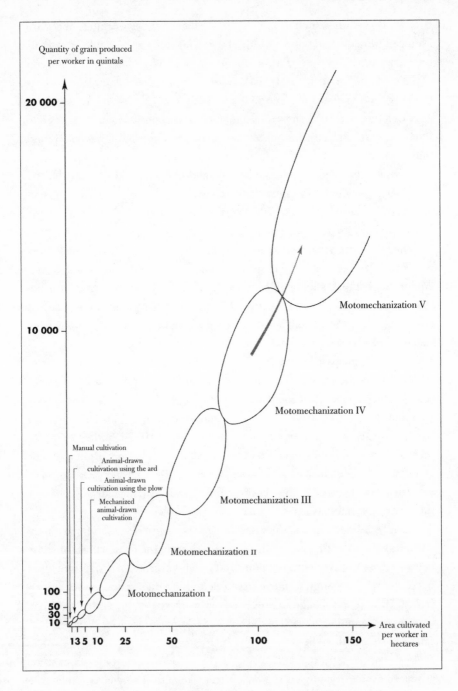

Figure 10.1 *Stages in the Development of Equipment and Motomechanization in Grain Cultivation*

fifty of them in a herringbone milking shed parlor with a milk tank, a hundred with a rotating milking stand (rotolactor) and more than 200 with a completely automated milking shed of the latest design.

## Advances in Agricultural Chemistry and in Selection

While each stage of motomechanization was expressed by a new increase in the farmed area or number of animals raised per worker, advances in agricultural chemistry and selection led to increased yields per hectare or per animal.

### Development of Fertilizer Use

From the nineteenth century, recall, synthetic (or chemical) fertilizers began to be used in Europe. At the beginning of the twentieth century, their use grew in the industrialized countries, but it exploded after the Second World War. While in 1900 the world consumption of the three principal mineral fertilizers, nitrogen (N), phosphoric acid ($P_2O_5$), and potassium ($K_2O$) did not reach 4 million tons of fertilizer units, in 1950 it was a little over 17 million tons, and, at the end of the 1980s, it reached 130 million tons.

As we already saw (chapter 8), this immense growth was made possible by an increase in extractive sources and by the development of industries for processing or synthesizing these fertilizers. At the beginning of the nineteenth century, there were guano from Peru and nitrates from Chile. At the end of the nineteenth century, there were superphosphates obtained by sparging natural phosphates in phosphoric acid; basic slag coming from the manufacture of steel and phosphoric pig iron; and potassium chloride extracted from the salt mines of Germany. At the beginning of the twentieth century, there were calcium cyanamide, obtained through fixation of atmospheric nitrogen in an electric furnace, and the synthesis of ammonia, from which most of the nitrogen fertilizers, such as urea, ammonium sulfate, and ammonium nitrate, were subsequently derived.

The biomass produced by a plant per unit of surface area is a function of the nutrient content of the soil solution (chapter 1). As figure 10.2 shows, beginning with no content, the increase of this mineral content is expressed first by weak increases in biomass production, and then become much greater (more than proportional). Beginning from a particular content (which corresponds to the point of inflexion on the production curve), increases of biomass production begin to slow down (less than proportional), and then they reach a maximum. Finally, with very high contents, which become toxic, production of biomass diminishes.

In practice, in cultivated soil, the initial nutrient content of the soil solution is already high enough, so that one is immediately situated in zone II of the curve in figure 10.2. If, then, one were to trace the curve for crop yields from a particular cultivated area as a function of the amount of fertilizers incorporated into the soil (all fertilizing elements mixed together in proper proportions), one obtains a curve that presents, first, increases of crop yields less than proportional to increases of the amount of fertilizer, then a maximum, and finally a diminishing yield (*Figure 10.3*).

The considerable increase in output per hectare of crops in the course of the last few decades results principally from increasing use of fertilizers, even if the improvement from treatments and from the mechanical work of preparing and maintaining cultivated lands also played a role in this increase. Take grains, for example. At the beginning of the century, yields of 10 quintals per hectare were obtained by manuring but without the use of mineral fertilizers in agricultural systems that did not use fallowing. In the 1950s, there was an average yield of 30 quintals with the use of quantities of fertilizer that included 100 kilograms of nitrogen (N) per hectare, plus phosphoric acid ($P_2O_5$) and potassium ($K_2O$) in the required proportion. Today, yields are approaching 100 quintals per hectare for quantitites of fertilizer that can exceed 200 kilograms of nitrogen per hectare. These amounts compensate not only for the large exports of minerals via the harvests, but also for the losses from leaching into the groundwater, which can represent several dozen kilograms of nitrogen per hectare (*Figure 10.4*).

## Selection of Cultivated Plants

In order to obtain such increases in yields, it is not sufficient to use such large quantities of fertilizers. It is also necessary to have varieties of plants that are capable of absorbing these increased quantities of minerals and making their use profitable. Such was not the case at the beginning of the twentieth century. Cultivated grain populations at that time could not have supported the amounts of nitrogen used today. The selection of more and more demanding and productive plant varieties was necessary in order to absorb the growing quantities of fertilizers produced by industry and make their use profitable. Certainly there was no sudden change from wheat populations capable of producing 20 quintals per hectare to varieties capable of producing more than 100. It was necessary to successively select several varieties with increasing potential, with the process taking place over many stages that conditioned the development of fertilizer use. With wheat, for example, varieties with increasingly shorter straw and rising grain yields were selected. Thus the grain's share of the total aboveground biomass went from 35 percent with varieties from the 1920s to 50 percent with varieties from the 1990s (*Figure 10.5*).

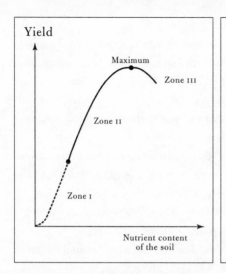

Figure 10.2 Crop Yield as a Function
of the Nutrient Content
of the Soil

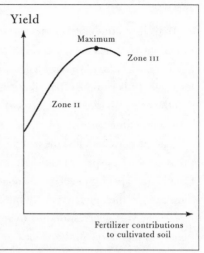

Figure 10.3 Crop Yield as a
Function of Fertilizer Contributions
to Cultivated Soil

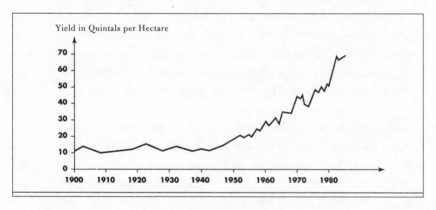

Figure 10.4 Evolution of Wheat Field Crops in France Since the Beginning
of the 20th Century

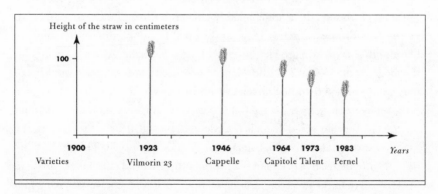

Figure 10.5 Reduction in the Height of Wheat Straw through Varietal Selection

In order to understand the economic mechanism that governed the adoption of varieties that are increasingly productive in relation to the use of growing amounts of fertilizer, we represent on the same graph the expense per hectare of fertilizers (it is assumed that all are mixed together in the proper proportions) and the gross product (yield x price) obtained per hectare for a given variety as a function of the amount of fertilizer used (*Figure 10.6*).

This graph shows that the margin M, that is, the difference between the gross product and the expense of the fertilizer, varies as a function of the quantity of fertilizer Q used per hectare. For a nil quantity of fertilizer, the margin has a value $M_0$. This margin then increases with the quantity of fertilizer used until reaching a maximum $M_{max}$ corresponding to an optimal quantity of fertilizer $Q_0$. Finally, with even higher quantities of fertilizer, the margin diminishes, even if the gross product per hectare continues to increase to its maximum $P_{max}$.

The *optimal* amount of fertilizer $Q_0$, that is, the one that procures the *highest margin* $M_{max}$ should not be confused with the amount of fertilizer that procures the maximum gross profit $P_{max}$. It is generally much smaller. Moreover, it is necessary to emphasize that if the price of fertilizers or of wheat varies, the optimal amount of fertilizer varies also. If the price of fertilizer increases, the optimal amount $Q_0$ and the maximum margin $M_{max}$ decrease, and conversely (*Figure 10.7*). If the price of wheat increases, the optimal amount of fertilizer $Q_0$ and the maximum margin $M_{max}$ increase and conversely (*Figure 10.8*).

Now we consider four varieties of wheat $v^1, v^2, v^3, v^4$, successively selected and increasingly productive. We represent, as before, the curves for the gross product of these four varieties as a function of growing amounts of fertilizer (*Figure 10.9*). This graph shows that the maximum margin attainable for the three varieties $v^1, v^2$, and $v^3$ grows increasingly larger ($M^1_{max} < M^2_{max} < M^3_{max}$). However, since the maximum gross product of variety $v^4$ is higher than all the others, the maximum margin attainable with this variety ($M^4_{max}$) is lower than that obtained with variety $v^3$, because variety $v^4$, more productive but too demanding, does not efficiently convert fertilizer into increased yield. In such conditions, varieties $v^1, v^2$ and $v^3$ will be adopted one after another because they procure a growing profit. On the other hand, variety $v^4$, although the most productive, will not be adopted because the profit that it would procure would be less than that obtained with variety $v^3$.

If someone suggests using the last variety with a high yield, he or she should not then be very surprised that it is not adopted. Indeed, it is the profitability of a variety that determines its diffusion and not its maximum yield. This profitability depends on the relative prices of the product (here wheat) and the inputs (here the fertilizers). In the preceding argument, in order to simplify, we are taking into account the expense of the fertilizer only. But quite obviously, in analyzing the

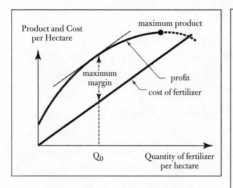

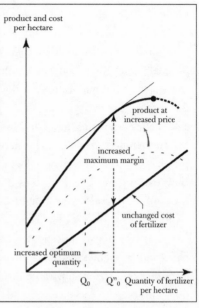

Figure 10.6 Optimal Quantity
of fertilizer (Q₀) per Hectare

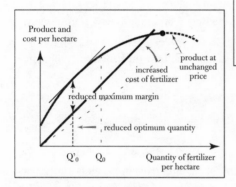

Figure 10.8 Increase in the
Optimal Quantity of Fertilizer
when the Price of the Product
Increases (Q"₀>Q₀)

Figure 10.7 Reduction in the Optimal
Quantity of Fertilizer when the Price
of Fertilizer Increases (Q'₀>Q₀)

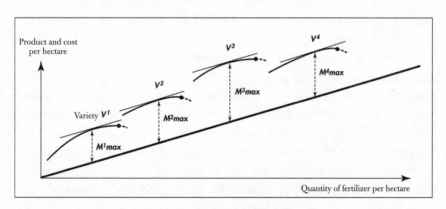

Figure 10.9 Gross Product, Fertilizer Expenses and Margins per Hectare as a
Function of the Quantity of Fertilizer Used for Four Varieties $V^1$, $V^2$, $V^3$, $V^4$

profitability of a variety it is appropriate to consider the combined cost of all the inputs determining the yield (fertilizers, pesticides, seeds).

But the objective of selection is not only to adapt plants to the growing use of fertilizers, it is also to adapt them to the use of new mechanical means. Thus populations of grains cultivated at the beginning of the century, which had relatively spread out maturations and were difficult to shell, were indeed suitable for harvesting with the scythe or the grain-binder and for transporting and storage in sheaves, which preceded the threshing by some time, but they would have been much less suitable for harvesting with the harvester-thresher. For that purpose, it was necessary to select the most homogeneous varieties in terms of their maturation date, and ones that were easier to thresh in the field, at all hours of the day and even at night.

Selection also aims to increase the resistance of crops to their enemies and to economize on use of pesticides. In addition, for many plants, and notably for fruits and vegetables, selection is increasingly a function of new requirements for industry, distribution, and consumers, requirements which go well beyond questions of yield and maturation date and which bear on size, form, color, and the taste qualities of the products.

## The Selection of Domestic Animals

The use of fertilizers and the selection of plants led to such an increase in the production of grains (rich in starch) and of legumes (rich in proteins), as well as other plant products and by-products, that more of these products could be devoted to feeding domestic animals. They served as raw material for a vast industry that manufactured livestock feed with a high nutritional value, called concentrated feed, intended principally for monogastric animals (pigs and poultry) but also for herbivores, notably dairy livestock (cows, ewes, goats), and for fattening livestock.

This large quantity of highly nutritional feed, combined with increases in the production of seeded pastures and other fodder products, made possible not only a strong increase in the number of domestic animals, but also a quantitative and qualitative improvement in their diet. In the same way that it was necessary to select plant varieties capable of valorizing increased mineral nutrition, it was necessary to select animal breeds capable of consuming increasingly nutritious feed rations and of generating a profitable return on the investment. A cow at the beginning of the century, which consumed 15 kilograms of hay per day and produced less than 2,000 liters of milk per year, would not be able to absorb the daily ration of a highly selected dairy cow of today, which produces more than 10,000 liters of milk per year and consumes 5 kilograms of hay and more than 15 kilograms of concentrated feed each day, double what it could ingest without digestive risk.

As with plants, selection of animal breeds also had the objective of adapting them to new mechanical means of production. The milking machine, for example, forced the elimination of cows whose teats were too large, too small, too long, too short, poorly formed—in brief, maladapted to the sizes of the teat cups——as well as the elimination of cows that hold back their milk or contracted diseases of the udder. The milking shed forced the elimination of cows that were too capricious to go along with the discipline of battery milking, and cows whose size and udder height did not conform to the standard of the new installations.

## Animal and Crop Protection

However, such carefully selected and richly fed animals represent substantial fixed capital as well as such a significant potential product heavily burdened with costs that losses of animals resulting from diseases or accidents are less and less tolerable. The risks from diseases are much stronger since the animals are concentrated in large numbers in huge buildings. That explains why rigorous sanitary precautions are taken in order to reduce losses and why, despite their high cost, a panoply of preventative treatments (vaccines) and curative treatments (serums, antibiotics), and even surgery in case of necessity (caesarians, setting of fractures), are called upon.

Annual crops certainly represent less significant fixed capital than animals or perennial plants. However, as a crop develops the expenses of seeds, fertilizers, labor, and fuel accumulate and often end up representing more than half of the expected revenue from the harvest. The margin between this revenue and these costs must then cover a portion of the fixed costs of the farm (amortization of the equipment and buildings, etc.). No losses of even an insignificant part of the harvest can be permitted. In order to limit the losses that could result from an abundance of weeds, from the proliferation of insects, from infestations of fungus, bacteria, or harmful viruses, large quantities of herbicides, insecticides, and other pesticides have to be used.

Finally, with crop losses unaffordable for the reasons outlined above, financial insurance is resorted to as much as possible in order to remedy other risks (hail, frost, various damages).

## Important Aspects of Specialization

While motorization revolutionized the agricultural means of production, it also revolutionized the means of transportation and thus the possibilities for exchange and specialization. Even farms in regions far from waterways and rail-ways could be supplied with all sorts of consumer and producers' goods by motorized road transport. These farms were thus freed from the necessity of practicing diversified production, which was necessary to satisfy the multiple needs of self-consumption and self-supplying. Instead, they could devote the largest part of their resources to a small number of the most advantageous prod-ucts, taking into account ecological conditions, conditions for shipping the products, and the know-how of farmers in the region. But this specialization of farms and of regions did not result only, as one would think, from improvement in the means of transport and trade. It was also greatly influenced by the devel-opment of motorization, agricultural chemistry, and selection.

With the appearance of tractors, farms were freed from the obligation to pro-duce fodder for maintenance of draft animals. Moreover, the use of fertilizers made it possible to increase not only harvested products but also the production of straw, tops, roots, and other crop residues. Beyond a certain level of fertilizer use, crop by-products became abundant enough to return organic material to the soil, thereby making it possible to maintain an acceptable level of soil organic matter in the soil content. Finally, the use of pesticide products freed farms from the old rules of rotation and plot allotment that they formerly had to respect in order to avoid an invasion of weeds, infestation of insects, and multiplication of plant diseases. For example, in order for a rapeseed crop to avoid infestations of parasitic insects (pollen beetles, weevils, flea beetles, aphids), it could only be cultivated on the same parcel of land every five to six years. Today, with the new treatments, rapeseed can be planted every three years on the same parcel.

From that moment on, farms could narrow their specializations. There was a spatial redistribution and reorganization of grain cropping, pastures, stockbreed-ing, wine growing, fruit and vegetable cultivation, etc. It would undoubtedly be tedious to follow in detail these alternating movements of delocalization and relo-calization for all agricultural activities, but it is possible to try to sketch the forma-tion of some important regional specializations.

## The Formation of Grain-Cropping Regions

Freed from the obligation to produce draft animals and manure, from the obli-gation to produce fodder, and from the old rules concerning rotation and plot allotment, farms in relatively flat regions, with fertile, easy-to-work soil, aban-doned fodder production and animal raising to devote themselves to motorized grain cultivation. Alluvial valleys and silt-laden plains and plateaus were cov-ered with simplified rotations consisting entirely of grains (maize-wheat or even continuous), or combining grains with other major crops such as rapeseed, sunflowers, sugar beets, or potatoes. Tractors and fertilizers even made it possible to convert to grain cropping some high lime heaths and grasslands (the squalid *savarts* of Champagne, for example) and heaths on acidic, poorly-drained sandy soils (Atlantic heaths), which were until then devoted to small-scale sheep-raising or softwood plantations.

By specializing in this way, these regions produced growing marketable surpluses quite cheaply. They were then able to export these surpluses to less-favored regions, which from then on would be better provided with such products.

## The Formation of Livestock-Breeding Regions

With the massive appearance of low-priced grains, dried legumes, oils, and potatoes coming from the preceding regions, farms located in hilly, wet regions with heavy or stony soil, which are more difficult to work with motorized and mechanized equipment, farms abandoned increasingly less profitable grain cultivation and turned to pasturage and stock breeding. In this way, the clayey plains with a mild, humid climate on the oceanic facade of northwestern Europe became large dairy areas. For their part, the distant mountain regions took advantage of their dairy cow breeds or mixed breeds that provided both milk and meat (Schwitz, brown Swiss, Siementhal, eastern red piebald, Salers, etc.) and their early tradition of producing long-lasting cheeses (gruyère, comté, cantal, tommes, fourmes, and bleus), in order to specialize in the production of quality dairy and cheese items.

Some wet, mid-elevation mountain regions, which formerly produced draft oxen for the plains, took advantage of their well-built cattle breeds (Limousine, Charolaise, Marchigiana, Aubrac, Pyrenees blonde) to specialize in the production of "lean" young cattle for the meat industry. The low-lying peripheral regions with heavy and poorly drained clay soils (Auxois, Bazois, Charolais) specialized in the fattening and finishing of these same cattle for meat. The dry mountains and calcarious plateaus of the south turned toward sheep-raising for meat and wool, or to dairy ewes for making special cheeses (roquefort, pecorino).

## The Strengthening of Wine-Growing Specialization and the Delocalization of Fruit and Vegetable Production

Farms in the most favorable wine growing regions frequently abandoned grain cultivation as well as animal raising in order to concentrate almost exclusively either on the production of quality wines or the production of large quantities of table wine. As a result, farms in other regions gradually abandoned the production of usually mediocre wine intended for local consumption, something that was still widespread throughout the Mediterranean and central Europe at the beginning of the twentieth century.

Many other specializations were formed as a function of the ecological conditions of each region. Due to rapid, refrigerated transportation, vegetable, fruit, and flower production, formerly located on the periphery of cities, spread to regions with light soil that was easy to work and quick to warm up (valleys of the Loire, Rhine, Garonne, Guadalquivir) and to the coasts with a mild climate (Brittany, Flanders). Production of early and off-season vegetables developed in hotter and sunnier southern regions (Lower Rhone, Valencian *huertas*, Sicily, Andalusia), while open field vegetables intended for the canning industry spread to regions with grain cultivation. Greenhouses and diverse procedures for preservation also made it possible to be freed from climatic constraints to a certain extent.

## Localization of Processing Units and Specialization

The main specializations were formed first as a function of regional ecological conditions. But they were also influenced by local economic conditions, notably by the localization of processing units. That is particularly true for agricultural raw materials that are rich in water, bulky, perishable, and difficult to transport, such as sugar beets, potatoes, vegetables intended for canning, and milk. Such products are necessarily produced in a limited radius around the sugar refineries, starch mills, canneries, and dairy processing plants. In fact, as effective as it is, transportation remains costly and, for this type of product, considerably lowers the price paid to distant producers.

In the mountains, where the collection zones for milk are stretched out and discontinuous, the cost of collecting the milk increases quickly with the size of the dairy processing plants. Only small farm or artisanal units making renowned cheeses can survive, and then on condition that these mountain cheeses are priced high enough to cover the inevitable additional costs of milk production and collection in the mountains. This assumes they are protected from low-cost industrial counterfeits by a label or by an *appellation contrôlée*.

Processing industries for dry agricultural products (grains, legumes, oilcakes, dehydrated fodder) are less sensitive to transportation costs than the preceding products. Nevertheless, large mills and livestock feed industries are not very far from regions of production, ports, and waterways. Also, pig and poultry breeding operations in which feed is brought to the animals are often established in nearby regions in order to benefit from lower feed delivery costs. Thus, in a region where the physical and economic conditions of production are relatively homogeneous, most of the farms tend to adopt the same combinations of production. They even tend to adopt the same equipment and same combination of inputs. In short, they tend to practice systems of production that are enough alike to be classed in the same category. That is why it is possible to speak of regional specialization.

## Relativity of Specialization and Diversity

But if this general movement of regional specialization is indeed real, it is neither as simple nor as absolute as one might think. Nothing prevents some farms from diverging from the dominant specialization in their region, whether because of their size (small farms, for example, interested in carrying on more labor intensive systems of production), their equipment, or their singular know-how, or whether because of particular microlocal physical or economic conditions (topography, land quality, microclimate, special clientele, etc.). As is well-known, particular conditions can determine the remarkable and unique qualities of some products, such as great vintages of wine, for example.

Let us make clear, finally, that specialization is not always as narrow as commonly thought. Monoproduction is, after all, quite rare. In many regions, farms carry on systems of production that can be qualified as mixed, combining, for example, grain cultivation, fodder production, and stockbreeding. There are also regions where, for reasons we will outline later, farms are divided up among two or more specializations having nearly the same profitability, which leaves to the farmers the possibility of choosing, without loss of earnings, their system of production according to their know-how and taste.

Finally, let us not forget that regions exist in which the abandonment of diversified production led to the disappearance of all types of agricultural activity and the return of natural vegetation. What specialization and what choices are then left for the farmers?

## 2. STRUCTURE AND FUNCTIONING OF AGRARIAN SYSTEMS ORIGINATING IN THE SECOND AGRICULTURAL REVOLUTION

Beyond the analysis of the development of motorization, mechanization, mineral fertilizers, selection, and specialization, it is necessary to comprehend the structure as well as the functional and developmental mechanisms of the vast agricultural, industrial, and food system that formed as a result of the second agricultural revolution. This system is one in which the social division of labor has taken on a truly global dimension.

# The New Division of Labor

## *Horizontal Division*

The specialization of farms and regions has led to the separation and regional grouping of different branches of plant and animal production that formerly were found together at the farm or village level. Specialization has given birth to regional agrarian systems, which contribute, each in their own way, to supplying the same national or international market. These specialized regional systems are complementary, interdependent subsystems, in which the landscape itself conveys the horizontal division of labor characteristic of the new multiregional agricultural and food system that has developed.

## *Vertical Division*

Complementary, interdependent subsystems supply agricultural raw materials to an extended network of agricultural industries made up of one, two, and sometimes even three stages of processing. Most of these industries produce consumer food goods. This is the case for flour milling, the dairy processing industry, sugar refineries, breweries, oil factories, etc. Some produce non-food consumer goods. This is the case for the textile industries, leather industries, perfume industry, and pharmaceutical industry. Others produce producers' goods intended for agriculture itself. Of particular importance in this category are the livestock feed industries. These industrial manufactures, some of which, such as textiles, began to develop at the beginning of the first industrial revolution, took on considerable importance in the twentieth century. Most often they were replacing manufactures formerly carried out on farms or in small artisanal units. This is the case, for example, with salted meats, cheeses, butter, canned foods, beer, etc. This tendency toward industrialization is pursued today in winemaking, candy, baking, and ready-made meals.

An analogous evolution also occurs upstream from agricultural production. An extended network of extractive industries and industries manufacturing new means of production (fertilizers, treatment products such as pesticides and antibiotics, motors, machines, fuel, and other supplies) takes the place of the old activities that supplied agriculture, be they artisanal (cartwrights, smiths, saddlers, builders) or agricultural (production of draft animals and manure, manufacture of farm implements).

Upstream as well as downstream, agricultural producers (and rural artisans) found that more and more of their activities and corresponding incomes had disappeared. They were gradually reduced to pursuing simple production of agricultural raw materials.

The vertical division of labor between these industries and agricultural production, properly speaking, has taken on such great importance that upstream and downstream industries represent today more than 10 percent of the national income of the industrialized countries, while agricultural production often represents less than 3 percent. Moreover, many other service activities (commerce, transportation, administration, consulting) are linked to the agricultural sector. If it is true that the working agricultural population represents less than 5 percent of the total working population, it should not be forgotten that agriculture and the set of activities linked to it employ two to three times more people. In other words, the productivity gains resulting from the second agricultural revolution are less important than they appear to be at first sight, because agriculture today has been relieved of a large number of tasks that were incumbent on agriculture yesterday, tasks that have been transferred to industry and the service sector.

## *Work of Design and Work of Execution*

As a result of this vast vertical division of labor, the design of new means of production (machines, fertilizers, treatment products, livestock feed, selected varieties and breeds) is largely outside the purview of agricultural producers. And, to a lesser extent, this is also true of their method of use, the ensuing work procedures and their diffusion. These different functions are henceforth the responsibility of new categories of intellectual workers, who operate in public or private centers for research, education, and popularization, are specialized by field of activity, and possess diverse levels of qualification. The effective use of new means of production also requires, on the part of the agricultural producers themselves, specialization and higher levels of skill, which must constantly be kept up to date. The horizontal and vertical divisions of labor are coupled, then, with an elaborate separation between tasks of designing and disseminating the new means of production and tasks involved in using them.

This division of labor is reflected in the specialized and hierarchical structure of the scientific, technical, and professional agricultural education system. It goes without saying that, in view of the number of specialties, the levels of qualification required and the rapidity of changes in equipment, it is hardly possible to foresee five or ten years in advance what will prove to be necessary for each type of activity and, consequently, train the correct number of qualified persons. In order to meet the extremely varied qualification needs effectively, while continually changing and enlarging them, it is necessary to have a flexible training system, making it possible to meet changing needs. But in order for this system to be effective, it is necessary that the initial training provide, at all levels, a sufficiently broad and high-quality scientific and cultural foundation that makes possible the rapid acquisition of new qualifications beyond the initial specialization.

Scientific and cultural training cannot be neglected, even at the so-called level of execution. In fact, the idea that the work of design and the work of execution should be entirely separate is an outmoded idea. No machine, no product, no procedure can be designed and developed without calling on the acquired experience and the active participation of technicians and practitioners themselves. The proper functioning of the chain of innovation requires that researchers, teachers, and students at all levels know the practice intimately, its conditions, its constraints, and its needs. Otherwise, many new inventions end up being inadequate, are rejected, and become an incredible waste of resources. When all is said and done, science and technique "propose," but it is practice and the economy that "dispose." Indeed, it is the farmers themselves who choose and combine the equipment, inputs, crops, and breeding activities that they use. It is they who develop the most advantageous systems of production, as a function of the particular conditions of the environment and pricing and of the constraints of land, labor, and financing, on their farms. It is precisely this design work that is the most difficult and, naturally, inseparable from practice.

That explains why downward centralized planning (going from the central planner to the agricultural production units) did not produce, in agriculture at least, very good results. (The same can be said, by the way, for systems of normative popularization that, in the colonial and postcolonial countries, pretended, and sometimes still pretend, to dictate to the "independent" producers what their equipment should be, what they should produce and the amount and nature of their investments. Happily, the farmers hardly ever obey injunctions of this type when they are contrary to their well-understood interests.) It is more than difficult, in the course of a process of rapid development, to redistribute equipment, inputs, crops, and breeding operations continually among all the regions and farms of a country in the most advantageous manner. And if, in order to facilitate their own task of central administration, the planners reduce the number of production units as much as possible, to the point of causing the remaining ones to grow beyond all good sense, this gigantism only complicates the management of each unit, rendering that management even more superficial and inadequate. Gigantism, technocratic omnipotence, and inadequate participation of the producers cause waste and shortages of all sorts. However, the technical efficiency of the new mechanical and chemical means of production is so large that it did not prevent some countries with planned economies from establishing an agriculture with great potential.

That said, it remains for us to understand how a multitude of dispersed and independent family farms in the industrialized countries with market economies, guided by their own interest and by their conditions of production and exchange, could carry out the second agricultural revolution. How could they achieve such an effective distribution of equipment, inputs, crops, and animal breeding activities in this manner, which is not itself devoid of disadvantages?

## The Mechanisms of Development of the Second Agricultural Revolution in Peasant Economies

What are the economic mechanisms by which a small fraction of the family production units at the beginning of the twentieth century succeeded in traversing all the stages of the second agricultural revolution, thereby transforming into heavily equipped production units that are dozens of times more productive and use large quantities of industrial inputs? How, then, did most of these farms end up disappearing? By what mechanisms did the farms of entire regions abandon their traditional activities in order to specialize so narrowly? How did some regions come to end all agricultural activities? How, in the interaction of alternating exchanges between regions, is the equilibrium of supply and demand in agricultural and food products achieved (and not always well, we might add)?

In order to answer these questions, it is necessary to analyze the conditions and methods of the economic reproduction of peasant units of production involved in the second agricultural revolution.

### *The Conditions of Economic Renewal for a Peasant Farm*

Let's begin with a family production unit, based on a single worker who also owns the means of production and who receives no positive transfer (subsidy) nor is subject to a negative one (tax, farm rent, interest on borrowed capital). In this particular case, the income of the farmer is equal to the net productivity of the farmer's labor.

*Productivity.* On the strictly economic level, in order for such a production unit to be able to renew itself, it is necessary and sufficient that the net output per worker, that is, the net productivity of labor $P_{nt}$ be above or equal to the income necessary to satisfy the needs of this worker and his family. Certainly, from one farm to another, and even from one period to another, these needs are variable, for objective reasons (a more or less numerous family, a celibate farmer, or one partnered with a person having an outside income) and also for subjective reasons (needs, of course, vary from one person to the next). But it remains the case that, in the long term, the income level considered satisfactory by an agricultural worker necessarily tends toward the income R that this worker would obtain on the labor market (wages and social benefits). Failing that, one day or another, this worker will change occupations or, if that doesn't happen, will probably not be replaced upon retirement.

Let us consider, then, for a given system of production (i.e., a particular combination of means of production and productive activities) and in a given system of prices, the economic variables that determine the productivity of agricultural labor in this type of production unit:

S: the area farmed *per worker*

p: the gross average product *per farmed hectare*, irrespective of the product

$P_{gr}$ = p x S : the gross output *per worker*, that is, the gross productivity of labor

c: the cost of variable inputs

a: the average cost *per hectare* of the amortization and maintenance of equipment and fixed assets, *proportional to the farmed area* (silos and livestock buildings, for example)

m = p - c - a : the margin *per hectare*

M = (p - c - a) x S : the margin *per worker*

A: the annual cost of amortization and maintenance of equipment and fixed assets, *not proportional to the farmed area*, necessary *for one worker* (tractor, cultivation equipment, harvest machine, equipment shed)

$S_{max}$: the maximum area that *one worker* can farm in this system, with this equipment

$P_{nt}$ = M - A = (p - c - a) x S - A, with S ≤ $S_{max}$ ; $P_{nt}$ is the net production of wealth *per worker*, that is, the net productivity of labor

R: the market price of low-skilled labor

*Threshold of Renewal and Threshold of Survival.* If the productivity $P_{nt}$ is greater than the income of labor R at market price, then the production unit has a net investment capacity per worker equal to I = $P_{nt}$ - R thanks to which it can develop, that is, increase its production capacity and its productivity.

If $P_{nt}$ = R then the production unit can renew all its equipment and remunerate its labor force at the market price, but it can make no new investments. The price of labor power R on the market constitutes then a *threshold of renewal* (or threshold of capitalization) for the production unit.

If $P_{nt}$ < R, then the production unit is even less able to make additional net investments, and it cannot even entirely renew its means of production and remunerate its labor power at the market price. Such a farm is in *crisis*, it can only *survive* by making sacrifices on one or the other of these two items. However, the possible sacrifices are not unlimited. In order for the production unit to be able to survive, it is necessary, all the same, that the productivity of labor be greater than a *threshold of survival* or *minimum income* r, below which the farmer can no longer meet his essential needs.

*Graphic Representation.* On a graph (*Figure 10.10*), we show the land area per worker and display the straight lines A (amortization and maintenance of capital necessary for a worker and that does not vary with the area which that worker cultivates), $P_{gr}$ (gross product per worker), M (margin per worker), $P_{nt}$ (net productivity per worker), and $S_{max}$ (maximum area that a worker can farm in the system thus defined).

To the threshold of survival r there corresponds a minimum area of survival $S_{min}$ below which any peasant farm cannot be maintained, even short term. To the threshold of renewal R corresponds an area of renewal $S_R$ above which a production unit can invest and develop, and below which it is in crisis and can only survive for a time, while regressing.

In a system of production in which the combination of cultivation and animal raising, the type of equipment, and the variable inputs are strictly defined, the net productivity of labor is, as a first approximation, a linear function of the area per worker and is represented by a straight line:

$P_{nt} = (p - c - a) \times S - A$

As shown by *figure 10.10*, when the cultivated area for a worker approaches its maximum Smax, the curve of productivity flattens out because it becomes increasingly difficult to accomplish all the productive tasks reasonably well and productivity is affected. (It would flatten out in an analogous way in the theoretical case in which the area per worker approaches zero, because then the cost per hectare of various inputs, the use of which entails an irreducible minimum, would increase in a significant manner.)

Moreover, for the same system and the same level of amortization (A and a) and variable inputs (c), the net productivity of farmers practicing the same system of production varies within a certain margin. In reality, from one farm to another, the technical paths are more or less well conceived and executed, and the environmental conditions, soil in particular, are not strictly identical. Ultimately, in practice, the farms of the same region engaged in the same specialization never practice the exact same system of production. From one farm to another, the combination of crops and animal-raising activities (and thus the gross product) varies somewhat, just as the equipment and variable inputs (and thus the costs) do.

All in all, because of all these variations, the productivity of labor falls between two extreme curves for the same type of production system. One higher (nearly) straight line represents the best productivity of labor available for farms practicing this type of system. A lower (nearly) straight line represents the lowest productivity to which some farms are reduced while practicing the same type of system. The quadrilateral formed by these two straight lines, the straight line of the maximum area per worker $S_{max}$ and the straight line of the threshold

of survival r, delimits the theoretical "space of existence" for farms practicing
this type of system (*Figure 10.11*).

## The Mechanism of Unequal Development of Farms Situated Above the Threshold of Renewal

Up to this point, the graph has been constructed in theoretical terms only. It
can be constructed in practical terms only through investigations focusing on
a systematic sample of production units. To be specific, a sample would be
taken from production units practicing the same type of system, these units
being both numerous and varied enough to reflect the different existing cases.
Such an investigation is useful for evaluating the particular parameters S, p, c,
a, and A that determine the specific productivity of each unit of production.
On a graph analogous to the preceding one (area per worker on the X axis,
productivity on the Y axis), the surveyed farms are each represented by a
point. A scatter of points is thus obtained (*Figure 10.12*). But the envelope of
this scatter of points does not suffice to define the theoretical space of existence of farms practicing this type of system. For that, it is still necessary to
evaluate, through investigations and specific calculations, the threshold of the
maximum area $S_{max}$ attainable with this type of system and the threshold of
survival r, just as the parameters p, c, a, and A determining the two upper and
lower lines of productivity defined above.

*Farms in Development and Farms in Crisis.* Graph 10.12 makes it possible to
distinguish farms in development, located above the threshold of renewal R,
and farms in crisis and temporarily surviving, located below this threshold. In
this last category, there are generally found undersized family farms, with run-
down equipment, without a plan, without a family successor or an outside
buyer, and whose land and other means of production will be acquired by other
farms, either completely or piecemeal, when it no longer carries on farming.
Note that there are no farms which employ wage labor in this category and for
good reason: as soon as the productivity of their wage laborers falls below their
cost—wages and related expenses—farms of this type quickly find themselves in
a position where wages cannot be paid.

Farms in development have an investment potential proportional to the
level of productivity above the threshold of renewal R. One can, moreover, verify that the most productive among these generally have plans, a family successor or, lacking that, an outside buyer. These are the farms that acquire the
spoils from farms that have no successors and can no longer carry on farming.

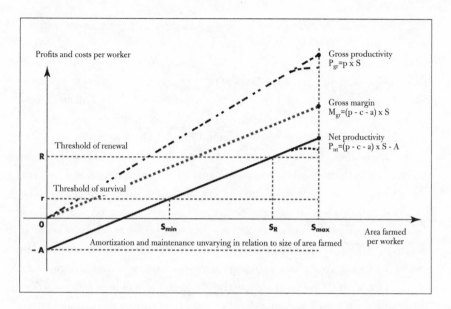

*Figure 10.10 Productivity of Labor as a Function of the Area Farmed per Worker*

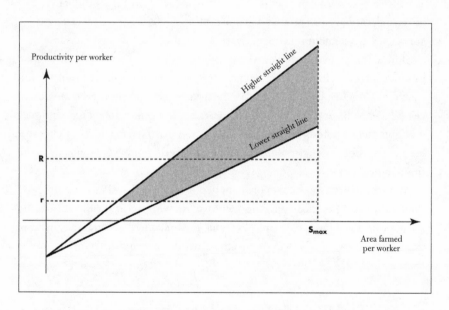

*Figure 10.11 Theoretical Space of Existence for Farms Practicing the Same Production System*

*Different Equipment Levels.* But research carried out at a given moment in the same region also shows that farms that have adopted the same specialization have different levels of equipment. Today in grain cropping there are generally three levels of equipment: a low level of equipment corresponding to a generation of old and obsolete equipment (motomechanization III); a middle level corresponding to an equipment generation not quite as old and with equipment still available for purchase (motomechanization IV); and a high level, still not widespread, corresponding to equipment recently put on the market (motomechanization V). If scatters of points and the quadrilaterals corresponding to these three levels of equipment are plotted on the same graph (*Figure 10.13*), it can be observed that nearly all of the farms practicing motomechanization system III are below the threshold of renewal R, and are thus incapable both of renewing their means of production and remunerating their labor force at the market price. In the end, this category tends to disappear. Some farms practicing motomechanization system IV are partly below threshold R and will be subjected to the same fate as the ones in the preceding category. Farms above this threshold both invest and develop. Finally, the few farms practicing motomechanization system V, which are the best equipped, the largest in size, and the most productive, are, in general, recently formed from the most successful farms in the preceding category. They are, without exception, clearly above threshold R.

This comparative analysis makes it possible to understand how farms that practice a successful system at a given moment (motomechanization IV, for example) are generally those that have the means to adopt a new, even more successful system made possible by the appearance of a generation of more powerful equipment (motomechanization V). On the contrary, farms practicing the least successful system (motomechanization III) do not have the means to go through this new stage.

But this analysis does not explain, in depth, why a series of increasingly productive systems (manual cultivation, cultivation with the ard, cultivation with the animal-drawn plow, mechanized animal-drawn cultivation, motomechanizations I and II) were eliminated one after the other since the beginning of the twentieth century.

## Processes Leading to the Crisis and Elimination of Farms Located Below the Threshold of Renewal

The double process of the development of new systems of production based on rising levels of equipment and the elimination of old systems based on lower levels of equipment has functioned without interruption since the end of the nineteenth century. In order to understand it, we will distinguish three principal stages:

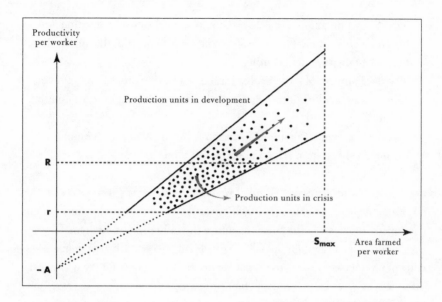

*Figure 10.12 Production Units in Development and Production Units in Crisis*

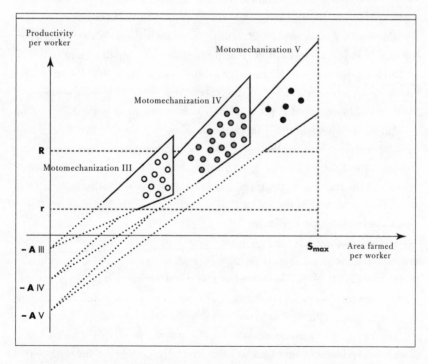

*Figure 10.13 Levels of Mechanization, Area Farmed per Worker, and Productivity*

1. The development of the mechanization of animal-drawn cultivation and the elimination of manual cultivation and cultivation with the ard, in the first half of the twentieth century
2. The development of motomechanizations I and II and the elimination of animal-drawn cultivation with the plow, whether mechanized or not, in the 1960s
3. The development of motomechanizations IV and V and the elimination of motomechanization III in the 1980s and 1990s.

*Productivity Gains for Some, Falling Prices and Lowering of Productivity for Others.* From the first half of the twentieth century, mechanization of animal traction developed sufficiently in the "new" countries and in Europe, on both medium and large farms, to entail a significant fall in agricultural prices and, consequently, a lowering of productivity (calculated at market prices) and income for the non-mechanized farms. This drop in income first pressured family workers to leave these farms, then led to the inability of these farms to renew their means of production, and finally resulted in their dismantlement when farming ceased. Figure 10.14 shows how this drop in prices had repercussions on the productivity of all systems.

*The Raising of the Threshold of Renewal.* At the same time, the development of the second industrial revolution created more and more employment in industrial and service sectors, and the large productivity gains attained in these sectors made possible an increase in real wages. As a result, a significant raising of the threshold of renewal R took place, which contributed to intensifying the agricultural exodus and the disappearance of small farms. Figure 10.15 shows how the raising of the threshold of renewal R had repercussions on the viability of different systems. In an analogous manner, in the 1960s the fall in agricultural prices and the increase in the threshold of renewal led to the disappearance of animal traction. And in the 1980s and 1990s the same mechanisms led to the elimination of motomechanization III.

By displaying these three significant stages in the development of mechanization on the same graph (*Figure 10.16*), it is possible to see how increasingly productive systems developed since the beginning of the century. It is also possible to see how, because of the successive drop in prices, on the one hand, and the progressive increase in the threshold of renewal on the other, manual cultivation, then animal-drawn cultivation systems using the ard, the plow, or the mechanized plow, followed by motomechanization systems I and II, fell below the threshold of renewal and were eliminated one after the other. Finally, it is possible to see that in the 1980s and 1990s, it was the turn of motomechanization III to be eliminated. Figure 10.16 shows that the only farms that remain today went through all the stages of development, one after another, since the beginning of the century, which is to say that they traversed at least one of these stages each generation.

This graph shows as well that within the same system, and for the same area per worker, the differences in productivity remain important. The gains in productivity attainable by a farmer who improves the choice of means of production, the particular combination of crops, and/or animal breeding activities and management of the farm's operations should not be underestimated. Lastly, this graph shows that the productivity of poorly equipped farms (motomechanization I, for example), if it is far below the current threshold of renewal, remains above the threshold of survival. That explains why this type of farm can last so long, above all in a period where jobs are in short supply and the conditions of life outside of agriculture are less attractive.

### Productivity of Labor, Income, and Investment Potential

Recall that the preceding analyses are valid only for farms of a well-defined type: farms with one worker who completely owns the means of production and that have no positive or negative transfer. In this simple case, the income of the farmer is equal to the net productivity of his labor, which makes the calculations for understanding them much easier.

But most often the income of a farmer working alone differs from his net productivity because it is necessary, depending on the case, to add or subtract from it certain transfers. If the farmer has to pay rent for all or part of the land he uses or interest on borrowed capital, the income will be cut back by that amount. Thus, to the inequalities of productivity resulting from unequal access to the means of production are added inequalities of costs resulting from the unequal distribution of the ownership of these same means, inequalities that increase the differences in income. Moreover, if a farmer receives subsidies or pays taxes, income will, consequently, be increased or decreased.

As for the potential of a farmer working alone to self-invest, we have previously considered it as equal to the difference between income and the threshold of renewal R, itself defined as the market price of low-skilled labor, which amounts to supposing that the consumption needs of this farmer and his family are equal to R. Now the consumption needs vary from one family to another (with the number, ages, and lifestyle of the family members) and, moreover, it often happens that a farmer has other sources of family income to invest. That is why the potential for self-investment is calculated beginning with the farmer's income, from which are subtracted the consumption needs of the farmer and dependents and to which are added the possible incomes received by particular members of the family from outside the farm and deposited in the latter's account.

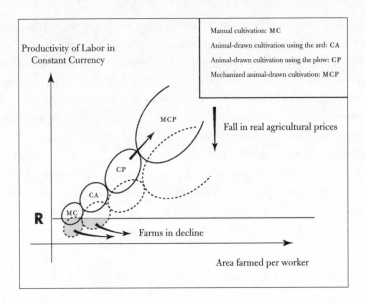

*Figure 10.14  Development of Mechanization and Fall in Real Agricultural
Prices in the First Half of the 20th Century*

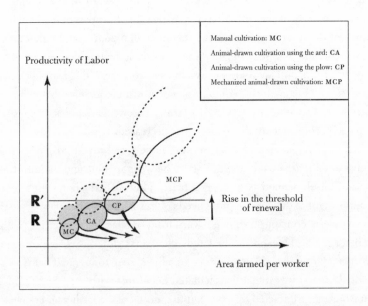

*Figure 10.15 Rise in the Threshold of Renewal in Real Terms
in the First Half of the 20th Century*

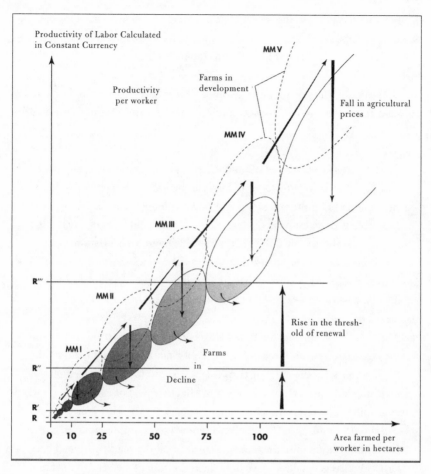

*Figure 10.16 Development of Motomechanization, Productivity Gains,*
*the Fall in Agricultural Prices and the Rise in the Threshold of Renewal,*
*Both in Real Terms, Since the Beginning of the 20th Century*

In the case of a family farm with two or several family members working together, the overall investment potential of the farm is calculated in an analogous manner (net production of all the family workers, plus or minus transfers, less the needs of the family, plus possible outside incomes). And in order to appreciate properly the development possibilities of such a farm, it is necessary to relate its overall investment potential to the number of its family workers. But, having done this, it should be remembered that the investment needs of a farm are not at all proportional to the number of workers. If some equipment must be purchased for each worker (tractors, for example), other large-sized equipment can be bought once for two, three, or four workers (a harvester-thresher or grape-harvesting machine, for example).

In order to calculate the investment potential of a family farm that includes some wage laborers (family or not), the wage expenses incurred by the farm must, naturally, be deducted from the overall net output.

In summary, the analysis of the development of specialized production systems through stages that are increasingly motomechanized and productive shows the following:

1. For a given agricultural specialization, there exist several systems of production, based on increasing levels of motomechanization, which developed in succession during the second agricultural revolution. The maximum yield attainable varies a lot from one system to another and is much greater the more recent the system and higher the level of motomechanization.

2. Inside each of these systems of production, the level of productivity also varies as a function of the farmed area per worker, the level of use of inputs, and the more or less favorable combination of products and means of production.

3. A threshold of renewal for farms always exists that corresponds to the income a farmer would receive on the labor market. Farms on which the income per worker is higher than this threshold have the possibility of investing, adopting a more costly and effective level of equipment, and expanding. The farms on which the income per worker is lower than this threshold cannot invest, not even to renew their equipment, and, at the same time, remunerate the family labor force at the market price. These farms in crisis generally survive until the retirement of the head of the farm.

4. The adoption of new equipment and higher levels of inputs by those farms that invest and the development of increasingly productive systems of production entail, in the long term, a fall in agricultural prices that is characterized by a lower productivity (calculated at market prices) for farms that could not invest. At the same time, the labor productivity gains in the industrial and service sectors make it possible to raise real wages and increase the threshold of renewal for farms. Consequently, the income from working small, underequipped, and relatively unproductive farms gradually drops far below the threshold of renewal and the crisis of these farms is made worse.

5. The labor productivity gains in agriculture and industry have led to the gradual elimination of the least-equipped and least-productive farms since the beginning of the twentieth century. The only farms that remain are those which, from generation to generation, have had the means to adopt the most productive systems of production, one after another.

## The Economic Mechanisms of Specialization of Farms and Regions

In order to explain why, in the same region, the majority of farms are engaged in a particular specialization, as well as why some farms occasionally practice different specializations, it is possible to represent on the same graph the productivities of every specialized system of production practicable in a given region.

### Grain-Growing Regions

Let's consider first of all one of the silt-laden plains of middle Europe that is quite amenable to the use of machinery. The soil and climate are favorable to varied crops and animal-breeding activities and the majority of farms developed, over the course of one century, increasingly successful grain-growing systems of production. Let's represent in a simplified manner the productivities of grain-growing systems practiced today (*Figure 10.17*), as well as the productivities of other practicable systems (which are, moreover, sometimes carried out by a few farms) in this same region: cultivated fodder-dairy cow systems, natural pasture-breeding cow systems, vine-growing and wine-producing systems.

This graph shows that today, in this type of region, the most productive specialization is indeed grain growing based on motomechanization systems IV and V. The former, with a cultivated area per worker between 70 and 120 hectares, has already been in existence for two decades, while the latter, in which the farmed area per worker can exceed 200 hectares, has only begun to develop in the past few years.

The graph also shows that for farms that have not been able to expand and have a farmed area per worker between 25 and 50 hectares, the dairy systems are more productive than the grain-growing systems, which explains the persistence of small farms entirely or partially given over to dairy production in regions of this type. And this graph shows that for farms having only 10 hectares per worker, the productivity of viticulture is higher than that of all the other systems, which are certainly no longer profitable on such small areas. But the productivity of viticulture on a cold, temperate plain, which can only produce wine of poor quality at low prices, is generally located below the threshold of renewal for farms. It is thus no longer practiced and has not been for a long time.

### Viticulture Regions

On the flanks or at the foot of some well-positioned hillsides located at the edge of a silt-laden plain (the mountains of Rheims dominating the plain of Champagne, the Vosgian foothills dominating the plain of Alsace, the Beaujolais hills dominating the plain of the Saone), viticulture that produces wines of high quality is very

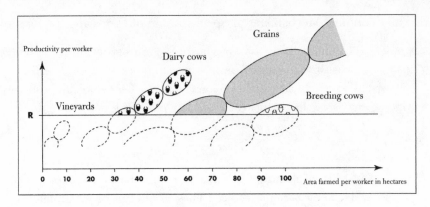

*Figure 10.17 Comparative Productivities of Different Systems of Production on a Silt-Laden Plateau with Average Rainfall*

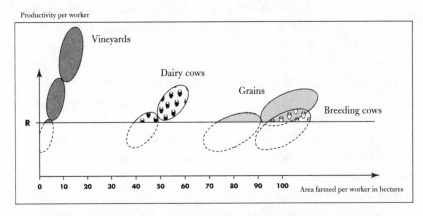

*Figure 10.18 Comparative Productivities of Different Systems of Production on a Sunny Hillside*

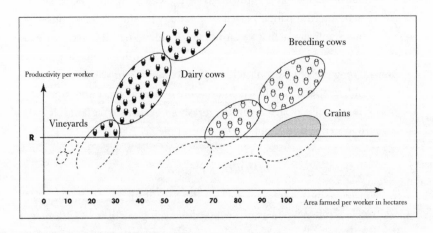

*Figure 10.19 Comparative Productivities of Different Systems of Production in a Humid Valley with Clay Soil*

profitable. Let's consider now one of these hilly regions, Beaujolais, for example, in which the majority of farms produce wine. As before, let's represent the productivity of the viticultural system, as well as that of other systems practicable in the region in a graph (*Figure 10.18*). It shows that the viticultural system is by far the most productive because of the quality and price of the wine. The other systems practicable (dairy cows, breeding cows, grains, conifer plantations) are carried out in this type of region only on lands incapable of producing quality wines. They are, moreover, less profitable than they would be on the plain.

## Dairy Regions

Lastly, let's consider an Atlantic region with heavy, difficult-to-work soil, a mild and humid climate, making it possible to limit winter stabling to a few months, and where improved natural pastures feed two mother cows per hectare. The graph (*Figure 10.19*) shows that, in this type of region, the productivity of dairy systems prevails over all other systems if the producer has a highly productive dairy breed that produces more than 6,000 liters of milk per cow per year.

## Sheep Raising and the Abandonment of Farmland

In addition to regions in which several coexisting systems of production make it possible for farms to stay above the threshold of renewal, there are regions in which only one system makes it possible to exceed this threshold. There are even regions in which no system makes it possible to reach this threshold.

For example, on the calcarious plateaus of the dry southern regions of Europe, characterized by thin soil and sparse natural pastures that barely feed one ewe per hectare, the only common system in which productivity exceeds the threshold of renewal is sheep raising for meat and wool. But for that it is necessary to have nearly 1,000 hectares of relatively flat and open land, where only one shepherd is able to manage a flock of some 1,000 head of small livestock (*Figure 10.20*).

On the other hand, on uneven, dry, scrub-covered mountains, where one shepherd can manage no more than 300 to 400 head, no system is viable (*Figure 10.21*). Regions of this type tend to be abandoned for agricultural purposes, unless the farms have an income linked to a particularly remunerative tradition of making sheep cheese (roquefort, for example).

## Inequalities of Income Between Regions

Comparison of graphs 10.17 to 10.21 shows that the maximum productivity attainable in these different specialized regions varies enormously from one region to another. But now if one compares similar specialized regions, it becomes clear that the maximum attainable productivity in areas practicing the

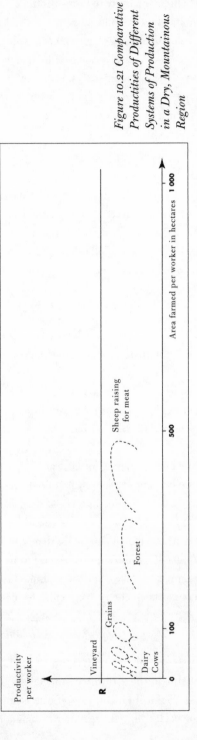

*Figure 10.20 Comparative Productivities of Different Systems of Production On a Dry Plateau*

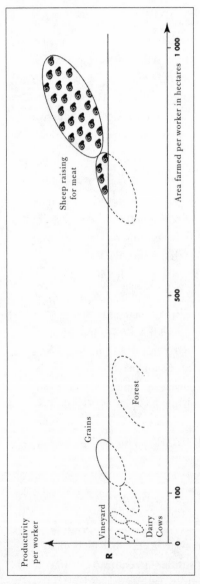

*Figure 10.21 Comparative Productivities of Different Systems of Production in a Dry, Mountainous Region*

same system also varies from one region to another. Let's consider all regions in which the most productive specialized system is the cereal-growing type, and let's represent on the same graph the productivities of these cereal-growing systems in the different regions (*Figure 10.22*). This graph shows that in the most favored regions (silt-laden plains of middle Europe), the maximum productivity is two to three times higher than the threshold of renewal. But in other regions (dry, southern), the maximum productivity does not reach the threshold of renewal and the surviving cereal-growing farms, if any remain, tend to disappear. These differences of productivity result from inequalities in fertility and yield from one region to another, but they also result from inequalities of maximum farmable area per worker, which is conditioned both by topography and how easy or difficult it is to work the soil. The same goes for any specialization practiced in different regions: for the same system, the maximum attainable productivity varies enormously from one region to another.

The comparative analysis of the productivity of different systems of specialized production that developed in different regions in the course of the second agricultural revolution shows the following:

1.  In every region, the productivity of agricultural labor varies from one specialized system of production to another, and there generally exists

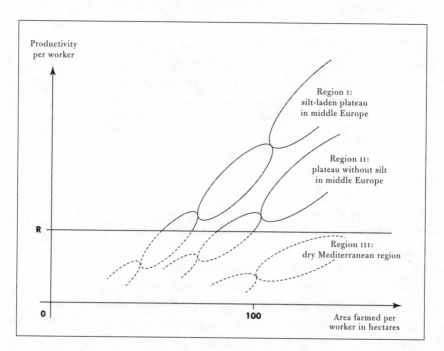

*Figure 10.22 Inequalities in Productivity Between Grain-Growing Regions*

one specialized system more productive than all the others. It is precisely this system that the majority of farms in the region tend to adopt over the long term. Consequently, it is the dominant system that ends up determining the size of the farmed area per worker, the size of the farms, and thus the density of the agricultural population of the region.

2.  For regions having analogous physical and economic characteristics, the most productive specialized systems are of the same nature. Since these analogous characteristics are not, however, identical, the yields, farmed areas per worker, and attainable levels of productivity of these systems vary greatly from one region to another.

3.  For regions having different physical and economic characteristics, the most productive specialized systems of production are different from one another and, of course, so are levels of productivity.

4.  For regions in which several specialized systems of production can attain levels of production higher than the threshold of renewal for farms, although the majority of farms tend to adopt the most productive system, there is a much larger choice. Small farms in particular can persist by practicing systems with a small farmed area per worker (for example, dairy farming, vegetable and fruit cultivation, possibly vineyards).

5.  In the case of regions where only one specialized system (vineyards or extensive sheep raising, for example) exceeds the threshold of renewal, the choice for the farms is clear.

6.  In regions in which no system of production can attain (at today's prices) the threshold of renewal, the long-run fall in agricultural prices and the raising of the threshold of renewal have already led, or will lead in the near future, to the elimination of all farms, the abandonment of agriculture and the expansion of uncultivated land and consequent return of natural vegetation.

From these analyses, one can conclude that a given system of prices (prices of products, equipment, inputs, and labor power) is able to control the adoption of the most productive specialized systems of production by a multitude of scattered agricultural production units, taking into account the physical and economic conditions in which they are found.

But let's go further. If we examine the farms engaged in the drive to be productive, as are all those that intend to go through the next stage in the development of the second agricultural revolution, and if we calculate, by adequate methods (linear programs), the optimum systems of production for these particular farms, we notice that the systems of production actually practiced on these farms are very close to those that could be determined by the calculation.

This is on condition, of course, that this calculation precisely takes into account the characteristics (of environment, size, equipment, labor potential, know-how, financial potential) unique to each farm and that it also takes into account the anticipated yields and prices pertaining to each farmer.[1] The determinative force of a system of prices over the production and investment decisions of a multitude of producers, even very small and scattered ones, as well as their adaptability, is measured in that way.

The art of the agricultural development advisor, then, is not to counsel the producers on the best manner of proceeding, taking into account the conditions in which they exist and the means and information at their disposal. It is, rather, to help them change these conditions (environmental planning, agricultural policies, the market, etc.), to put at their disposal new means of production (tools, varieties, breeds, and other inputs), to help them acquire these items (credit) and also to help them educate and inform themselves.

From this analysis of the mechanisms of development and specialization specific to the second agricultural revolution in peasant agriculture, it is possible to conclude that the long-run fall in real agricultural prices and the raising of the threshold of renewal for farms are capable of leading, in the long term, to the development of increasingly capitalized, specialized, and productive systems by eliminating the oldest and least productive systems, one after the other. The control of peasant agriculture by prices goes well beyond the immediate choices and the medium-term strategies of the producers. It controls the spatial and social accumulation and the distribution of agricultural capital and products, the disappearance of many categories of farms, the exclusion of entire regions, the exodus from agricultural areas, and, finally, the density of the agricultural population in different regions. In brief, it goes so far as to control the spatial distribution of capital, products, and people.

## Economies and Diseconomies of Scale

Up to now, our study of the developmental mechanisms of the second agricultural revolution has consisted of an analysis of the differential development and specialization of peasant farms, and for a good reason: in the developed countries, farms with one or two family workers are by far the most numerous.

However, there also exists in these countries, aside from the majority peasant farms, other categories of farms. There are wage labor farms using, beyond the family labor force, a few wage laborers. There are also organizations of family farms that buy their supplies or sell their products together, or even use agricultural equipment bought in common. Some of these groups even go so far as to merge in order to form a cooperative or associated unit of production. Lastly, in some countries there exist large cooperative, capitalist, or state agricultural enterprises, employing a large number of members or wage laborers. We thus also must analyze, even if only briefly, the effect that the size or scale of these units of production can have on their economic results, their competitiveness, and their development.

Not so long ago, many economists still thought that, in agriculture as in large industry, a strong increase in the size of production units (up to thousands of hectares and hundreds of workers) would make it possible to realize significant "economies of scale," in other words, greatly reduce the cost of production per unit of product (i.e. the unit cost of production). According to this presupposition, these economies of scale should confer on very large farms a strong competitiveness that should lead them, in the end, to triumph over peasant farms, farms employing same wage labor, and small cooperatives. Now, contrary to this prognostication, the large "industrial size" enterprises of agricultural production, whether they be "cooperative" (kolkhozes), capitalist, or state, have encountered many difficulties and even experienced failures. In general, they have lost ground in the countries with planned economies as much as in the countries with market economies.

In fact, very large units of agricultural production have developed and persisted only on lands benefiting from high differential rents (certain vineyards of high quality, for example) or in countries where the near monopoly of land (countries characterized by the latifundia-minifundia system) protected the large estates from competition from peasant agriculture and assured them of a labor force at a very low price. They also have developed in countries where they benefited from all sorts of supports and privileges from the state (former socialist countries, in particular). Finally, industrial-sized units of production have also been formed in various types of stockbreeding operations, where manufactured livestock feed is brought into the enterprise from the outside, which allows the feeding of thousands of animals. But in this case, paradoxically, advanced automation and robotization of livestock breeding operations has made it possible for units of large size to function with only one worker or even with one part-time worker.

Should we conclude from the general lack of success achieved by large agricultural enterprises that there is no, or only a very small, economy of scale in agriculture? Not really. The answer is a little more complicated. Indeed, throughout the development of the second agricultural revolution there has been, recall, a contradictory development of different types of farms. There has been the continual elimination of the smallest and least productive peasant farms. In addition, there has been the step-by-step progress of the most productive medium and large peasant farms, as well as farms employing same wage labor and small organizations of producers. Finally, apart from the exceptions that we just saw, the very large, industrial-size agricultural units of production have experienced difficulties.

In order to make sense of this mixed development, the following hypotheses are essential:

1. In agriculture, the increase in the size of units of production certainly makes it possible to achieve significant economies, but these economies of scale emerge only up to a relatively low threshold of size. This threshold is generally below a few workers (three to seven), which, depending on the current specialized systems of production, corresponds to a farmed area of a few hectares (horticulture), of several dozen hectares (vineyards), of several hundred hectares (large-scale farming), and sometimes several thousand hectares (extensive raising of herbivores).
2. Beyond this threshold, the increase in size of agricultural units of production no longer achieves any significant economies. To the contrary, it gives rise to additional costs and an increase in the total unitary cost of production, which seriously harms the profitability of large units of production.

In order to verify the validity of these hypotheses, it is necessary to analyze more precisely how the different categories of production costs per unit of product vary, as a function of the size of farms practicing the same system of production in identical ecological and economic conditions. In other words, it is necessary to analyze the variation in costs of production per unit as a function of the size of farms, "all other things remaining equal."

In order to do that, let's consider first of all a farm with only one worker who is not using means of production extraneous to the farm and possesses, consequently, the whole range of necessary equipment to carry out the system of production under consideration. In this case, the costs of amortizing the equipment per unit produced (or per hectare) gets smaller and smaller when the farmed area per worker increases. These costs diminish up to the point where the maximum farmable area for only one worker ($S_{max}$) is reached.

This maximum area is reached precisely when the full employment of this worker and/or the full use of one (or several) piece(s) of equipment is attained during one (or several) nondeferable seasonal work period(s). In order to go beyond the maximum area thus attained, it would be necessary to resort to additional labor and/or equipment.

In the case of a farm with one worker, outfitted with all the necessary equipment to implement a given system of production, the cost of the labor and the cost of amortizing the equipment per unit produced, then, are at minimum levels when the area per worker reaches its maximum level. Moreover, as we have seen, when the other so-called proportional costs of production (variable inputs costs, proportional amortizations) remain practically constant per unit produced, the *total* unitary cost of production is at a minimum when the area per worker is at a maximum.

## Real Economies of Scale

However, even when the maximum farmed area per worker is attained, it remains the case that, in a farm with only one worker, most of the necessary equipment is not completely used. That is why in larger farms, whose size (and consequently number of workers) is 2, 3, 4 ..., n times higher, there is generally no need for each worker to have the complete range of equipment necessary to practice the system of production under consideration. In large-scale farming, for example, it is adequate to have one large harvester-thresher for two or three tractor workers, and in viticulture, it is sufficient to have one grape-picking machine for five to ten vineyard workers. There is also other equipment for working the soil, transport, and processing, which can number less than the number of workers.

On a farm with several workers, a surplus of equipment can then be reduced and the unit cost of amortizing equipment is, consequently, much lower than on a farm with only one worker who owns the whole range of necessary equipment. These are economies of fixed capital that constitute the bulk of the economies of scale attainable in agriculture. Moreover, large units of production benefit from discounts on purchasing their supplies and from incentives on their sales, since the quantities bought or sold are large.

However, these economies and commercial advantages are not exclusively reserved for large farms. In fact, farms with one or two workers can also achieve significant economies of fixed capital by participating in organizations for buying and common use of the most underutilized and most costly agricultural equipment, or by resorting to custom agricultural work companies, or even by buying used equipment. They can also benefit from advantageous commercial conditions by participating in bulk buying and selling organizations. Even very small, "part-time" farms, employing less than one permanent worker, can use these means to obtain a sufficiently high productivity per hour of actual work and sufficiently low unit costs of production. That explains why farms of this type are so numerous in the developed countries.

Small farms can, then, limit overequipping and the extra costs that ensue and benefit, to a certain extent, from the commercial advantages bound up with the volume of transactions. It is nevertheless necessary to recognize that grouping together small units of production, whatever purpose it may serve, just like using custom agricultural work and service companies, is not always easy. Such a choice can sometimes lead to some losses, such as a machine unavailable at the requisite time or work poorly executed. But that also happens to the large farms. It quite often remains true that the strong competitiveness of peasant farms is based on underpaid family work.

## Economies of Scale of Limited Scope

It is important to note that economies of fixed capital linked to the size of production units have a limited scope. These economies become insignificant as soon as the maximum area farmable by a small team of workers (three to seven workers depending upon the system in question), organized around a well-balanced combination of all the machines necessary to implement a given system of production, is exceeded. Indeed, in much larger units of production, counting not one but several teams of this type (or a total of several dozen workers), it is, in practice, necessary to have available for each of these teams the same well-balanced combination of all necessary equipment, which means that, beyond several workers, there is no longer any economy of fixed capital in practice.

## Significant Diseconomies of Scale

Moreover, in a production unit based on a small team of workers, there is no need for supervisory personnel who do not directly participate in productive work. In this case, the workers, whether they are family, wage laborers, or members of a cooperative, can perfectly well coordinate their tasks, or work under the direction of a manager or owner who participates in the agricultural work. On the other hand, in large units of production, consisting of several work teams, it is necessary to have administrative and managerial personnel, without which the quantity and quality of the work diminish and squandered inputs and mounting production losses proliferate. The more the unit of production expands the farmed area, the more the hierarchy of personnel not participating directly in agricultural tasks increases (director, department heads, foremen, storekeepers, guards, secretaries, drivers, etc.).

We should also note that when the size of a farm reaches several thousand hectares, the time involved in moving the labor force and equipment weighs heavily on productivity and costs of production. In the same way, in large animal-raising operations, the cost of transporting and manuring animal dung becomes prohibitive. Moreover, when agricultural labor is applied in a standardized manner to excessively large cultivated parcels or stockbreeding establishments, micro-local ecological variations and the particular needs of each animal are not adequately taken into account, which leads to waste and loss of earnings. Thus large agricultural units of production necessarily bear either significant administrative costs or losses or both at once. That is, they basically entail diseconomies of scale.

In the end, one can say that in agriculture significant economies of scale are only attainable up to a modest threshold, corresponding to an autonomous work team of several persons. Beyond this threshold, diseconomies of scale make their appearance, which increase in proportion to the size of the production unit. For most of the agricultural systems of production practiced today, the most favorable size for the economic efficiency of the production unit (whether this unit employs same wage labor, is part of a cooperative, or is a family operation) is commensurate with a small number of workers (between three and seven), it being understood that the farmed area per worker must be close to its possible maximum ($S_{max}$) in the system in question.

Basically, this characteristic of the farm economy is due to the fact that in the current state of development of agricultural machinery most of the existing agricultural equipment can be used by only one worker or by a small team. There does not exist at this time any large machinery (such as blast furnaces and assembly lines in industry) that both requires and controls the work of dozens, indeed hundreds of workers.

This does not mean that farms based on a small team of workers will prevail in the developed countries over the course of the coming decades. As we have seen, the competitiveness and staying power of individual farms having only one worker, full- or part-time, should not be underestimated. From another perspective, it is not impossible that systems of large agricultural machines will develop in the future (for example, remote-controlled automatic machines, traveling cranes carrying equipment and running on automated rails). But it must be noted that the cost of amortizing fixed capital and the cost of agricultural labor per unit produced are already low enough in the most successful farms. It becomes increasingly difficult to reduce them even more. Lastly, never underestimate the waste and losses that can result from the application of standardized labor to an environment and to plant and animal populations whose heterogeneity increases greatly with the size of the operation. As dependent as it is on industry, agriculture itself is not an industry.

### 3. DIFFICULTIES, DISADVANTAGES, AND FAILURES OF THE SECOND AGRICULTURAL REVOLUTION AND ITS AGRICULTURAL POLICIES

The development mechanisms of the second agricultural revolution in a peasant agriculture governed by prices appear to be particularly effective. Nevertheless, as we have already discussed, this type of development is neither easy nor harmonious nor entirely positive. It encounters many difficulties, gives rise to disadvantages and excesses of all sorts and can even lead to actual failures: disequilibria of markets and fluctuations in prices; inequalities between farms and between regions; unequal development of some farms, crisis, poverty, and elimination for others; massive exodus, abandonment of whole regions, and unemployment; attacks on the environment and on the quality of products; genetic degeneration of some domestic species and reduction in the biological diversity of ecosystems, etc. That is why, throughout this vast transformation that has generally been encouraged by the governments of the developed countries, these same governments have also implemented diverse policies aimed at removing the difficulties, limiting the disadvantages, and avoiding or correcting the excesses and failures.

We will not study these policies in detail (their means of action, their effects, or the competing influences that affect them) here. Rather, we will try to show that these policies are social choices that are rooted deeply in the historical, geopolitical, and cultural conditions of the country and that their significance goes well beyond the immediate economic objectives and results often attributed to them.

## Fluctuations in Agricultural Prices and Their Tendency to Fall

The first difficulty in a developing peasant agriculture is that market prices, which guide the choices of the producers at each moment and govern the transformations in agriculture over the long term, are unstable if there is no organized regulation of prices. The quantity and price of an agricultural commodity, on which a producer and a buyer or a group of producers and a group of buyers, brought together in the same market, agree at a given moment, varies from one instant to another. However, it is possible to calculate the average price of a commodity, weighted by the quantities exchanged at each transaction, on one or several markets, for a day, a week, a month, or an entire agricultural year. This is how the average annual price of a commodity for a country and average world price for all the exchanges realized among all the countries in the course of a year are calculated. These annual prices mask, then, a multitude of variations linked to irregularities of supply due to climatic, biological (diseases), or political (wars) setbacks and as a function of the evolution of demand. Even these annual prices themselves also vary from one year to another.

However, these variations are not completely erratic because if one traces the changing curve, in constant currency,[2] of the annual prices of an agricultural commodity over a long period of several dozen years, one can observe large-scale, long-term movements of price. These movements are of two types. First, note that for most of these commodities, there is a more or less regular succession of periods of high prices and periods of low prices, whose rhythm can vary from several years to several decades depending upon the products in question. These oscillations are called cycles or fluctuations. Beyond these fluctuations, there is also, usually, a general tendency for real prices of agricultural commodities to fall (in constant currency), a fall that results, as is well-known, from gains in productivity due to the agricultural revolution.

The curve of wheat prices in the United States since 1860 clearly illustrates these large fluctuations and the tendency of prices to fall (*Figure 10.23*). While the real price of wheat tended to fall by a factor of nearly five in a little more than one century, this price also varied by as much as 100 percent over the course of twenty to thirty years. Certainly, this cycle was disrupted by the two world wars, which caused strong price increases. But it is also clear that the low prices of the 1960s, for example, which led to a reduction in reserves, was followed by a strong rise in prices (accentuated by speculation) in the mid-1970s, then by a new and strong fall in prices in the 1980s.

The magnitude and duration of these fluctuations vary from one product to the next. The pig's cycle, for example, archetype of agricultural economics manuals, has a length of three years, which is governed by the time it takes the size of the herd of mothers to adjust and by the time it takes to raise the piglets and fatten the meat-trade pigs. But such a short and regular cycle does not deter well-informed stockbreeders from managing their breeding in a countercycle, thereby modifying the original cycle. Another example that stands out, the beef cattle cycle lasts seven to eight years. Much longer cycles, lasting several decades, can also affect vineyards and fruit trees, whose lead-in time for production lasts from five to ten years and whose length of time in production generally exceeds twenty years.

## Origin and Consequences of Fluctuations

In order to explain these fluctuations and analyze their effect on the evolution of production, let's place ourselves in a period of relative scarcity and high prices for a particular agricultural commodity on the market. Let's consider all the regions $(R_1, R_2..., R_n)$ that, to a greater or lesser degree, produce this commodity and that take part in supplying the market. Let $P_1, P_2, P_3 ..., P_n$ be the maximum productivities attainable in each of these regions, calculated in the price system of the time and arranged in descending order. Let's represent side by side on the same graph (*Figure 10.24*) these productivities, as well as the quantities $(Q_1, Q_2, Q_3 ..., Q_n)$ of this commodity that each region can produce. For each of these regions, let's represent side by side in the same manner the productivities and quantities produced by each farm. Finally. let's trace on this graph the threshold of renewal R of the farms (valid for every system and every region).

Thus, when a period of high prices begins, a large number of farms, including those in the less advantaged regions, are clearly above the threshold of renewal. It is in the interest of these farms to pursue their productive investments further and they have the means of doing so, but, for diverse reasons (the time to reestablish a financial situation compromised by earlier low prices or to be convinced of the solidity of the new prices), investment decisions are not

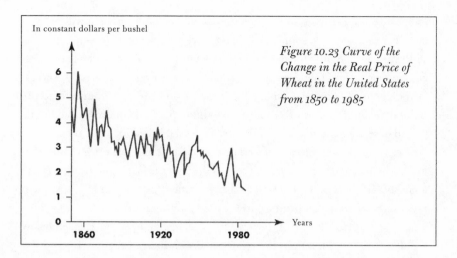

*In constant dollars per bushel*

*Figure 10.23 Curve of the Change in the Real Price of Wheat in the United States from 1850 to 1985*

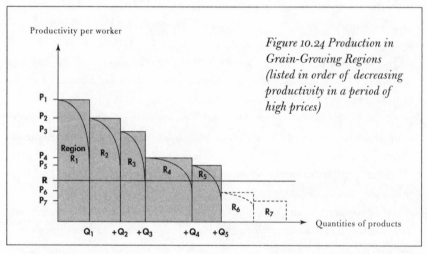

Productivity per worker

*Figure 10.24 Production in Grain-Growing Regions (listed in order of decreasing productivity in a period of high prices)*

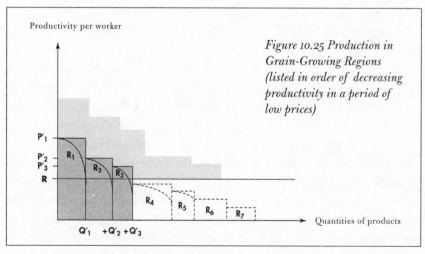

Productivity per worker

*Figure 10.25 Production in Grain-Growing Regions (listed in order of decreasing productivity in a period of low prices)*

immediate and increases in production that will result from them will be even later. For several years, supply remains relatively weak and high prices persist, all the more so because the demand for agricultural and food products is generally not elastic and therefore hardly weakens. However, at the end of several years, the productive investments bear their fruits and they continue, to the point that supply becomes in the end too high, prices drop excessively, and a period of relative abundance and low prices is established for some time.

Let's now look at the case of a relative abundance and low price of the commodity in question. Let's represent on a new graph the maximum productivities attainable ($P'_1$, $P'_2$ ..., $P'_n$) as well as the quantities ($Q'_1$, $Q'_2$ ..., $Q'_n$) that the n regions in question can produce in the new price system (*Figure 10.25*). This graph shows that, because of the low prices, a large number of farms fall below the threshold of renewal. In the relatively disadvantaged regions $R_4$ and $R_5$, since the maximum attainable productivities $P'_4$ and $P'_5$ are now lower than the threshold of renewal, all farms are in crisis. The graph also shows that the quantities produced per region tend to decrease ($Q'_1 < Q_1$, $Q'_2 < Q_2$, $Q'_3 < Q_3$, $Q'_4 < Q_4$). The decline in production is not immediate, because farms in crisis can survive until their equipment is completely worn out or until the head of the farm retires. As a consequence the relative abundance of supply and the low prices persist for several years. Demand barely increases. The low prices discourage investments to the point that supply will become in the end completely insufficient, prices will shoot up excessively, and, since investments and their effects on production will have to wait, a period of relative scarcity and high prices will be established for some time.

In a general manner, the cyclical fluctuations of a commodity's price result from the fact that the reaction of agricultural supply to variations in price (elasticity of supply in relation to prices) is weak in the short or even medium term, while, after a certain delay, it is abrupt and exaggerated. However, demand for basic agricultural and food products is not sensitive to prices, except among consumers with low incomes.

Concretely, this delay in the reaction of agricultural supply to variations in price results, essentially, from the inertia of the production apparatus and the discrepancies between the predicted prices that govern the decisions of farmers and the real movement of these prices. When the prices rise, it takes time to decide to invest, to gather the means to invest, to implement those investments, and to harvest the fruits of those investments. Conversely, when prices fall, it also takes time to decide to stop investing in the products in question and to complete harvesting, at lower cost, the fruits of earlier investments. Moreover, farmers' investment potential follows in large part from the profits realized over the course of preceding years, so that the productive investments they make at a given moment often depend more on profits obtained in a past conjuncture than from profits expected from the coming conjuncture.

The magnitude of fluctuations in agricultural prices is because, in agriculture, the variation in a commodity's supply depends on the geographic expanse and production potential of the regions ($R_4$ and $R_5$ in our example) that go into or out of production on the occasion of price changes. Allowing for exceptions, the production capacity thus mobilized or suspended largely exceeds the variation in supply that would be required to reestablish an "average" equilibrium price, and the resulting overabundance or scarcity causes prices either to plummet or to soar. That is why, once the fluctuations of agricultural prices have begun—and for that to happen all it takes is several bad or, on the contrary, several good harvests—they are forcefully stimulated with each alteration in the level of prices, and tend to get worse rather than not. Moreover, it is not uncommon for the average annual prices of an agricultural commodity to vary from one to four, unless stabilized by an appropriate policy.

There is no doubt that such enormous price fluctuations pose grave problems both for producers and consumers. Periods of high prices are the source of suffering for the most deprived consumers, and since food needs are essential and cannot be postponed, this suffering, which can include shortages and even famine, has no price and it cannot be compensated for by low prices later. That is why food security is not reducible to security of supply as in some other inessential consumer good. Food security is an inalienable human right, and it should be considered as a public categorical imperative.

Conversely, periods of low prices harm producers, whose incomes fall excessively, thereby holding back the necessary development of some, ruining the efforts of others, aggravating the difficulties and the crisis of many. Prolonged low prices cause the untimely ruin of farms that would have remained viable without such wide fluctuations. They accelerate the agricultural exodus in proportions that are unrelated to the labor needs of other sectors and can consequently be the source of unemployment. Marginal farms and regions excluded from production during the periods of low agricultural prices are no longer there to take advantage of the periods of high prices that follow, and it is the surviving farms and regions that profit from their elimination by investing and conquering additional parts of the market. Fluctuations in price aggravate the crisis and accelerate the exclusion of disadvantaged farms and regions. Further, price fluctuations accentuate the concentration of production in an increasingly reduced number of farms and regions.

Markets for basic agricultural commodities are not chaotic nor are they uncertain. They are generally marked by a regular alternation between periods of relative abundance and low prices and periods of relative shortages and high prices, as well as by a tendency for real prices to fall. What is uncertain and unpredictable are the dates at which the next reversal of the tendency will take place, and the magnitude of each price fluctuation. What is chaotic is less the fluctuations themselves than their destructive effects: prices so low that they can destroy entire areas of production or prices so high that they can starve and even kill part of the consuming population.

## Policies of Correcting Price Fluctuations

Markets of agricultural commodities are far from functioning, then, in the most efficient and harmonious manner. That is why public policies and professional initiatives aimed at reducing the price fluctuations and limiting the disadvantages for consumers and producers have been implemented in numerous countries for a long time with some degree of success.

In antiquity, Athens and Rome tried to protect consumers against shortages, speculation, and high prices by prohibiting exports, favoring imports, limiting the accumulation of supplies for speculation, fixing the price of wheat, flour, and bread, and even subsidizing them if needed (see chapter 6).

In medieval England, the laws on grains (the Corn Laws) aimed to limit high prices unfavorable to consumers by combating speculation practiced by merchant guilds and, if necessary, by limiting exports. Beginning in 1660, these laws also aimed to maintain a price level favorable to agricultural producers and landowners, by taxing imports as much as necessary. Beginning in 1815, imports of grains were even prohibited each time their price fell below a level fixed by law. It was no longer a question, then, of measures that aimed to limit price variations, but indeed of measures that were clearly protectionist, aiming to sustain the internal prices of agricultural commodities for the greatest profit of the landowners and employers of English agriculture. But, on the other hand, the high level of food prices was unfavorable to consumers and to industrialists obliged to pay their wage laborers much more in order for them to be able to feed themselves.[3] But, as we have seen in chapter 8, the Corn Laws were abolished in 1846, under pressure from industrial circles.

Beginning from the end of the nineteenth century, most industrialized countries had recourse, to a greater or lesser degree, to price stabilization policies for numerous agricultural products. Management of imports and exports (fixing export/import quotas, taxing), as well as management of supplies, made it possible to maintain prices at levels or reference brackets fixed by the administration or professionals concerned. These means of intervening, effective for reducing price fluctuations, could also be used for protectionist ends.

## Agricultural Protection Policies

A policy of agricultural protection can, then, attempt to maintain internal prices above world prices in order to favor national producers to the detriment of foreign producers. It can also attempt to reduce a country's expenses in foreign-currency, by limiting its imports. But it can also have another objective.

In a country in which agriculture is not very competitive and in which other resources and foreign currency receipts are insufficient to pay a heavy agricultural and food bill, a policy of protection against imports aims above all to avoid the brutal impoverishment and elimination of the small peasantry, as well as the abandonment of entire relatively disadvantaged regions. It aims at avoiding a massive agricultural exodus that would quickly exceed the creation of nonagricultural employment and lead to unemployment and emigration. It also aims at maintaining high enough agricultural incomes to make it possible for at least part of the farming population to invest, to advance, and catch up with the productivity gains of their foreign competitors. Finally, it aims at avoiding a long-lasting disequilibrium in the external balance of payments and in the indebtedness of the country.

Harshly confronted with competition from the new countries since the second half of the nineteenth century, several European countries, such as France and Germany, independently of one another at first, then by grouping together in the European Economic Community, have carried out protectionist policies of this type for decades. These policies have been relatively well planned and adapted. They have been carried out with differing levels of determination, depending upon the countries and time periods in question, but have had unequal results. There is hardly any doubt, however, that they are a part of a mode of economic regulation that has made it possible for these countries to move their agriculture and economy to the first ranks of the developed countries, despite two deadly wars.

In some developed countries, such as Switzerland and Japan, agricultural protectionism has been pushed much further. In order to maintain an adequate level of self-supplying to guarantee their food security in all circumstances, and in order to avoid the human abandonment of entire, relatively unfavored, areas of their territory, these countries have protected themselves to the point of maintaining internal agricultural prices at a level several times higher than international prices. Certainly, such high agricultural prices push the producers to use all inputs, thereby allowing them to increase yields per hectare. But since they contribute to keeping many relatively inefficient small and medium farms in business, they slow down the activity of freeing the lands, expanding farms, and increasing agricultural productivity. Since the income of small farms is relatively good, the nonagricultural sectors are obliged to pay high enough wages to attract the labor force they need, which consequently constrains them, more than in other countries, from investing to increase their productivity.

Undoubtedly the high level of agricultural protection adopted by Switzerland and Japan was at the beginning more a strategic choice, linked to a relatively unfavorable agricultural geography, than it was an economic choice. One could even assume that such a policy was going to slow down the development of these countries. It must be recognized, however, that this high level of agricultural protection has not prevented these two countries from being among the most competitive in the world, while having the highest income per inhabitant and one of the lowest rates of unemployment.

## Speculation and the Food Weapon

When a country has an export monopoly of an essential agricultural commodity, or even a dominant position on the market for such a commodity, it can use its position as a means of managing foreign trade and supplies for speculative purposes, indeed even to exert political pressures on countries that import this essential commodity. In 1973, for example, the United States, which then had a monopoly of exports on soy and its derivatives, profited from the low level of supplies by instituting harsh measures to limit exports of these products. These measures caused a strong increase in prices, which benefited the American balance of trade for several years.

As for the food weapon, this deadly power that one or several grain-exporting countries can have of putting an overly dependent importing country under embargo and condemning it to famine has nothing to do with economic policy. It generally has as its aim to force the government of the importing country to submit to a particular political demand of the exporting country or countries. However, the food weapon has a negative influence on the development of international trade in agricultural commodities. As long as it is a threat, numerous importing countries will continue to protect their food-producing agriculture in order to maintain their ability to supply themselves at a level sufficient to guarantee their food security.

But whatever the reasons for regulatory policies or agricultural price supports, it is difficult for the responsible institutions to fix and make changes to these prices so that they reflect established productivity gains and control investments, without these measures creating significant disequilibria between supply and demand for products or between the elimination of agricultural labor and the creation of nonagricultural employment. Moreover, it has to be recognized that it is difficult for makers of policies to foresee all the effects of their decisions. Consequently, while correcting the worst effects of fluctuations, policies of price regulation can also have unforeseen, and not always desirable, consequences. Certainly, price fluctuations on a free market present many disadvantages, but administered prices, without definite and coherent long-term objectives, present others just as formidable

## Policies for Accelerating the Development of the Second Agricultural Revolution

Just after World War Two, the major concern for the governments of the industrialized countries was to encourage and accelerate an anticipated agricultural revolution that would take part in the improvement of diet and general well-being, would free the most labor possible, which the expanding industrial and service sectors needed, and would offer outlets and raw materials necessary to the rapid development of upstream and downstream industries.[4] At the time, politicians and government officials, agronomists and economists were worried above all about the inertia of farms that did not adopt the new means of production quickly enough. They cursed the survival ability of the small peasantry that continued to "block" a good portion of the lands. Some went so far as to dream of the rapid formation of a large capitalist or collectivist agriculture, much more capable of implementing the advances to come according to them. Others even dreamed of planning the allocation of capital, products, and people.

Most of the industrialized countries of western Europe, impressed by American agriculture, then adopted policies aimed at accelerating the development of the second agricultural revolution.[5] In broad outline, these policies consisted of facilitating the selling of products (organization of transparent markets, setting up interprofessional offices by product) and guaranteeing to producers prices sufficiently stable and remunerative to stimulate production and provide farms likely to develop with self-investment potential. In order to enlarge these farms' investment possibilities, government-subsidized loan systems were also set up. Moreover, to facilitate the acquisition of new means of production, machines, fertilizers, treatment products, buildings, and land planning were not only exempted from taxes but sometimes even subsidized. In addition, laws guaranteeing long-term, regularly renewed leases to tenant farmers, and limiting the level of farm rents greatly influenced the effectiveness of all these measures.

At the same time, national research and development systems were strengthened or created. These systems were composed of central organizations for agronomical research, taken over by specialized technical institutes, themselves relying on an extended network of local centers of experimentation, information, and agricultural popularization. This is in addition to the corresponding hierarchical and specialized system of education.

These policies to provide incentives to development increased the number of farms becoming involved in development and increased the total amount of their investments, to the point that many farms ended up overequipped in relation to their farmed area. As a result, agricultural development policies also sought to increase the flow of vacated lands by accelerating the disappearance of farms that were having difficulty, and direct and facilitate the takeover of these lands by farms that were developing.

In order to increase the supply of lands, various kinds of pensions for farmers were established (the lifelong annuity for leaving the land dates from 1962 in France and from 1972 in the European Economic Community), which made it possible for the heads of farms to bring forward their retirement dates and thereby accelerate the freeing of lands. On the other hand, the so-called antiaccumulation laws prohibited farmers who already had a sufficient farmable area to secure a return on the new equipment and fully employ the family labor force from growing larger. These measures made it possible, then, to reserve available lands for medium-size farms. Small farms whose farmed area was below a minimum threshold generally had no access to certain subsidies, low-interest loans, and freed lands. In particular, young farmers who settled on land areas that were too small (below the "minimum settlement area") did not receive the settlement grant. These arrangements consequently reduced the demand for land on the part of small farmers and facilitated the growth of others. By excluding small farms from development assistance, these measures accelerated their disappearance and the freeing of lands.

Basically, all these laws facilitated the development of medium and large family farms and prevented, to a certain extent, the development of large capitalist farms employing numerous wage laborers. These same laws barely helped the small farmers to develop or even to survive, but they did not force them to disappear suddenly, either. All these measures strengthened the mechanisms of unequal development between the medium and large farms, which were pushed forward, and the small farms, which survived for one generation.

## Disadvantages and Failures of Development

From the end of the 1960s, the disadvantages of this type of development became clear and increasingly less accepted in public opinion. In particular, the inequalities between farms and between regions became too glaring.

## Cumulative Unequal Development and Crisis of Disadvantaged Farms and Regions

At each stage of this unequal development, only those farms whose productivity was above the threshold of renewal invested, and the higher their productivity the more they invested. At each stage of the development, the initial inequalities were magnified by additional inequalities, which were themselves a function of these initial inequalities. Farms and regions favored at the beginning invested and progressed more than the others, and they gradually ended up even more favored. It is not sufficient, then, to speak of unequal development between farms and between regions. Rather, this is about *cumulative unequal development.*

In the course of this process, the farms that were the least well located, the least capitalized, the least appropriately sized, and the least productive one day or another found themselves unable to invest sufficiently to traverse a new stage of development. They were no longer in the running, if one may say so, and because of the tendency of prices to fall, they were relegated below the threshold of renewal. These farms in crisis generally survived until the retirement of the head of the farm, and then they disappeared. Since the beginning of the twentieth century, nine-tenths of the farms disappeared in a little more than three generations in most industrialized countries, and only one family farm out of ten benefited from start to finish from all the stages of the second agricultural revolution. In some disadvantaged regions, the whole agricultural economy was obliterated because all the farms disappeared.

## Inequitable Distribution of the Fruits of Agricultural Labor

In a peasant economy, the income from labor and the fate of the farmer and the farmer's family vary enormously from one region to another and from one farm to another. Essentially, these immense economic and social inequalities result from the quantity of capital and the size and quality of the lands inherited by each farmer. Certainly, the inequality of inherited means (and of good fortune) does not prevent the quantity and quality of family labor and the relevance of the farmer's choices from having a significant effect on the farm's productivity, income, and evolution. The inequalities between farms of the same size from the same region having the same level of capital show this well. But it would be absurd to conclude from that that the results and development of each farm are uniquely the fruit of the labor, enterprising mind, and personal "dynamism" of the farmer. It is just as absurd to believe that the stagnation and regression of small farms result from the laziness and conservatism of the small farmers. For that to be true, all the farmers from regions in crisis and in the process of being abandoned would be unsuccessful, backward people.

The natural or acquired inequalities between farms and regions are more heavily determinative. Whatever their qualities may be, farmers of the dry and uneven mountainous region of southern Europe can have neither the results nor a future comparable to those of farmers in the large silt-laden plains of middle Europe. And whatever his qualities may be, a young farmer taking over a grain-growing family farm of sixty hectares, with the obligation to pay back two-thirds of it to his brothers and sisters, can have neither the income nor the same life nor the same farm to leave to his successor that a neighbor inheriting the entire property of a 300-hectare farm can have. In a peasant economy, the rule governing the distribution of the fruits of labor is closer to "to each according to his/her inheritance" than to "to each according to his/her labor." In these conditions, one understands that the income inequalities from agricultural labor appear, in the eyes of many, as particularly unjust.

It must be emphasized that in an agriculture made up of large capitalist farms employing wage labor the inequalities in labor income that are so significant between farms and regions are inconceivable in peasant agriculture. In this case, agricultural labor is remunerated at the market price and the differences in productivity resulting from more or less advantageous natural or economic conditions do not have repercussions (or not many) on wages. Mainly, they have repercussions, as the classical theorists of economics explain quite well, on farm rents paid to owners, which vary from one region to another (the theory of differential rent of David Ricardo, Karl Marx, and J. H. von Thünen). As for differences in productivity resulting from assets acquired (level of capital, know-how, etc.) by agricultural enterprises, these are essentially encountered in the profit level of these enterprises. Let's note, however, that a tenant farmer having carried out productive investments appropriate to the property he leases does not easily find an equivalent property if his lease is not renewed. As a result, unless there is particular legislation protecting the tenant farmers against such an eventuality, the property owners are in a situation where they can extract from their tenant farmers a portion of the profits resulting from the investments of the latter. An additional rent, which is linked to the quality of the land, is thus added to the differential rent.

On the other hand, for a capitalist entrepreneur to maintain a business, he must not only pay market prices for the wage laborers and the lands that are farmed but also must extract from the capital invested in agriculture a profit rate higher than or equal to the profit rate attainable in the rest of the economy. Otherwise he is going to invest the money elsewhere. Now, that is not necessarily the case for a family farmer, who generally has neither satisfactory investment opportunities nor employment opportunities outside of the farm and who prefers, consequently, to invest in it in order to maintain the business and allow one of his descendants to carry it on, even if it means accepting remuneration for his labor, capital, and land below the market price.

The threshold of renewal for capitalist enterprises is therefore much higher than that for family farms, and the laws of development of the second agricultural revolution are not the same in the one case as they are in the other. In capitalist agriculture, production units cannot exist on a long-term basis below the threshold of productivity that makes it possible to pay for land, capital, and labor at market prices. In peasant agriculture, however, one-third or one-half of the farms, or even more, are commonly below this threshold. These farms manage to survive for one generation. They sometimes even manage to develop by paying for the factors of production that belong to them at a lower rate than market prices.

## Other Negative Consequences: Pollution, Desertification, Unemployment

Many other negative consequences come to be added to the large inequalities in income and survival potential between peasant farms over the course of the second agricultural revolution, such as: *regional* concentration of plant production, *local* concentration of a high number of animals in factory livestock operations, abusive use of fertilizers and pesticides and animal pharmaceutical products, difficulty of maintaining adequate public services and an acceptable social life in regions where the farmed area per worker exceeds 100 hectares, desertification of abandoned regions, etc.

Beginning in the 1970s, the agricultural exodus, resulting principally from the disappearance of small farms and the abandonment of whole regions, continued at an accelerated pace, even though general economic growth slowed down and job creation outside of agriculture declined. Unemployment began to expand well beyond the proportions ordinarily necessary to ensure the mobility of the labor force.

## Corrective Policies

From the 1970s, various measures were implemented to limit the worsening of inequalities and disequilibria and avoid pollution or remedy its effects.

### Targeted Development Plans for Farms

Among these measures, the "development plans," established in 1972 in the EEC, had the objective of helping low-income small and medium-sized farms make a set of comprehensive investments, using subsidies and subsidized loans, in order to bring the farmer's income in four or five years to a level higher than or equal to a so-called parity income (defined as being the average regional income attainable outside agriculture). But such an objective was not attainable by farms located far below the threshold of renewal, particularly numerous in the disadvantaged regions undergoing abandonment, which were, consequently, largely excluded from the benefits of this policy.[6]

That is why it was necessary to take specific actions that aimed at compensating for the loss in earnings and the additional costs of farming in regions suffering from serious natural handicaps and very slow investments. Compensatory allowances for these natural handicaps were accordingly given to farms from these regions, allowances calculated as a function of the size of the herd or farmed area, and differentiated according to areas (high mountain, mountain, piedmont, other disadvantaged areas and dry areas). Assistance for farm mechanization and farm equipment, particularly costly in hilly areas, as well as assistance to offset the additional expense of collecting milk, were also established. Research and development, which had until then concentrated the main part of its efforts on the needs of advantaged regions, began to redirect a part of its efforts to assist farms and regions in difficulty.

These measures certainly had positive effects, but they generally proved to be too late and insufficient to restore parity of income between regions and prevent the increasing abandonment of farmland. Policies of more balanced development, intended to prevent unequal development between farms and regions, which began to be applied much earlier in some countries such as Switzerland, Austria, the Netherlands, and to a certain extent Germany, in the end produced more significant results.

## Preservation of the Environment and of Product Quality

In the 1980s and 1990s, the first European measures intended to preserve both the environment and product quality began to be applied. Essentially, these are regional and local programs and long-term development plans directed at individual farms. These plans gave monetary assistance to committed farmers, which made up for the loss of earnings resulting from less polluting agricultural practices or which paid for maintenance work on the landscape (roads, hedges, canals, ditches). Restrictive regulations concerning livestock buildings and applying of animal excrement were also enacted. Efforts to ensure that these buildings and practices complied with these regulations were partially subsidized. Finally, protected labels of origin and various other labels made it possible to distinguish, enhance the status of, and indeed improve the quality of certain products. But there exist neither general regulations limiting quantities of fertilizers, pesticides, or livestock gathered in the same place to below their pollution level nor taxation of potentially polluting inputs directed at diminishing the level (economically optimal) of their use. Despite these inadequacies, it may be that a comprehensive policy is in preparation, aimed at promoting a well-planned ecological agriculture and quality food, thereby meeting the aspirations of the greatest number.

## Surpluses and Fixing Quotas

Policies aimed at accelerating the development of the second agricultural revolution, and the lack of systematic measures actually capable of preventing the excessive use of some productive inputs, also had the effect of accentuating the disequilibria of markets for vegetable and animal products. In the 1970s, in order to profit from high world prices for basic commodities, conquer additional market shares, and improve the balance of trade, many developed countries, spurred on by the nascent crisis, again strengthened their policy of aiding agricultural development. The effects of this assistance, added to the stimulating effects of the high prices, accentuated the tendency to create surpluses and contributed to the collapse of prices that occurred at the end of the 1970s. One can expect that, conversely, the Malthusian policies deployed in the 1980s and 1990s to reduce the surpluses of taking land out of production, production quotas, and reducing agricultural subsidies will accentuate the increase in prices of agricultural commodities, which will not fail to appear one day or another.

Thus a policy of regulating production that, instead of preventing future fluctuations, only responds to the conjuncture of the moment ends up accentuating this fluctuation. We do not know if measures that consist of paying highly productive farmers not to produce or paying small farmers, whose productivity and agricultural income are negative, to continue to play their role as peasants in the landscape constitutes a "policy" comprehensible enough to be acceptable to farmers and taxpayers. Do not such measures appear, rather, as a series of complicated and costly tactical expedients that attempt to limit the damage caused by the shortcomings and lack of strategic direction in an agricultural policy increasingly disoriented by outside political pressures and its own trajectories and counter-trajectories?

Contemporary agricultural policies should not, indeed, be reduced to a series of interventions intended to facilitate the development of the agricultural revolution and correct its shortcomings. Beyond these more or less explicit technical and economic objectives, every agricultural policy also arbitrates among the interests of different social classes. Every price policy arbitrates between the interests of farmers and landowners and those of industrialists and consumers. It arbitrates among the interests of different groups of producers, grain growers and livestock breeders, for example. Every foreign trade policy arbitrates between the interests of national producers and those of their foreign competitors. Every policy of assisting agricultural development through subsidies, government-subsidized loans, priority allocation of freed lands, and the direction of research and development arbitrates between the groups of farms and regions that benefit from them and those that do not benefit. In particular, a policy of development can aggravate or, quite the opposite, reduce the unequal development between farms and regions.

Every agricultural policy, just like every economic policy, is a major social contest. And consequently it is the object of all types of demands, pressures, negotiations, representations, and influence games, which express the interests of different groups: national producers of all types (grain growers, animal breeders, vine growers, well-off farmers or those in difficulty, farmers from the plains or those from the mountains, foreign producers, industrialists, consumers, ecologists, regionalists, etc). This is as much as to say that the prices of products and of agricultural means of production, which control the development of contemporary agriculture, are not simply the result of commercial negotiations between buyers and sellers. They are also the result of ongoing social and political negotiations. Agricultural prices are not formed only on the grain exchange and in the livestock markets; they are also negotiated in interprofessional organizations, government ministries, meetings of the World Trade Organization (formerly GATT), counsels of ministers of the European Union. Agricultural prices are social relations and subjected to a much more complex regulation than the interaction of supply and demand alone.

But above all it must be remembered that every agricultural policy appears as first and foremost a choice of, or at least a national preference for, definite property and farm structures: peasant family farms and small associations of producers as is the case in the countries that we have considered, or capitalist enterprises employing wage labor (southern and eastern Europe) or even large production cooperatives and state farms (former socialist countries).

Finally, let's not forget that measures of general economic policy have at least as significant an impact on the development of agriculture as the measures of agricultural policy proper. Monetary policy, which has a strong effect on inflation and the exchange rate, and foreign trade policy, which controls quotas as well as taxes and subsidies for imports and exports, have a significant influence on the international competitiveness of a country's agricultural products. Monetary policy, through the interest rate, and price policies have a strong effect on the profitability of investments. Budgetary policy influences the total amount of public funds allocated to agriculture. Industrial and wage policies strongly influence the agricultural exodus.

## 4. CONCLUSION

In the space of several decades, the new means of production and transportation, of unprecedented power and produced by large, concentrated industry in the twentieth century revolutionized the conditions of agricultural production and trade in the developed countries. The biological conditions were changed with the selection of better and more productive plants and domestic animals, as were the ecological conditions, with the simplification of specialized cultivated ecosystems. Also affected were the conditions of work, with high-performance motors and machines that made the recourse to animal power unnecessary and reduced the need for labor drastically. There were significant changes in economic and social conditions, with the continual growth of an increasingly smaller number of farms, which largely turned over production of their means of labor and the processing of their products to upstream and downstream industries, and the progressive exclusion of the large majority of the others.

Increasing production by tenfold and the productivity of labor by fifty-fold, this new agricultural and food system, composed of specialized subsystems, exploiting specially selected biological materials consonant with its requirements, proved more than capable of feeding the whole population, only a tiny fraction of which still had to be involved in agriculture. That is, it made possible at the end of the twentieth century a whole range of nonagricultural activities— the most useful, but sometimes also the most useless, most ridiculous, or most harmful. In the developed countries "modern" agriculture has triumphed beyond all expectations.

But the greatest triumphs, as long as they are poorly controlled, always lead to excess. Modern agriculture will be dangerous, as were all new agricultures before it, as long as the use of new means and new methods of production is not tempered by preventing abuses and negative consequences. Even though axes of polished stone were welcome agricultural tools, used wildly they were also dangerous tools of deforestation. Overused on erodible lands or inadequately manured lands, plows often became formidable instruments of soil degradation. When manure was carelessly piled up close to sources of potable water, it became a veritable agent of death on a great many occasions, despite its usefulness. If it were applied too late or in quantities that were too large, its effects were lost to the seeds. Carried too far, the large clearings of the central Middle Ages had to slow down, which contributed to the large food crisis of the fourteenth century. And the tremendous expansion of the railroads and agricultural colonies in the nineteenth century plunged the world into the first great crisis of agricultural overproduction.

How much greater still will be the damage caused by the use of such powerful means and such extraordinary methods of production today, if this use is not consciously and socially controlled, i.e., if the most immediate dangers and the most intolerable distant consequences are not kept at a reasonable distance? Without curbing their use, fertilizers, pesticides, and animal pharmaceuticals continue to be employed up to their profitability level, i.e., sometimes well beyond their level of harm. Without strict bans, dangerous but profitable products will be used. Lacking an absolute ban, questionable raw materials will be used by the animal feed industry. The most irreplaceable sites will be cultivated. The rarest species will be destroyed.

Too much ignorance of and contempt for the past, too much haste and presumption, too few human, ecological, and qualitative precautions inevitably lead, in the long run, to an excessive concentration of cultivation and animal-raising activities, excessive number of empty regions, excessive exodus from agriculture, and excessive unemployment. Where do such outcomes come from if not from the mechanisms of competitive development, mechanisms that turn out to be so effective in pushing the means, methods, and organization of production to abundance, but can also end up just as effectively carrying them beyond the well-understood bounds of usefulness to excess?

There is, then, very little sense in believing that it would be possible without risk to do without prohibitions, rules of production, and draconian controls, even if true that regulation must be simple in order to be effective and that it will never suffice alone to make production more ethical and create labor processes and products of perfect quality. Moreover, in an open world economy, rules of use, prohibitions, and codes of good conduct must be shared and strictly applied by the producers of all countries, without which those who respect them will be penalized by the unfair competition of the others. A well-planned, ecological agriculture and quality food will exist at this price. It is illusory to pretend that generalized deregulation leads to the best of all possible worlds and that the free market is capable of avoiding disequilibria, the fluctuating actions and reactions of the conjuncture, excesses, waste, poverty, and abandonments, which are in fact the counterpart of the impetuous competitive development of the agricultural revolution itself.

But the contemporary agricultural revolution and its effects do not stop at the frontiers of the developed countries. Looking further, as far as the most distant regions of the developing countries, let's now find out in what limited and deformed way this agricultural revolution has advanced in these countries and at what point the consequences of unequal development, crisis, and exclusion become overwhelming. The crisis that today strikes the majority of the peasantry in the developing countries is the essential source of the growing poverty that affects one-half of humanity, a poverty that is at the origin of the current crisis of the world economy.

# 11

# Agrarian Crisis and General Crisis

The political problem of humanity consists in combining three things: economic efficiency, social justice, and political liberty.

—JOHN MAYNARD KEYNES, The Collected Writings

At the end of the nineteenth century, after ten thousand years of agricultural evolution and differentiation, the world's peoples were heirs to multiple forms of agriculture. They were as different from one another as the agriculture of the intertropical forests and savannas, the irrigated agricultures of the arid and semi-arid regions, the wet rice-growing agriculture of the humid tropical regions, the agricultures closely linked to animal raising in the temperate regions and in some tropical regions, not to mention the multiple forms of pastoralism of the cold or semi-arid grassland regions.

These forms of agriculture, which had been formed thousands of kilometers and thousands of years from each other, were already at that time quite unequally productive. The average gross yields per hectare, measured in grain-equivalent, were about 10 quintals for rainfed agricultures and 20 quintals for irrigated or aquatic agricultures, while the farmed area per agricultural laborer ranged from less than one hectare in manual cultivation to 10 hectares in mechanized animal-drawn cultivation. Thus, a hundred years ago, the gap in productivity between the least productive and the most productive agricultures of the world ranged from ten quintals per agricultural worker to a hundred, or a ratio of 1 to 10.

In less than one century, the contemporary agricultural revolution increased the labor productivity of agriculture in the industrialized countries and in some limited agricultural sectors of the developing countries by several dozen times. Consequently, the ratio of gross productivity between the least productive manual agriculture and the most productive motorized agriculture is today on the order of more than 1 to 1000!

This tremendous advance in a certain form of modern agriculture was not in itself harmful to the development of others. But at the same time the transportation revolution opened up and put all of the world's agricultures into competition. Thus the relatively unproductive manual agricultures, dominant in the developing countries, were confronted one after the other with low prices for grains and other basic agricultural commodities coming from the most developed agricultures. Over time, they were subjected to the tendency of agricultural prices to fall in real terms, a fall that results from constant gains in productivity due to the agricultural revolution. For example, the real price of wheat in production in the United States fell by a factor of nearly four since the beginning of the twentieth century. This should provide some idea of the magnitude of the fall in agricultural prices that has had such a strong impact on the agricultures of the developing countries.

Disadvantaged by their low level of equipment, the producers of the developing countries tried to take advantage of their natural advantages by specializing, at least partially, in tropical export crops, for which there was initially less competition. Ultimately, many of these crops were also subjected to the competition of crops from the developed countries (sugar beets versus sugarcane, soya versus peanuts and other tropical oilseeds, cotton from the southern United States, tobacco, etc.), while others were subjected to the competition of replacement industrial products (synthetic rubber versus rubber tree cultivation, synthetic textiles, etc.).

Moreover, in the developing countries, export crops have been affected one after the other by the progress of the second agricultural revolution. But if selection, fertilizers, and pesticides benefited the peasantry in the developing countries to a large extent, motorization and large-scale mechanization benefited only large capitalist or state farms and a small fraction of the well-off peasantry. Thus, even though the development of this agricultural revolution may have been limited in this part of the world, it nevertheless contributed to dragging the prices of most of the tropical export commodities downward.

Progressively deprived of profitable activities, the immense majority of the underequipped and relatively unproductive peasantry of the developing countries ended up with inadequate incomes to invest and develop, that is, with incomes below the threshold of renewal (or threshold of capitalization). Consequently, even today more than 80 percent of African farmers and 40 to 60 percent of those in Asia and Latin America continue to work with strictly manual tools.

As relatively unsuccessful and poorly paid as their work may be, most of these underequipped peasants have had to continue producing for the export market in order to renew their meager tools, obtain a few special consumer goods, and, if need be, pay taxes and other fees. Innumerable poor peasants contribute to the increase in the supply of export products and the lowering of prices that results. They continue to do so up to the point where the labor income they obtain from export crops becomes equal to the income that they can make from food-producing crops. The price of export crops is thus linked to that of basic food commodities and the tendency of grain prices to fall inevitably leads, in the long run, to lower prices for agricultural export commodities.

This lowering of real agricultural prices forces the peasantry to devote an increasing share of its forces to products intended for sale and, consequently, to reduce production for self-consumption and maintenance work for the cultivated ecosystem. Weakening of labor power and degradation of fertility lead in turn to a lowering of production. That combines with the lower prices to reduce the already pathetic income of this peasantry even further. The moment soon comes when the already undernourished peasantry can no longer even renew its seeds and tools. The peasant population falls below the threshold of survival and has no other option than to leave for the shantytowns or refugee camps, that is, if no economic, climatic, biological, or political disaster occurs in the meantime to make the situation worse and lead to famine on the spot.

Certainly, to date, this process of impoverishment and exodus has not affected all of the peasantry carrying out manual cultivation. It has affected above all the most destitute peasants from the most disadvantaged regions. But so long as the tendency for grain prices to fall continues, which entails behind it the lowering of prices for other agricultural commodities, the huge agricultural exodus and the enormous swelling of the shantytown populations will also continue. Because of the lack of urban infrastructure and adequate employment in the industrial and service sectors, the exodus of the poor peasantry is transformed into unemployment or into underpaid jobs, i.e., into urban poverty. In the end, the wages of the unqualified labor force are established on a level barely above the cost of reproducing labor power—a level close to the threshold of survival for poor peasant agriculture.

Therefore the fall in agricultural prices and in the incomes of the poor peasantry leads to rising unemployment and the lowering of basic wages in all branches of employment in the underindustrialized developing countries, and it drags down the prices of all goods and services supplied by these countries as well.

The object of this chapter is, first, to explain the enormous explosion of inequalities of productivity and income between different agricultures in the world, an explosion that occurred in the twentieth century because of the agricultural and transportation revolutions. We show that in the existing system of international trade the tendency for agricultural prices (in real terms) to fall, which has continued for decades, plunges entire and increasingly broader sections of the poor peasantry of the developing countries into crisis. Further, we show that this immense agrarian crisis is the cause of mass poverty and the failure of the poor agricultural countries to modernize.

Second, we will establish that the insufficiency of demand in the developed countries lies in the insolvency of the needs of this other half of the world. This insufficiency of demand is the essential cause of the general slowdown in growth, the trend toward speculative investment, and the world economic crisis that has developed since the beginning of the 1970s.

Based on this analysis, we will try to show that the solution to this crisis cannot come from exacerbating competition between countries and from the currently common national deflationist policies, which reduce employment and income. As we see it, the revival of the world economy will occur through a decisive enlargement of effective demand on a world scale, which can only result from a significant, gradual, and sustained increase in agricultural prices paid to the poor peasantry of the developing countries and from a massive increase in incomes and buying power in this part of the world.

Finally, we will show that in order to implement a world anti-crisis policy of this type, able to lead the world toward a balanced and long-lasting development, a new, more equitable international system of trade, and a new, more stable world monetary and financial system will be necessary.

## 1. ORIGINS AND EXPANSION OF THE AGRARIAN CRISIS IN THE DEVELOPING COUNTRIES

Between 10,000 and 5,000 years ago, the Neolithic agrarian societies, that is, the pastoral societies of the steppes, prairies, and savannas and the slash-and-burn agricultural societies of the forested environments, using manual tools that were neither very diversified nor very effective (axes and adzes of polished stone, digging sticks, harvest knives, sickles with microliths) had still not conquered more than half of the exploitable environments of the planet. However, beginning 6,000 years ago, the forested environments that had been cultivated the longest and the least resistant to ax and fire began to be deforested and give way to various ecosystems, offering different possibilities for agricultural use.

### Differentiated Agrarian Systems Inherited from the Past

In the Bronze and Iron Ages, between 5,000 and 1,000 B.C.E., as this deforestation progressed, diverse post-forest agrarian systems emerged, thousands of kilometers and thousands of years distant from one another: systems of floodwater and irrigated cultivation in the arid regions of the Sahara and southwest Asia (Mesopotamia, Nile and Indus Valleys, a little more than 5,000 years ago), of America (Olmecs, more than 3,000 years ago; Teotihuacán and the Mayas, more than 1,000 years ago; pre-Inca civilizations, more than 2,000 years ago); *wet rice-growing systems* in the monsoon regions of Asia (China and India, more than 3,000 years ago); rainfed grain-growing systems with fallowing and associated animal breeding in the temperate regions (Mediterranean region, more than 2,500 years ago; northwest Europe, more than 2,000 years ago).

In connection with this vast movement of diversification in cultivated ecosystems, significant inequalities of equipment developed. Beginning in the Bronze Age, in a few societies of Eurasia, Africa, and America, some metallic hand tools (knives, small axes, points of digging sticks) were made, though they were still not very effective. But beginning with the Iron Age, new, much more powerful tools (axes, hoes, spades, iron-tipped sickles) were made and used more and more widely in the agriculture of the "Old World." Moreover, thanks to the progress of artisanship in iron and wood, new equipment (ard, packsaddle, cart) made it possible to use animal energy. At the end of antiquity, cultivation with the ard was used in hydroagricultures and in systems based on fallowing in the most advanced regions of the Near East, North Africa, Asia, and Europe, while carts, wagons, and chariots, originating in Asian pastoral societies, were already used for transportation and war. Finally, in the Middle Ages, in northwestern Europe and in some areas of Asia, new progress in the artisanship of iron and wood made it possible to traverse a new stage in the use of animal energy (animal-drawn cultivation using the plow, harrow, and wagon) and water and wind energies (growth in water and windmills).

At the dawn of modern times, highly differentiated and unequally evolved agrarian societies were already in existence. Animal-drawn cultivation based on the plow had developed only in northwest Europe and in a few deltas and valleys of monsoon Asia, while in the Mediterranean region, the Near East, and other regions of Asia and Africa, the dominant agrarian system was, at best, animal-drawn cultivation using the ard. In most of the world, manual cultivation was either largely predominant or exclusive. Such was the case particularly in the intertropical forests and savannas of Africa, Asia, and South America. In some of these regions, the manual tools were still made from polished stone. At this time period, only hunter-gatherers frequented the vast northern and equatorial forests.

Some of these societies, poorly equipped for production and poorly armed for defense, were subjected for several centuries to the continual ravages of colonization—such as the partial destruction of the intertropical Amerindian civilizations, which were repressed and subordinated to the construction of agro-exporting latifundia-minifundia economies; the centuries-long exhaustion of the intertropical African populations, largely through the slave trade; the formation of enclaves of colonial plantation economies throughout the whole tropical world; the nearly total destruction of precolonial societies in the temperate regions of the two Americas, Australia, and New Zealand, and the transplanting to these regions of entire sections of European agrarian societies, with their people, tools, plants, animals, and methods of cultivation and animal raising. While colonization transported to America the domestic plants and animals of the Old World (wheat, rice, sugarcane, cattle, sheep, goats, etc.), American plants (maize, potatoes, cassava, tobacco, tomatoes, sunflowers) moved in the opposite direction.

During this time, from the sixteenth to the nineteenth century, the farms of Asia and northwest Europe continued to develop. In Asia, double annual rice cropping and animal traction gained ground, while in northwest Europe a new agricultural revolution developed, in close relation with the first industrial revolution. This agricultural revolution made it possible to double agricultural production and productivity once again by replacing fallowing with the cultivation of fodder and row crops.

So, in the middle of the nineteenth century, after thousands of years of the differentiated evolutions and interactions of agrarian systems, the peoples of the world ended up the inheritors of quite different and unequally productive agricultures. To appreciate these inequalities better, let's represent side by side on a graph the maximum net labor productivity attainable in the main agricultures in existence at that time (*Figure 11.1*). Estimated in quintals of grain-equivalent, this productivity is calculated as follows: maximum cultivable area per worker multiplied by the yield per hectare attainable in good conditions of fertility, after deducting seeds, losses, and the quantity of grain necessary to cover the cost (small enough in these systems) of inputs and the amortization of equipment. For each main type of system, the maximum area per worker and the maximum yield per hectare varies as a function of the different regions in question, which explains why the attainable net productivity varies to a certain extent.

As Figure 11.1 shows, the systems existing at that time can be classed in order of increasing net productivity in the following manner:

— *Systems of manual cultivation*, in which maximum net yield is on the order of 10 quintals per worker. These include part of the rainfed cultivation systems in the intertropical forests and savannas and some temperate forests in America and Asia as well as irrigated and wet rice-growing cultivation systems with one harvest per year.
— *Systems of animal-drawn cultivation based on fallowing and use of the ard*, in which maximum net productivity is on the order of 20 quintals per worker. These include grain-growing systems based on fallowing of the Mediterranean regions and some regions of Asia and South America and irrigated and wet rice-growing cultivation systems with one harvest per year.

— *Systems of animal-drawn cultivation based on fallowing and use of the plow,* which persisted in some temperate regions of Europe and America, and in which maximum net productivity is on the order of 35 quintals per worker and irrigated and wet rice-growing cultivation systems with two harvests per year using animal traction, in which productivity is on the same order of magnitude.

— *Grain-growing systems based on animal-drawn plows and without fallowing in the temperate regions,* in which yield is on the order of 50 quintals.

In the middle of the nineteenth century, the ratio of net productivity between the least effective manual cultivation systems in the intertropical regions and the most effective systems of cultivation using animal-drawn plows without fallowing in the temperate regions was on the order of 1 to 5. As Figure 11.1 shows, this disparity in yields between systems was already higher than disparities in yields existing inside each system.

## *The End of the Nineteenth Century: A Productivity Ratio of 1 to 10*

In the second half of the nineteenth century, the disparities in productivity widened even more. In northwestern Europe and North America, rapidly expanding industry began to supply farmers with new animal-drawn mechanical equipment (Brabant plows, reapers, harvesters), which made it possible for them to double the farmed area per worker and the productivity of agricultural labor, while the transportation revolution (railroads, steamships) made it possible for them to obtain amendments and fertilizers from distant locations and begin to sell their products in distant markets and specialize their production. At the end of the nineteenth century, the new grain-growing systems using mechanized animal traction and no fallowing in the temperate regions reached a net productivity on the order of 100 quintals (10 hectares/worker x 10 quintals/hectare = 100 quintals/worker), or around ten times more than the productivity of manual cultivation (*Figure 11.1*).

Note that at this time manual cultivation still existed in the most advanced areas of Europe, America, and Asia. In all the villages of these areas, next to the rich farmers who were well provided with draft animals and equipment and produced several dozen quintals per worker, there existed numerous peasants using manual cultivation, whose yield barely exceeded 10 quintals. Also let's note that in some deltas of East and Southeast Asia, the maximum attainable productivity by the best-equipped rice growers, resulting in two harvests per year, was not far below that of the most advanced farmers in the most advanced cold temperate countries, though it was undoubtedly a bit higher than that of farmers using the ard in the perimeter of the Mediterranean region.

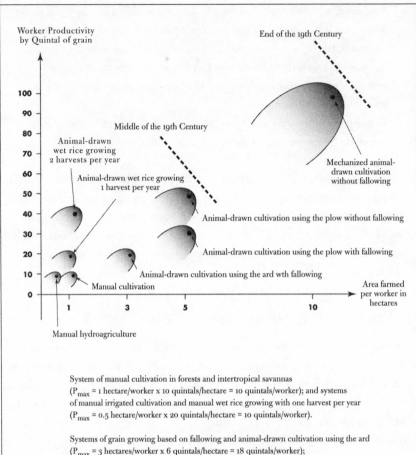

Worker Productivity
by Quintal of grain

End of the 19th Century

100
90
80 — Middle of the 19th Century
70 — Animal-drawn
        wet rice growing
        2 harvests per year
60 —
        Animal-drawn wet rice growing
50 —        1 harvest per year
40 —
30 —                                    Animal-drawn cultivation using the plow without fallowing
20 —                                    Animal-drawn cultivation using the plow with fallowing
10 —
        Manual cultivation          Animal-drawn cultivation using the ard wth fallowing
0

                Mechanized animal-
                drawn cultivation
                without fallowing

Area farmed
per worker in
hectares

1      3      5          10

Manual hydroagriculture

System of manual cultivation in forests and intertropical savannas
($P_{max}$ = 1 hectare/worker x 10 quintals/hectare = 10 quintals/worker); and systems
of manual irrigated cultivation and manual wet rice growing with one harvest per year
($P_{max}$ = 0.5 hectare/worker x 20 quintals/hectare = 10 quintals/worker).

Systems of grain growing based on fallowing and animal-drawn cultivation using the ard
($P_{max}$ = 3 hectares/worker x 6 quintals/hectare = 18 quintals/worker);
and systems of animal-drawn irrigated cultivation and wet rice growing with one harvest
per year using the ard
($P_{max}$ = 1 hectare/worker x 20 quintals/hectare = 20 quintals/worker).

Systems based on animal-drawn cultivation using the plow with fallowing
($P_{max}$ = 5 hectares/worker x 6 quintals/hectare = 30 quintals/worker);
and systems of animal-drawn irrigated cultivation and wet rice growing
with two harvests per year using the plow
($P_{max}$ = 1 hectare/worker x 2 harvests per year x 20 quintals/worker = 40 quintals/worker).

Systems of grain growing based on animal-drawn cultivation using the plow
without fallowing
($P_{max}$ = 5 hectares/worker x 10 quintals/hectare = 50 quintals/worker).

Systems of grain growing based on mechanized animal-drawn cultivation without fallowing
($P_{max}$ = 10 hectares/worker x 10 quintals/hectare = 100 quintals/worker).

*Figure 11.1 Comparative Productivities of the Major Agricultural Systems Existing
in the World in the Middle of and at the End of the 19th Century*

# The Weak Penetration of the Contemporary Agricultural Revolution into the Developing Countries and the Explosion of Global Inequalities in Agricultural Productivity

As significant as they were at the end of the nineteenth century, the disparities in productivity between the different agricultures of the world still pale compared to those that have developed since. Indeed, in the twentieth century, these disparities literally exploded. In several decades, the second agricultural revolution (motorization, selection, mineral fertilizers, treatment products, specialization) spread to the developed countries, vigorously increasing agricultural productivity dozens of times, while the largest part of the agricultures in the developing countries remained isolated from this movement. In fact, only a small fraction of these underequipped agricultures were affected by this agricultural revolution, and then in an often incomplete and deformed manner.

## Limited Motomechanization and the Persistence of a Manual Agriculture Among the Largest Number of People

The large grain, cotton, and sugar-growing estates in Latin America adopted motorization, large-scale mechanization, and mineral fertilizers somewhat later than their North American equivalents. Some of the large and medium-sized farms of Latin America and the Near East took the same road. In these regions of the world today, the tractor is used on more than one-third of the farms. But in Africa and the Far East, tractors are found on fewer than 10 percent of the farms.

The small farms practicing manual cultivation—the great majority in Africa, Asia, and Latin America—have never had the means to acquire any form of motomechanization, even on a small scale. The large majority among them have never even had the means to attain animal traction that, even today, is present in fewer than 15 percent of the farms in intertropical Africa, fewer than 20 percent in Latin America and the Near East, and fewer than 30 percent in the Far East. Strictly manual cultivation, which is not very productive, continues to be predominant in the developing countries. More than 80 percent of the farmers in Africa, and 40 to 60 percent of the farmers in Asia and Latin America use manual cultivation. It should be added that many of them have benefited very little from selection of crops and the use of agricultural chemicals.

## Selection and Synthetic Mineral Fertilizers—The Green Revolution Pulls Up to the Gates of Poor Agriculture

During the colonial period, and sometimes well after, tropical agronomical research focused most of its efforts on export crops, selecting improved varieties and advocating cultivation methods better adapted to the production conditions of large plantations than to the needs and means of peasant farms. Food crops were often neglected.

After World War Two, international centers of agricultural research, financed by large American private foundations (Ford, Rockefeller), selected high-yield varieties of rice, wheat, maize, and soya requiring high inputs in fertilizers and treatment products and developed appropriate cultivation methods on experimental stations. In the 1960s and 1970s, the diffusion of these varieties and cultivation methods made it possible to increase yields and seed production in many countries of Asia, Latin America, and, to a lesser degree, Africa. This large-scale expansion of some elements of the second agricultural revolution (plant and animal selection, mineral and synthetic fertilizers, treatment products, pure culture of genetically homogeneous populations, partial mechanization, strict control of water) to three main grains widely grown in the developing countries was called the "Green Revolution." But the gains in yield and production connected to the Green Revolution, as significant as they were, benefited above all the fertile regions most able to get a return on the necessary costly inputs and benefited the farmers with adequate means to buy those inputs and apply the corresponding technical advice. Marginal regions and the poor peasantry, once again, remained largely outside of this movement.

Moreover, many food crops considered as secondary (millet, sorghum, native peas, taro, sweet potato, cassava, yam, plantain) were not the objects of significant research. Most of the local species and breeds of large and small livestock (zebus, yaks, buffalos, donkeys, sheep, goats, native pigs) were also neglected, as were multiple varieties of vegetables and fruits, however significant in the diet.

The research effort above all focused on the most specialized production systems and on standardized methods of cultivation (the famous "technology packages"), conforming to the conditions of relatively well-equipped farms. Complex production systems (associated crops, mixed systems combining crops, animal raising, arboriculture, or fish farming), with flexible and diversified cultivation methods that were less risky, used fewer inputs, were more labor intensive, and  much more appropriate to the needs and possibilities of small underequipped farms, were neglected.

Although the Green Revolution made it possible to greatly increase production in numerous countries, it could hardly, at the beginning anyway, contribute to saving and developing poor peasant agriculture from less favored regions of the developing countries.

## The End of the Twentieth Century: An Agricultural Net Productivity Ratio of 1 to 500

All things considered, at the end of the twentieth century and beginning of the twenty-first century, the advanced sectors of agriculture in the developed countries and in some limited sectors of agriculture in the developing countries are at a level of capitalization that allows them to attain a net productivity (variable inputs and amortization deducted) on the order of 5,000 quintals of grain equivalent per worker (100 hectares/worker x 50 quintals net/hectare). At the same time, in the developing countries manual cultivation, producing on the order of 10 quintals net of grain-equivalent per worker, continues to be overwhelmingly predominant. Hence the ratio of net productivity between the least productive and the most productive agriculture in the world, which was 1 to 10 at the beginning of the century, is today 1 to 500. In a little less than one century, this ratio has increased by 50 (*Figure 11.2*).

## Motorization of Transportation, International Competition, and the Tendency for Agricultural Prices to Fall

The second industrial revolution has not only produced the means to increase the disparity of productivity between manual agriculture and the most successful motorized agriculture by *fifty times*, it also has provided the means to put them in actual competition with one another. Since the Second World War, the motorization of road transportation, extending that of marine, rail, and air transportation, has gradually penetrated into every region of the world, including the most remote regions of the developing countries. The capacity and speed of this transportation has increased and its costs have diminished to such a point that most of the world's agricultures are no longer sheltered from competition from the most productive agricultures, which continue, moreover, to advance. Certainly, distances have not been abolished, transportation costs have not vanished, and institutional obstacles to international trade (taxes, quotas) remain quite real. But as a result of the reduction in transportation costs and the liberalization of international trade, the prices of basic food commodities, particularly grains, are today more and more the same in most of the countries of the world. They are determined by the low-priced exports from the surplus countries with high agricultural productivity in North America, South America (Argentina, Brazil), Europe, and Oceania (Australia, New Zealand).

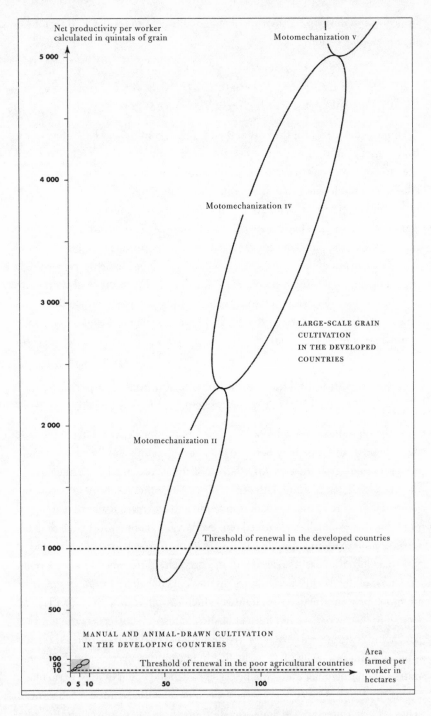

*Figure 11.2 Gaps in Net Productivity of Labor Between Motomechanized Grain-Growing Systems that Use Agrichemicals and Systems of Manual or Animal-Drawn Cultivation in the Developing Countries*

## The Fall in the Prices of Basic Food Commodities

Over the last decades, the appearance of low-priced grains has caused a significant fall, in real terms, in the domestic prices of grains and substitute food commodities in most of the developing countries. The first consequence of this tendency for uniformity in grain prices and substitute commodities has been to make apparent the enormous disparity in productivities existing between farmers using a hoe, producing on the order of 10 quintals net per worker, and high-equipped farmers, who produce several thousand quintals. As soon as food commodities are priced approximately the same for workers from both systems, the disparities in productivity per worker are expressed purely and simply as disparities in incomes.

At $20 per quintal of grain, for example, a well-equipped European grain grower, working alone, producing 5,000 quintals net (variable inputs and amortizations deducted), receives $100,000. This $100,000 represents a net creation of wealth (that is, net value added for the grower's country) that this grain grower must possibly share with the landowner if he is a tenant farmer, with the banker if he is indebted, and with the tax department if he is subjected to taxes. After that, there will remain between $20,000 and $50,000 per year to pay for his own labor and to invest.

Paid at the same rate, $20 per quintal of grain, a Sudanese, Andean, or Indian manual farmer producing 10 quintals net would receive $200 if the entire production were sold. But since 7 quintals of grain must be saved to feed the farmer and the farmer's family, the monetary income is barely more than $60 per year. Again, that is on condition that there is no farm rent, interest on borrowed capital, or taxes to pay.

At the rate of $20 per quintal of grain, it would require one life of labor (33 years) for a manual farmer having a monetary income of $60 per year to acquire a pair of oxen and small animal-drawn equipment costing $2000, supposing that this farmer can devote all of the monetary income to this purchase. It would require 100 years to acquire sophisticated animal-drawn cultivation equipment. It would require 300 years of labor to buy a small tractor at $20,000, and it would require 3,000 years to buy the complete set of motomechanized equipment, with a value of $200,000, comparable to that of a European or American farmer.

*Food Dependence.* Grains, and the other food products replaceable by grains, were the first affected by the competition with the agriculture of the developed countries and by the lower prices that resulted from that competition. Without abandoning the food crops used for self-consumption, the farmers in the developing countries then reduced or abandoned cultivation of food crops intended for sale in order to devote an increasing part of their resources to tropical export products, which encountered less competition. In doing this, they chose the most advantageous products, taking into account the physical conditions of each region. Thus large agro-exporting specializations were formed or strengthened: coffee, tea, cacao, tobacco, peanuts, cotton, pineapples, bananas, etc. These export crops developed, in successive waves, during periods of rapid growth in world demand and high prices.

The relative decline of food crops intended for sale, at the same time that urban demand increased, plunged many developing countries into growing food dependence. In intertropical Africa, from 1965 to 1985, grain imports (wheat and flour, rice and the main secondary grains) per inhabitant more than tripled, going from 10 kilograms to around 35 kilograms. At the same time, production fell from 135 to less than 100 kilograms of grains per inhabitant.[1] Consumption per head thus fell (close to 10 percent) and this despite or because of the low-priced imports.

*Agro-Export Specialization.* Naturally, the large agro-export plantations were the first to take advantage of periods of high prices. They profited from those prices by adding to their equipment and sometimes by expanding their operation, to the detriment of the peasant economy. However, in countries where the land was not monopolized by the large estates, export crops also brought higher incomes to the peasantry than those it would have obtained with grains or other substitute food commodities. Those higher incomes made it possible for part of this peasantry to invest and develop. Hence, peasant plantations of coffee, cacao, rubber trees, etc., expanded in Africa, Asia, and Latin America, in regions having forested land reserves. In the years 1950–1960, the least deprived fraction of cotton and peanut producers in West Africa could adopt animal traction.

## Lower Prices Spread to Export Commodities

But many tropical export crops were also affected by competition from identical or substitute commodities produced by the highly productive agriculture of the developed countries. Tropical sugarcane, for example, has for a long time had strong competition from sugar beets, a crop pivotal to the agricultural modernization of temperate Europe. Peanut oil and peanut oilcake was subjected and is still subjected to competition from large American production of soya. According to the World Bank, from 1950 to 1984, the price for oils and fats relative to the price of manufactured goods imported by the developing countries fell 1.29 percent per year.[2]

Moreover, some export products important for the developing countries, such as natural rubber and cotton, saw their prices greatly reduced by competition from competitive substitute industrial products. Most tropical export crops were also affected one after another by the progress of the second agricultural revolution: selection of varieties that both require more fertilizers and are more productive as well as the development of synthetic fertilizers and specific pesticides, guidelines for cultivation, and sometimes even large, specialized machines, for harvesting or other activities.

These new and costly means of production were only adopted in their totality by large capitalist or state production units and by the wealthiest strata of peasant agriculture employing same wage labor, even though, for want of means, the large majority of the peasantry went no further than manual cultivation or animal-drawn cultivation with the ard and had only limited access to improved varieties and fertilizers. The fact remains that the second agricultural revolution, after having led to the fall in prices for basic food commodities, also led to a fall in the prices of most tropical export commodities. From 1950 to 1984, the average price of tea, coffee, and cacao, tropical commodities not affected by competition from countries in the north, fell 1.13 percent per year in relation to the price of manufactured products imported by the developing countries.[3] The same causes produced the same effects. The producers of export agricultural commodities were affected, in turn, by the fall in their incomes, as were the producers of grains and other basic food commodities.

*Competition Among the Poor.* Since the range of relatively profitable products continued to narrow in many regions, increasingly numerous peasants turned toward the few still profitable products. In order to do that, millions of peasants around the world abandoned their homes and moved thousands of kilometers. Sahelian and Sudanese peasants abandoned peanut and cotton crops to grow coffee and cacao in the equatorial forest zone. Rice growers in Southeast Asia colonized the last forests of the region to cultivate rubber trees. Peasants of the Andes came down from the mountains to cultivate vegetables on the periphery of coastal cities or to cultivate coffee or coca on the Amazonian slopes.

Any export product still profitable at a given moment attracts such a large number of deprived producers, ready to accept poverty wages, that supply increases and prices fall, even for products not affected by the second agricultural revolution and without competition from synthetic products. The prices of these commodities fall, then, up to the point where the income they obtain becomes equal to the income obtained by the sale of devalorized food commodities. Moreover, it should be noted that when the price of the last practicable export crop falls below this level, peasants abandon this export crop and return to food crops, however poorly remunerative.

Consequently, over a long period of time a system of relative prices for food and export products is formed, such that the income per worker obtained by the peasantry from these different products tends to become more equal, conforming to the lowest income. Certainly, it is a question of a general law as tendency, which cannot be verified at every moment in all places because the patterns of price fluctuations of different commodities are not the same. But if the curves representing the evolution of real prices of some important agricultural commodities, such as wheat, sugar, rice, maize, and rubber, are placed on a graph, it is possible to take into account simultaneously the enormity of the fluctuations, the extent of the tendency of the prices of these commodities to fall, and the close correlation of these downward trends over the long term (*Figure 11.3*).

## The Development of "Naturally Protected" Products

The only commodities that escape, to a certain extent, the competition of imported products are quickly perishable commodities intended for the domestic market, such as certain fruits and vegetables, fresh dairy products and eggs, or commodities of low value such as firewood. These "naturally protected" products benefit from growing urban outlets. Truck farming, fruit growing, and small-scale livestock breeding in the interior and on the periphery of the cities of Africa, Asia, and Latin America have all undergone a strong development to meet the growing demand. There has also been an expansion of woodcutting with the consequent deforestation occurring in an increasingly larger ring around urban centers.

But the growth of urban and peri-urban agriculture is nevertheless limited by the weak buying power of the majority of city populations and by the imports of replacement products (deep-frozen products, powdered milk, various kinds of fuel). Moreover, urban agriculture is thwarted by the new construction that continually whittles away at its territory, regardless of how much labor went into its development, and peri-urban agriculture is always pushed back further by land speculation. The time and the delivery costs for products increase continually, which accordingly reduces the income of the producers.

While basic food and export crops, which encounter too much competition and pay too little, languish, this strong development of urban and peri-urban agricultural activities shows that the adaptability and courage of the poor peasantry are quite high. All it takes to be convinced is to see the truck farmers of Kenskof, who supply Port-au-Prince from the heights above the city, carrying heavy loads at night on top of their heads over many kilometers, or the armies of porters with palanches converging at dawn on the large cities of Asia, or the bicycles overloaded with bananas tearing down the hills surrounding Bujumbura at breakneck speed, or lines of carts and donkeys and dromedaries with packsaddles carrying, from more than a 50-kilometer radius, their daily wood delivery to the households of Niamey.

# The Crisis of the Poor Peasantry

## *The Mechanism of the Crisis*

For the mass of peasants using manual cultivation in the developing countries, the fall in real agricultural prices over more than a half-century led first to a fall in their purchasing power. The majority of them have quickly found themselves unable to invest in more effective tools and even sometimes unable to buy improved seeds, fertilizers, and pesticides. In other words, the fall in agricultural prices has been expressed by a true blocking of development for the vast majority of the least well-equipped and least well-situated peasants.

With the continuing fall in agricultural prices, the peasants who have not been able to invest and achieve gains in productivity have clearly fallen below the threshold of renewal. In other words, their monetary income has become inadequate to renew their tools and inputs, buy consumer goods they have not produced themselves (such as sheet metal for their roof, salt, cloth, shoes, kerosene, medicines, pencils, paper) and, if need be, pay taxes, all at the same time.

In these conditions, in order to renew the minimum set of tools necessary to continue working, these peasants have had to make sacrifices of all kinds: selling livestock, cutting back on purchases of consumer goods, etc. At the same time, they have had to expand as much as possible the crops grown for sale, but since their production capacity is strictly limited by the inadequacy of their tools, in order to do that they have had to reduce the area allocated to food crops grown for self-consumption. In other words, the survival of a peasant farm where income falls below the threshold of renewal is possible only at the price of reducing its capital (sale of livestock, poorly maintained and limited tools), underconsumption (peasants in rags and barefoot), and undernourishment.

*Ecological and Health Crisis.* Peasants are increasingly unable to work because they are more poorly equipped, undernourished, and poorly cared for than ever before. They are obliged to concentrate their efforts on tasks that are immediately productive and neglect maintenance work for the cultivated ecosystem. In hydraulic systems, poorly maintained installations deteriorate. In slash-and-burn systems, in order to reduce the difficulty of clearing, peasants attack regenerating forest lands that are younger and less remote, which accelerates deforestation and the deterioration of fertility. In systems combining cultivation with animal breeding, the reduced number of livestock leads to diminishing transfers of fertility to the cultivated lands. Generally, cultivated lands that are poorly weeded become "messy," and the cultivated plants, deficient in minerals and poorly maintained, are increasingly subject to diseases.

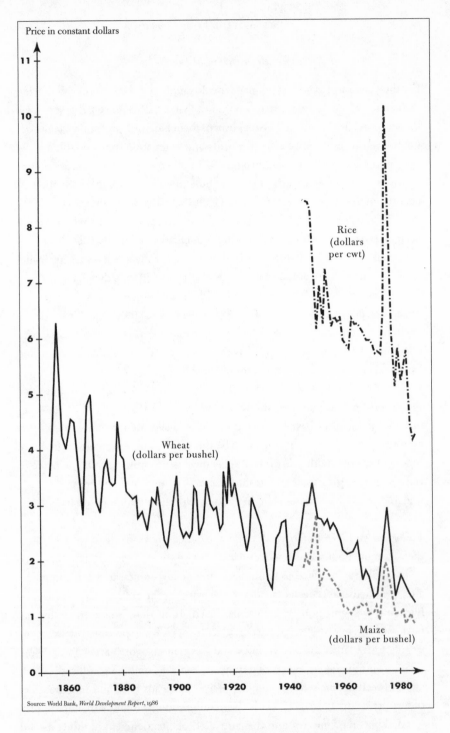

Source: World Bank, *World Development Report*, 1986

*Figure 11.3 Tendency of Real Prices to Fall and Fluctuations in those Prices for Some Major Agricultural Commodities in the United States*

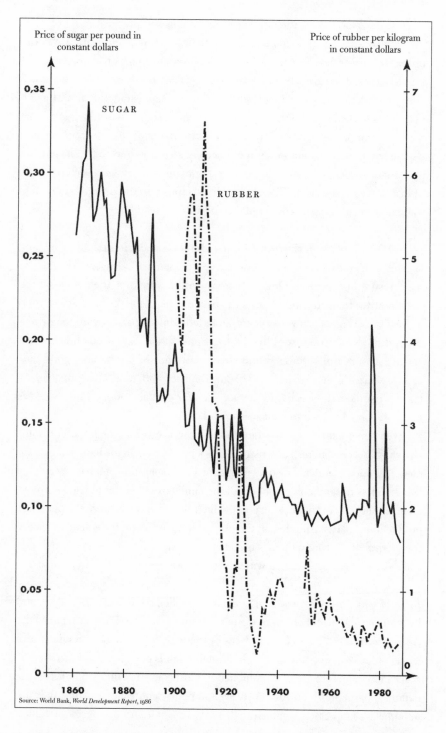

Source: World Bank, *World Development Report*, 1986

*Figure 11.3 (continued) Tendency of Real Prices to Fall and Fluctuations in those Prices for Some Major Agricultural Commodities in the United States*

The deterioration of the cultivated ecosystem and the weakening of labor power also lead the peasants to simplify their systems of cultivation. "Poor" crops, requiring less fertility, water, and labor, supplant crops with higher requirements. The diversity and quality of vegetable products intended for self-consumption declines which, added to the near disappearance of animal products, leads to dietary deficiencies in proteins, minerals, and vitamins.

Thus the farm crisis spreads to all components of the agrarian system: declining equipment, deterioration of the ecosystem and lowering of its fertility, undernourishment of plants, animals, and humans, and a general deterioration of health (see chapter 7). The economic instability of the productive system leads to the ecological instability of the cultivated ecosystem.

*Indebtedness and the Agricultural Exodus.* Impoverished, undernourished, and farming a degraded environment, weakened peasants are dangerously close to the threshold of survival. One bad harvest suffices to force them into debt, if only in order to eat during the months between harvests. At this stage, a good harvest can even make it possible for an indebted peasant to pay off the principal and the heavy interest of his debt, eat poorly, and save seeds for the next sowing season. But with his diminished conditions of production, good harvests are rare, the average harvest becomes smaller, and, most often, after repayment of the debt, there is hardly any food left for several months. The peasant is therefore forced to go into debt sooner and deeper.

Even being deprived of food up to the limit of survival, the possibilities for repaying the debt diminish and there comes a moment when the indebted peasant no longer finds a lender. The only remaining course of action, if it has not already happened, is to send able members of the family to search for outside employment, either temporary or permanent, which decreases even more the peasant's production capacity. And then, if the outside incomes are not adequate to ensure the survival of the family, there is no other recourse than to leave the countryside and move to the urban shantytowns, unless the peasant resorts to illegal crops.

*Illegal Crops.* In some remote and poorly controlled regions of Africa, Asia, and Latin America, illegal crops of opium poppies (Golden Triangle), coca (the Andes), and Indian hemp (Africa, Near East) are both possible and remunerative enough to make it possible for hundreds of thousands of poor peasants to survive. Since these crops are actually prohibited in most countries, they suffer less international competition. Furthermore, even in poorly controlled regions, they are nevertheless suppressed, so much so that they benefit from a sort of risk premium.

*Famine.* Although a peasantry having significant surpluses can support one or even several bad harvests, a peasantry that is reduced to the limits of survival is at the mercy of the least accident that could suddenly reduce the volume of its harvests or revenues. Whether this accident be climatic (flood, drought), biological (plant, animal, or human disease, invasion of predators), economic (drop in product sales, downward price fluctuation), or political (civil war, troops passing through), the peasants are condemned to famine on the spot or to refugee camps, if any exist nearby.

After more than a half-century, continually renewed strata of poor peasants in the developing countries were blocked in their development by competition and impoverished by the tendency of prices to fall. One after the other, they were excluded from agricultural production and forced to leave the country for the shantytowns or emigrate, or, in extreme cases, reduced to famine on the spot. This process of exclusion, which reduces agricultural labor power, has not yet affected all peasants using manual cultivation. It has, however, affected the most deprived peasants in the most disadvantaged regions. It is clear that if the fall in agricultural prices continues, new strata of the peasantry will also be excluded from production.

## Aggravating Circumstances of the Crisis of the Poor Peasantry

Our intention is not to add to the exposé of the world's miseries and still less to go one better than the apocalyptic visions which appeared at the beginning of this millennium. In the long run, pity and fear lead more to despair, indifference, and abandonment than to lucidity and supportive and lasting commitment, which are necessary to go beyond the simple continuation of sporadic assistance and eliminate the sources of mass poverty on a long-term basis. But in order to find the way, it is necessary that we examine, beyond the general mechanisms of the peasantry's impoverishment and exclusion we have just explained, the particular circumstances that still aggravate the crisis of the poor peasantry in the developing countries. These peasants are unequally affected; some suffer more on a long-term basis than others from particular disadvantages, be they natural or infrastructural, economic or political. The world economy of today is a game of comparative advantages. But some countries, some regions, some categories of peasants reap only disadvantages.

*Natural Handicaps.* Some of these disadvantages can be considered natural. Tropical regions having only one rainy season (Sahelian and Sudanese regions, for example) are disadvantaged in relation to equatorial regions with two rainy seasons, where it is possible to make two harvests per year. Sahelian regions with one small rainy season are more disadvantaged in relation to wetter Sudanese regions. In an analogous manner, cold, high-altitude regions of Central Asia and the Andes have a significantly smaller production potential. These dry or cold regions are sometimes even so handicapped that no production intended for export or the domestic market allows the producers to attain the threshold of renewal. The populations concerned support themselves just above the threshold of survival. They are at the mercy of the least climatic or biological accident. They are plagued by shortages and, quite often, political troubles that also contribute to and aggravate the crisis in these regions (Ethiopia, Somalia, Sudan, Chad, Colombia, Bolivia, the Peruvian Andes, Yemen, Afghanistan).

*Deficiencies in Hydraulic Infrastructures.* But if, in areas with rainfed cultivation, it is possible to consider the lack or excess of water as a natural handicap that aggravates the crisis of the poor peasantry and can precipitate famine, it is not the same in areas with hydraulic agriculture. There the lack or excess of water results from hydroagricultural installations inherited from the past and from the ability of current hydraulic institutions to maintain and extend these installations when needed. As we know, in this type of society the squandering of potential investment surpluses and the decline of institutions have particularly catastrophic consequences for the peasantry.

In China and Egypt at different times, because of the shortcomings of the state and hydraulic institutions, the developed lands were not extensive enough in relation to the number of people and their needs. Today, many rice-growing valleys, deltas, and coastal basins in Asia and Africa do not have the hydraulic infrastructures necessary to deal with prolonged droughts or deadly floods, whether the latter come from the land or the sea. Bangladesh, for example, lacking protective dikes, is periodically ravaged by devastating floods. There are also countries where the hydraulic infrastructures, while remaining extensive and strong enough, are neither maintained regularly nor repaired quickly enough in case of deterioration.

*The Smallest Farms.* Outside of these natural and infrastructural handicaps, one of the worst things that can happen to an underequipped peasantry is to have insufficient land to employ the family labor force completely and ensure its survival. While a manual farmer can cultivate from 0.5 to 2 hectares depending on the system practiced, there are many regions in the world where the majority of the peasantry does not have half, or even a quarter, of this area. These very small farms, or minifundia, are the result either of unequal land distribution among the farms or of overpopulation and excessive subdivision of farms, or of both simultaneously.

*Latifundia and Minifundia.* The most extreme case of unequal division of the land is the combination of excessively large and excessively small farms, i.e., *lati-minifundism*, a widespread agrarian social structure in the Latin American countryside. In this region of the world, large agricultural estates of several thousand, indeed even of several tens of thousands of hectares, often underexploited, monopolize the largest part of the agricultural lands, while the poor peasantry is confined on ridiculously small farms, which do not even produce enough to cover the minimum dietary needs of the peasant families.

These families are forced, in order to obtain the necessary additional income, to sell their surplus labor to the large estate owners, who are often the only employers in the countryside. This underpaid labor force comes to be added to the mass of "landless peasants," sometimes homeless, who go to look for work from region to region according to the agricultural seasons. For the large estate owner, this landed property structure presents the double advantage of avoiding competition from a true peasant economy and having available a numerous labor force, at the lowest price possible.

*Unequal Land Distribution and Minifundia.* The largest part of the land does not have to be concentrated on a few large estates for a significant fraction of the peasantry to be confined on minifundia or totally deprived of land. In regions with hydraulic agriculture, in particular, the organized farmable area is often hardly sufficient for all peasant families to have a farm in keeping with their means and their needs. In such conditions, it suffices that there be just a slightly unequal distribution of the land to reduce a portion of the peasantry to minifundist status. In many rice-growing valleys and deltas of Asia, in the Nile Valley, etc., it suffices that a minority of "rich" peasants (peasants who are often just a little less poor than the others) hold more than half of the lands for a majority of the peasantry to be deprived of land. But that can also be the case in regions with rainfed cultivation where the cultivable lands cannot be expanded at will.

*Overpopulation and Minifundia.* Demographic pressure can be, alone, a cause of the proliferation of minifundia. In any agricultural system, when the population density increases, there inevitably arrives a moment when all the cultivable lands have been taken advantage of and the cultivated area per farm is reduced. In order to maintain productivity and income, the peasants therefore increase the quantity of labor and production per unit of area. They "intensify," as it is called, by increasing the number of crops (associated crops, accelerated crop sequences, fruit plantations) and the care and attention given to them. But as can be seen in many regions of the world (Rwanda, Burundi, overpopulated deltas), this type of gardening has limits. Beyond a certain threshold, the additional labor does not yield very much. Consequently, if this peasantry does not have additional means that allow it to adopt a new, more productive system, the increase in population is expressed purely and simply by a growing underemployment of the labor force, a fall in income per worker, and impoverishment.

It is exceptional that this process of proliferating minifundia due to overpopulation is not worsened by inequalities arising from distribution of the land. But it remains the case that in some regions overpopulation is indeed the essential cause of the proliferation of minifundia. Thus, in the delta of the Red River, the Vietnamese government conducted at the end of the 1980s a redistribution of the lands of the former cooperatives to peasant families. This relatively egalitarian redistribution was made in proportion to the number of people to feed in each family. However, the area of the farms formed in this way does not exceed one half hectare, and it is often less than half of the area that could be cultivated by each of these families.

*Policies Unfavorable to Agriculture.* But beyond these natural, infrastructural and land ownership disadvantages, many countries have also carried out economic and agricultural policies that are unfavorable to agriculture in general and to the poor peasantry in particular. In this regard, the costly policies of infrastructural and administrative modernization, the overvaluation of currencies and the protection of industry have been particularly harmful for agriculture.

*Modernization, Overvaluation of Currency, and Protection of Industry.* Ruinous investments, oversized in relation to the needs and financial capacities of countries, and to a great extent not too productive or even unproductive, have abounded. They not only have taken capital away from agricultural production, but they have also attracted a significant fraction of the young labor force away from agriculture, and that all the more so since the minimum legal wage observed in the administration and in public works has often been much higher than the income attainable for a peasant. Insofar as this reduction in the agricultural labor force has not been compensated for by investments that make an increase in productivity possible, the effect has been a reduction in agricultural production per inhabitant. In Congo, for example, in thirty years half of the working population of the country went from the countryside to the cities. The number of mouths to feed per agricultural worker doubled, going from 4 to 1 at the end of the 1950s to 8 to 1 in the 1980s. Since the productivity of manual cultivation of forests and savanna did not increase one iota during the same period, the country depended on food imports for nearly half of its food needs.[4] In order to finance all these modernization expenses, the poor states have resorted to large-scale borrowing, both domestic and foreign, and to creating money, which in turn generated inflation. This inflation has been much higher in the developing countries than in their commercial partners among the industrialized countries, which has led to a loss in the relative value of their currencies. Rather than devalue to compensate for this loss in relation to foreign currencies, governments generally have preferred to maintain the overvaluation of their currencies, which comes down to subsidizing imports and taxing exports and is particularly unfavorable to agricultural producers in the developing countries.

Naturally, the overvaluation of national currencies could be harmful to industrial production, too. But because of the high priority accorded to industrialization in most of these countries, the industrial sector has not only benefited from all sorts of tax exemptions, low-interest loans, subsidies, and a significant part of the public investments, it has also often been protected from foreign competition by all kinds of measures (high import taxes, quotas, etc.). By limiting imports, this industrial protectionism has contributed to the overvaluation of national currencies, and by causing the internal prices of manufactured products bought by farmers to rise, it has degraded the terms of exchange even a bit more, to the detriment of agricultural products. According to a study conducted in seventeen countries in Africa, Asia, and Latin America, the protection of industry has been the economic policy measure that, from 1960 to 1985, has exerted the most influence on the relative fall in agricultural prices in relation to other prices.[5]

*Agricultural Price Policies.* Generally, agricultural price policies have only reinforced this tendency, because in many developing countries, the poor urban population has become excessive, indeed even the majority, and its political influence, as in Rome of the past, has become much greater than that of the peasantry. To respond to the pressure of urban consumers and limit wage increases in industry and the administration, governments have attempted to supply cities with low-priced food commodities. The most commonly implemented measures to lower food prices, and thus agricultural prices, have included resorting to food aid, commercial imports at the lowest possible prices, subsidies for the consumption of imported food products (grains and flour, in particular), and sometimes obliging peasants to deliver specific quantities of products at low prices.

But in order to have the means to pay for the growing imports of all kinds, governments have often heavily taxed exports of agricultural products. These taxes have formed in many cases one of the principal sources of revenue for the state budget. The prices paid to the producers have then been cut back drastically in proportion. In some countries, this taxation, by being added to other factors lowering agricultural prices, has even ended up leading to a decline in production. This is what happened in several countries of Africa (Togo, Congo) when the coffee producers first stopped planting, then stopped maintaining the plantations, and finally, even refrained from harvesting.

*The Pillage of Agriculture in the Developing Countries.* In many poor countries, policies of increasing investments of dubious value and unproductive employment, protecting industry, overvaluing the national currency, taxing agricultural exports, subsidizing food imports, and compulsory deliveries at low prices have been combined to devalorize yet again the fruits of agricultural labor. The study cited earlier demonstrated that for the seventeen countries of Africa, Asia, and Latin America considered in the period 1960–1985, the cumulative effect of these policies was generally a significant deduction from prices paid to farmers, which led to a fall even more significant in proportion in their net income.[6] All things considered, this deduction took effect as an enormous transfer of income toward the state, industry, and urban consumers, a transfer so significant that the authors of the study do not hesitate to view it as a veritable "pillage of agriculture in the developing countries."

This study demonstrates moreover that the "taxation" of agriculture harmed agricultural development. The countries that strongly taxed their agriculture had a rate of agricultural growth less than half that of countries that weakly taxed theirs. It also demonstrates that the countries that strongly taxed their agriculture had a much lower rate of general economic growth than the others. It shows, in addition, that countries like South Korea that protected their agriculture instead of taxing it had the highest rates of economic growth.

Not every developing country has carried out policies unfavorable to its agriculture. But above all, let us not forget, in cases where policies unfavorable to agriculture existed, the cumulative effect of these policies on the prices paid to farmers, as significant as it was, generally remained much less than the effect on prices from competition from more productive agricultures. It is necessary to recognize that the economic and agricultural policies carried out in the developing countries have had the advantage of not completely passing along the effects of strong fluctuations in world prices for agricultural products to domestic prices and thus generally stabilizing production prices. In fact, in countries where the great majority of producers and consumers are poor, the negative effects of fluctuations in agricultural and food prices are disastrous.

## The Disastrous Effects of Price Fluctuations

As for export crops, periods of low prices reduce the monetary income of peasants in a dramatic manner. The most poorly situated and most destitute of the producers in the developing countries are, by the millions, plunged below the threshold of survival and condemned to leave the countryside or even to suffer famine on the spot. Then, in the periods of high prices that follow, since most of the producers previously excluded from production do not have the means to return to the land, their market shares are partly taken over by better-equipped producers in more favored regions and countries.

As far as grains are concerned, when the world market is saturated and the prices are low (as was the case at the end of the 1960s and in the 1980s), food aid is abundant and the large producer countries even subsidize their commercial exports. Grains imported at low prices increase their share of the markets and consumption in the poor countries, so that the local producers of other food commodities (millet, sorghum, local rice, yams, cassava, sweet potatoes, taro, plantains, etc.) are plunged into crisis, and food dependence is widened. Some years later, when world production and reserves of grains are insufficient (as was the case in the 1970s), high prices return. But the producers excluded during the previous period are no longer there to take advantage of the price increases and the revival of domestic production in the poor countries is weakened, while the needs of the cities continually increase. In this situation, food aid becomes scarce, the bill for food imports becomes overwhelming, and, unless basic food commodities are subsidized, consumption by the poorest is reduced and shortages and famines reappear.

Natural or infrastructural handicaps, proliferation of small farms and harmful policies up to and including the "pillage of agriculture" contribute in no small measure to the agrarian and food crisis of the poorest agricultural countries. In countries and regions where several of these particularly unfavorable circumstances intersect, the multiple factors combine to take a devestating toll. This is what happened in northeast Brazil, where the aridity of the climate, lati-minifundism, and the predominance of one crop, sugarcane, which has suffered many vicissitudes, combined. Such is also the case with Bangladesh, which suffers from the shortcomings of an inadequate hydraulic structure and a minifundism resulting from both the unequal distribution of land and overpopulation. Such is still the case in many Sahelian countries.

But as unfavorable as they may be and as dramatic as their consequences may sometimes be, these worsening circumstances should not mask the fact that the essential cause of the agrarian crisis and of the rural and urban poverty that strikes the poor agricultural countries lies elsewhere. This crisis and poverty were inevitable from the moment that the underequipped and poorly productive agricultures of these countries were confronted with competition from forms of agriculture several hundred times more productive and with the resulting fall in agricultural prices. And there is no doubt that if the tendency of grain prices to fall and the subsequent fall in the prices of all other agricultural commodities continues, the huge agricultural exodus—the excessive increase in the population of shantytowns, and emigration—will also continue.

In the same way that the contemporary agricultural revolution and the revolution in transportation led to the elimination of the underequipped and poorly productive small peasantry in the developed countries, the expansion of the agricultural revolution to tropical branches of production and the extension of the transportation revolution to the developing countries are leading to the impoverishment and massive elimination of the underequipped peasantry in these countries.

But the analogy stops there. In the developed countries, tens of millions of workers excluded from agriculture since the beginning of the twentieth century were, except during crisis periods in the 1930s and since 1975, gradually absorbed by the development of industry and services, without reducing the production potential of agriculture, which was actually more productive. On the other hand, in the developing countries, there are not tens but hundreds of millions of poor peasants who, in only a few decades, were condemned to an agricultural exodus. As noted, in most of these countries, this massive exodus was not compensated for by gains in agricultural productivity, and investments from the whole world have not sufficed and will not suffice, not by a long shot, to absorb this uninterrupted stream of the rural poor looking for a new means of existence.

## 2. FROM THE AGRARIAN CRISIS TO THE CRISIS OF THE DEVELOPING COUNTRIES

### From Rural Poverty to Urban Poverty

In the course of the last decades the exodus of hundreds of millions of poor peasants from the countryside has led to an excessive increase in the size of the cities of Latin America, Asia, and Africa, an excessive increase in the sense that these cities did not have the necessary infrastructure to accommodate them nor the industrial and service businesses able to employ them.

From that moment on, these immense migrations have led to the formation of megalopolises split into two distinct parts. In one part, an urbanized core, the so-called *formal* administrative and economic activities are concentrated along with the social groups having jobs and regular incomes. In the other part, proliferating shantytowns, populated with a growing mass of poor people. These are poor who have come directly from the countryside, as well as descendants of those who had taken the road to the city a generation earlier.

Among the masses of these job seekers, only a minority of the able-bodied and well-adapted is in a position to gain the skilled and regularly paid jobs offered by the government and by well-established national or foreign businesses. These few, stable jobs are often reserved to persons coming from well-off circles (large landowners, planters and wealthy peasants, merchants and businessmen, officials and other regularly paid wage earners), when they are not given to "expatriates" coming from the developed countries.

## Visible Unemployment and Hidden Unemployment

The immense majority of the poor in the cities is doomed either to unemployment or to insecure, thankless, and underpaid employment in businesses in the so-called *informal sector*, or even small individual jobs produced by the profusion and endless subdividing of service activities: the last resellers of cigarettes sell them one at a time, lit or not, at the corner of each street, to buyers who do not have the means to pay for a whole pack. This is not to mention prostitution, criminality, and begging that, because they transgress public morality, are the origin of other forms of exclusion and suffering.

Many informal activities (resellers, shoe-shiners, windshield washers, guards for cars and houses, occasional porters) require availability for long periods of time everyday (eighteen to twenty hours) for insignificant actual work time. By occupying a large number of people as much as possible, they contain and conceal more unemployment and poverty than they actually supply jobs and income. Low levels of capitalization, productivity, and labor income generally characterize the unregistered businesses of the informal sector, which escape all labor legislation. These levels are, in fact, hardly above those of poor agriculture but, as minimal as they are, this tiny advantage explains why the shantytowns are nevertheless attractive for the most destitute country people, in particular for the young without family responsibilities.

Since the agricultural exodus greatly exceeds the labor needs of the formal and informal sectors, significant structural unemployment appears, and this is only the visible part of an immense, hidden underemployment.

## The General Depreciation of the Fruits of Labor in the Developing Countries

In this context of massive unemployment, officially recognized or not, the wages of the unskilled labor force are established in the following manner. The daily wage of an occasional worker is hardly above the price of his daily food. The annual wage of an unskilled employee, occupying a relatively undemanding job, is set in the range of the price of ten quintals of grains, or of what inadequately feeds a family of four or five persons. The annual wage of a relatively unskilled or completely unskilled employee occupying a job requiring daily attendance, sustained attention, and reliability covers the dietary needs of a family as well as a minimum of other essential needs.

This is why the basic wage paid by a business, whether national or foreign, even were it the most modern in the world, is without any relationship to the productivity of labor in that business. The wage is in line with the market price of the local labor force. In other words, in countries that are already relatively industrialized and that protect their agriculture, like South Korea, the basic wage approaches that paid in the developed countries, but in the still largely agricultural poor countries, like China or Vietnam, it is thirty to forty times lower.

The extremely low price of the relatively unskilled labor force reduces to next to nothing the costs of production and the prices of goods and services produced and consumed locally, which leads to a lowering of the wages of other categories of employees. Thus, with equal skills and work, the expert or interpreter originally from a developing country, employed in his country by a firm or an international organization, is paid up to ten times less than his counterpart in a developed country.

The integration into the same market of less industrialized countries, heirs of a relatively unproductive agriculture, and of industrialized countries, heirs of a highly productive agriculture, is characterized by the establishment of an exchange relation seriously unfavorable to the former. It takes dozens of years of labor for a peasant or wage laborer in a developing country to buy the product of one year of work in a developed country. Conversely, several days of labor for a wage laborer or a farmer in a developed country are adequate to buy the product of one year of labor in a developing country.

## Deterioration in the Terms of Trade

Most of the economic studies concerning the prices of different categories of commodities show that over the course of the last decades, the prices of agricultural and nonagricultural raw materials fell in relation to the prices of manufactured products.[7] According to the World Bank, between 1950 and 1984, the average weighted prices of grains, in relation to the average prices of manufactured products imported by the developing countries, fell 1.3 percent per year, while the prices of all agricultural products and of raw materials respectively fell 1.03 percent and 1.08 percent per year.[8]

As a result, many developing countries that are essentially exporters of raw materials and importers of manufactured products were subjected, in the long term, to a strong "deterioration in the terms of foreign trade." But in this context it is necessary to point out that many developing countries became importers of grains and other basic food commodities, that some even became *net* importers of agricultural products, and that a handful of new industrialized countries are already principally exporters of manufactured products. It is then possible that the differential evolution of commodity prices became less unfavorable than in the past for some developing countries, indeed even became favorable in some cases. But these hypotheses, difficult to verify, are still controversial.

# The Failure of Modernization Policies in the Poor Agricultural Countries

Except for some Asian and Latin American countries, which had a productive enough agricultural subsector to be able to extract a non-negligible surplus while continuing to advance, and except for some large petroleum-exporting countries, which had and still have very significant incomes, tax revenues, and exports, most developing countries are poor agricultural countries, which inherited a completely underequipped agriculture and possibly some mineral resources. Now, despite their mediocre incomes and receipts, almost all these countries, like the others, have embarked on policies aimed at rapidly modernizing their infrastructure and their state apparatus, in order to create, at least that was their hope, the necessary conditions for industrial takeoff and growth.

The rulers of the poor countries, like most of their advisors from West and East, equated underdevelopment with a simple infrastructural, industrial, institutional, and educational backwardness, and they had the ambition of catching up, in a historically short time period, with the level of development and income of the developed countries. They also expected that, following the example of nineteenth-century Europe, North America, and Japan, the agricultural sector of their countries would be able to free the necessary capital and labor for the development of industry, infrastructure, and services. But by doing this, they underrated the notable weaknesses of their agriculture and, certainly, were unaware of the lowering of agricultural prices (in real terms) that was going to swoop down on their economy during the following decades.

## Public and Foreign Deficits

The heavy public expenditures for urban infrastructure, communication (ports, railroads, highways, airports, electrification, telecommunications), education, health, general administration, defense, etc., not only greatly exceeded the meager tax revenues of the poor countries but, more seriously in most cases, did not engender the expected dynamic of investment and agricultural and industrial development. Despite an often advantageous investment code, modernization was not enough to retain or attract the mass of capital that would have been necessary to lead to a true economic takeoff. The available capital was oriented first toward the developed countries and their vast market, then toward a small number of Latin American and Asian countries that offered it maximum economic advantages and political guarantees, countries that today have become or are becoming the "new industrialized countries."

Taking note of the insufficient productive private investments, but fearing the grip of foreign capital on their economy and undoubtedly impressed by the industrial progress recorded in the 1930s and 1950s by the Soviet Union, a number of governments committed their countries on the path of more or less extensive state control of mining, agricultural, industrial, and commercial activities. The additional heavy expenditures in inconsistently profitable productive investments came to be added to the public expenses of general modernization.

Whether their rulers claimed to adhere to liberalism or interventionism, the poor countries were plunged into significant chronic deficits. From 1972 to 1982, for all the non-petroleum-producing developing countries, the public budget deficit went from 3.5 to 6.3 percent of their gross domestic product.

On the other hand, since modernization was conceived on the model of the industrialized countries, it required numerous imports of goods and services from them and heavy expenditures in foreign currencies. Except in transient periods of high prices for raw materials, these expenditures largely exceeded the foreign currency receipts of the poor agricultural countries, leading to a chronic deficit in the foreign balance of payments. From the end of the 1960s to the beginning of the 1980s, in fifteen years, according to the statistics of the International Monetary Fund, the deficit in the current balance of payments for all of the non-petroleum-producing developing countries went from some $6 billion to nearly $100 billion.

## *Overindebtedness*

The foreign deficits, a good part of which were created from public deficits, were then covered by massive borrowing from foreign states (developed countries or petroleum-exporting countries), commercial banks, and international financial institutions. In the 1970s, the corresponding loans were granted all the more easily since a lot of capital was poorly invested and the ephemeral high prices of raw materials made it possible to overestimate the potential of the developing countries to repay them. In many poor countries, the foreign debt came to represent a significant part of the gross national product, and sometimes even came to exceed it (Ivory Coast, Costa Rica).

At the end of the 1970s, after the collapse of raw materials' prices, these heavily indebted countries found themselves unable to pay back their debts, at the same time that they continued to pay for their imports of manufactured goods, food products, and petroleum. As had already happened in the nineteenth century in countries like Egypt and Peru (see chapters 4 and 5), the desire to catch up quickly with Western modernity was transformed into a veritable financial trap and therefore a political one.

## Policies of Stabilization and Structural Adjustment

Except for opting not to pay the debt service, which would have consequently led to the loss of all international credit, the inevitable collapse of their imports, and de facto harsh austerity, the overindebted countries had no other alternative than to seek a rescheduling of their debts as well as additional loans. Since it is acceptable in such circumstances, these "generous gifts" were granted to requesting states only on condition that they implement "stabilization" policies, that is, austerity policies aimed at reducing in the short term public budget deficits and foreign payments, while continuing to pay back old and new debts.

These stabilization policies, carried out under the aegis of the International Monetary Fund, consist of reducing investments and consumption in a draconian manner. In order to do that, various measures are applied: reduction in wages and in the number of officials, reduction in the costs of operating the administration, reduction of public subsidies and welfare payments, increase in taxes, generalized wage austerity, increase in the interest rates, restriction of credit granted to the state, to businesses and to households, and currency devaluation. But if these policies indeed contribute to reducing budget and foreign deficits, without always succeeding in reestablishing the corresponding equilibria, they also inevitably have negative effects on economic growth and on the purchasing power of countries that implement them.

This is why these short-term stabilization policies are accompanied with "structural adjustment" policies in the medium term, which aim, under the aegis of the World Bank, at stimulating production and trade. These policies are based on the premise that free enterprise and free trade are, in all circumstances, the best way possible to promote economic development and social well-being and, consequently, advocate liberalization of prices and foreign trade, deregulation of markets, particularly the labor market, the organization of financial markets, and, where these exist, the improvement of their effectiveness, the disengagement of the state from all economic activity, and privatizations. But this premise, which is advanced by only some neoliberal economists, is far from being accepted by the majority of economists.

## The 1980s: A "Lost Decade for Development"

After fifteen years of a more or less strict implementation of these policies, the least that can be said is that they have not had all the expected effects relative to the stimulation of economic growth. In the 1980s, the average per capita income fell 10 percent in Latin America, 25 percent in sub-Saharan Africa, and, in some countries, real incomes were reduced by more than 50 percent.[9] However, this "lost decade for development," according to the expression used by the United Nations, was not lost for everyone. In these same years, the average per capita income increased by 50 percent in South and East Asia. There are many reasons to think that these quite contradictory developments are the result of the unequal economic and social legacies of different regions of the world, and their relative positions in the world economy, rather than of the more or less strict application of adjustment policies.

## The Case of the Petroleum Countries and the New Industrialized Countries

In the system of international trade established over the last few decades, the developing countries that have had sufficient capital to invest and significantly increase their population's income are the exception. Some exceptions are a few large petroleum-exporting countries such as Saudi Arabia, the United Arab Emirates, or Brunei, whose export receipts are so large in relation to their population that imported modernity has been pursued quite far and poverty has almost disappeared. These countries even attract a numerous labor force from poor countries and, in addition, realize savings, a large part of which is deposited in foreign countries. But most of the petroleum-exporting countries, such as Mexico, Venezuela, Algeria, and Nigeria, do not have such high per capita export receipts, which does not prevent the capture and redistribution of petroleum income from exercising a multiplier effect on imports and a sort of eviction effect on activities directly productive of goods and services. Thus these countries are far from having reduced poverty and unemployment and are even today among the most heavily indebted countries, their repayment potential having been widely overestimated following the two "oil crises."

Otherwise, some countries of Southeast Asia and Latin America have been able to accumulate capital, ensure that all or part of their infrastructure and administration comply with "international standards," and create favorable enough conditions to attract large amounts of international capital. Some of these countries have even found the way to strong and long-lasting growth. In commercial and financial centers without hinterlands, such as Hong Kong and Singapore, and in small territories such as Malaysia, unemployment has almost disappeared. However, even in countries such as Taiwan and South Korea, which have been able to base their industrialization in part on their own agricultural surpluses and where there is a clear tendency to increase wages and enlarge the internal market, unemployment and poverty have not been eliminated. As for the large, partially industrialized countries, such as Indonesia, Thailand, India, China, Brazil, Mexico, Argentina, Chile, etc., unemployment and rural and urban poverty remain immense, and industrialization has only a limited effect on the basic income level, which remains very low.

Beyond the developed countries already industrialized at the beginning of the twentieth century and setting aside a handful of developing countries endowed with relatively productive agricultural activities or large enough commercial or petroleum rents, the large majority of the world's countries have inherited only an underequipped, largely unproductive agricultural sector, incapable of financing a rapid and costly modernization imported from the developed countries.

Integrated into a system of international trade that has put them in competition with developed countries having a much more productive agriculture, these underequipped and relatively unproductive agricultural countries have suffered a severe depreciation of the fruits of their labor. This depreciation has been worsened by the tendency of agricultural prices to fall resulting from the sustained progress of the most productive agricultures. These so-called countries in development have in fact become impoverished agricultural countries, that is, countries with very low agricultural incomes and very low wages, with little or no industrialization, producing little, having low public revenues and low foreign currency revenues. These are indebted countries that do not have the means to accumulate capital in order to begin a real development, or even the means to modernize sufficiently in order to attract foreign capital. These are countries long in crisis, where unemployment and mass rural and urban poverty reach unsustainable proportions; countries in which hunger and massacres are not uncommon, where the impotence and disintegration of the state are increasingly manifest.

In these conditions, it is illusory to think that national policies exist that would make it possible for each of these states to lift their countries out of poverty. Not that the policies that they carry out are unimportant, as we will see. But it is clear that in order to raise all of the incomes of the poor agricultural countries significantly, allow them to accumulate capital, develop, and modernize, it would be necessary to establish first a totally different system of international trade. Not a system in which agricultural prices tend to be standardized and fall in real terms, thereby reducing the revenues of these countries to the poverty level of their agrarian heritage, but a more equitable system of international trade in which prices would be differentiated and raised so as to compensate for the formidable handicaps in equipment and productivity from which these countries suffer.

## 3. FROM THE CRISIS OF THE DEVELOPING
## COUNTRIES TO THE WORLD CRISIS

### The Twenty-five "Glorious" Years of Sustained Growth

From the end of World War Two to the beginning of the 1970s, the world, in particular the developed countries with market economies, experienced nearly three decades of strong and sustained economic growth. Supported particularly by technology and the high financial capabilities of an American economy that had ended the war greatly enlarged, the countries of Western Europe and Japan were first rapidly reconstructed. Then, learning from the underconsumption crisis of the 1930s and the demonstrated successes of Keynesian policies of stimulating production by stimulating demand, the developed countries carried out policies aimed at maintaining demand at a sufficiently high level to stimulate production and achieving full employment of the labor force: public investments, public expenditures in the general interest (defense, education, health), wages indexed in practice on productivity gains, extended social welfare, steady agricultural prices, investment assistance, etc.

These policies were facilitated by the international monetary, financial, and commercial system put in place by the Bretton Woods Accords in 1944. This system rested in the first place on the stability of exchange rates, in order to secure the forecasts and estimates of investors and avoid a series of competitive currency devaluations, like those that had been carried out by most countries in the 1930s. The exchange rates were nevertheless adjustable. A country confronted with a fundamental disequilibrium in its economy (a rate of unemployment or inflation considered to be unsupportable) could decide to devalue its currency. Moreover, in this system the monetary and budgetary policies of each country remained independent and capital movements were controlled.

But the Bretton Woods system, like previous systems for that matter, left to the country experiencing a long-lasting trade deficit the responsibility to reestablish the equilibrium of its external balance by carrying out policies to reduce internal demand (consumption and investment). This arrangement ran counter to the proposals of John Maynard Keynes, who had recommended that trade balances between countries be reestablished by increasing demand in the surplus countries rather than reducing it in the deficit countries. From this perspective, J. M. Keynes had also proposed to create an international currency that was not convertible into gold, and not to use for that purpose the U.S. dollar guaranteed by gold, as imposed by the American government. Moreover, during this whole period of exceptional growth, the cold war, the armaments race, and a few hot wars (Korea, Algeria, Vietnam) also contributed to keeping economic activity at a high level.

Thus, from 1950 to 1973, the wealth produced in the world increased an average of 4.7 percent per year, and world trade advanced at the rate of 7.2 percent per year.[10] Further, it should be noted that from 1965 to 1973 the average annual rate of growth of developing countries exceeded that of developed countries: 6.2 percent versus 4.4 percent. But since these rates of growth corresponded to initial, inordinately unequal levels of wealth, the increase in the quantity of wealth produced and consumed in absolute value was much larger in the developed countries than in the developing countries. In 1973, at the end of these twenty-five glorious years of sustained expansion of the world economy, the global purchasing power of all the developing countries remained very low in comparison to that of the developed countries.

## Insufficiency of Effective Demand and the Slowdown in Growth

From 1973 to 1990, the growth in production and world trade slowed down considerably. The annual rate of growth for production fell to 2.5 percent and that of trade to 3.9 percent, or a fall of nearly half in relation to the previous period. From the beginning of the 1970s, the development of world production potential began to come up against the limits of planetary buying power. In the 1960s, medium and long-term studies of the market, carried out by large economic research departments (Rand Corporation in the United States, Société d'économie et de mathématiques appliquées and Metra International in France and Europe) or by specialized services of large businesses and banks, had already demonstrated that, for numerous goods and services, expected demand for the beginning of the 1970s was going to be far below expected supply according to the investment and development plans of the industries concerned. Armed with these predictions, large investors then revised downward their development plans, taking into account the predicted limits to effective demand.

Note that in the 1920s, relatively dispersed businesses did not have the means to make studies and effective economic predictions, which would have made it possible for them to adjust their investments as a function of expected demand. That explains why in the 1970s, contrary to what happened in the 1930s, the weakness of demand did not result in the formation of excessive production capacities, the accumulation of unmarketable stocks, the collapse of prices, the multiplication of failures, mass business closures, massive layoffs, an enormous stock market crash, and a financial debacle.

## Unemployment, Speculation, and Stagflation

If, in the 1970s, the crisis resulting from the insufficiency of demand relative to productive capacity did not take the catastrophic turn of an overproduction crisis, as in the 1930s, it nonetheless remains the case that the curbing of productive investments was still characterized in the developed countries by a net slowing in growth, the development of unemployment, and the appearance of a mass of capital in search of profitable investments. A growing part of this "floating" capital was directed, depending on circumstances, toward all kinds of speculation: currencies, raw materials (oil crises, doubling of grain and soya prices in the middle of the 1970s), gold, real estate, securities (shares, bonds, holdings), and derivatives (hedging in the futures market, options).

These multiple types of speculation were made possible by the dismantling of the system of fixed exchange rates (1973) and by the financial deregulation (suppression of controls on international movements of capital by the United States in 1974, then by most other countries in the 1980s). In the 1970s, these various types of speculation contributed to making goods and services more expensive, while, in the OECD (Organization for Economic Cooperation and Development) countries, policies of supporting demand and creating money fed inflation, without succeeding in stimulating economic activity. This paradoxical combination of two phenomena, inflation and stagnation, until then considered incompatible, was simply baptized with the name *stagflation*, instead of being clearly explained.

## Living on Credit

In order to find uses for an abundant savings with few profitable investment outlets, public, private, and international financial institutions embarked on extensive campaigns to make loans to the governments of the developing countries, as well as to the governments of the socialist countries and the developed countries. From the beginning of the crisis, almost every country in the world ended up living largely on credit. Further, if this credit contributed to enlarging consumption and to stimulating production a bit in the short term, its repayment was necessarily a burden on purchasing power in the medium and long term, barring, of course, cancellation of the debts. Lacking the potential to invest immediately in production, a growing part of world savings took refuge in investments that lived on profits from speculation and interest on loans.

## Modernizations, Relocations, and Reduction
## of World Effective Demand

In this context of a weak expansion of demand, businesses able to invest could hardly develop other than to increase their share of the market to the detriment of competing businesses. In order to accomplish this, they had to lower their costs of production and prices and search for a significant improvement in quality, marketing, and after-sales service for their products.

In branches where significant productivity gains were possible, businesses that had the means (that is, those with financial reserves, credit, and possibly government aid) carried out heavy investments in modernization, which made it possible to reduce their labor force greatly. In the automobile industry, for example, Japanese, then American and European businesses automated their manufacturing from 1970 to 1990 to an extent that the necessary labor time to assemble an average car was reduced by around half. These modernizations led to the elimination of numerous jobs in the branches concerned, which, because of insufficient growth elsewhere, was characterized by significantly increased unemployment and thus reduced demand for consumer goods.

In branches using a large labor force that would be difficult to reduce (textiles, shoes, etc.), businesses that had the means reduced the costs of production by subcontracting their manufacturing, or building new factories in low-wage countries that offered attractive conditions for effective investments and political guarantees—that is, in a few Asian and Latin American countries in the process of industrializing and in several formerly socialist countries with low wages.

Insofar as the relocated factories are substituted for factories in the developed countries, they lead, in some of the latter, to reductions in employment and income, not only in the activity directly concerned but also in upstream and downstream activities and in all other activities that are linked to it. Since these reductions in employment and income are not compensated for by creating employment and income in other sectors, this type of movement is characterized by a distribution of incomes to the developing countries far below the incomes eliminated in the developed countries, and therefore by a reduction in world demand for consumer goods.

Certainly, the relocation drives in the 1970s entailed additional demand in capital goods, which had some stimulating effect in the developed countries. But this phenomenon was considerably less in the 1980s, in part because some newly industrialized countries began to produce their own producers' goods, which they even exported to the developed countries, and in part because many developing countries had to reduce their imports of capital goods in order to repay their debts.

In the final analysis, in a world economy with few outlets, the relocation of industrial activities from a country with high wages to a country with low wages has the effect of restricting growth in world demand for consumer goods. That is particularly true when the relocation takes place in countries that are only just beginning their industrialization process, such as China and Vietnam where the wages are thirty to forty times lower than in the developed countries. That is less true for countries already largely industrialized, such as South Korea and Taiwan where wages are relatively higher.

Of course, it is not the industrialization of the developing countries as such that is at issue. Any productive investment in a low-wage country that meets an increase in purchasing power and is expressed by a net creation of income on a world scale is welcome, because it contributes greatly to the expansion of global demand. However, industrialization of the developing countries on the basis of absurdly low wages and at the price of deindustrialization of countries with higher wages poses a problem. In sum, for industrialization of the developing countries to give rise to a global net creation of jobs and incomes, it must not rest on the very low wages existing in these countries and aim principally to export products to the high-income countries. It must be founded on a growth in local purchasing power that, to be significant, must involve the mass of poor people in the countryside and cities, which necessarily presupposes, as we have seen, a *preliminary* raising of the incomes of the peasantry.

### *Growing Unemployment and Lowered Wages in Developed Countries*

As a result of the slowdown of investment in industry and the services, modernization and relocation, and the continuing exodus of millions of farmers, unemployment has increased considerably in the developed countries from the middle of the 1970s. Some branches of mining (coal, iron ore), primary processing (steel metallurgy), and manufacturing (textiles, shoes, watches) were partially dismantled, cities and whole regions (Liverpool, Lorraine) were deindustrialized. In the OECD countries, between 1975 and 1995, employment in the manufacturing sector fell 8 percent. It fell 20 percent in the European Union and 35 percent in the United Kingdom.[11] In the OECD countries again, the number of unemployed passed 30 million at the beginning of the 1980s and reached 35 million in 1994. Moreover, more than 10 million persons now work part-time, against their will, and many more, having given up looking for work, no longer even appear in the unemployment statistics.

Unemployment and increasingly stiffer competition from modernized or relocated companies exercises a strong downward pressure on wages, in particular on incomes of less skilled workers. In some countries, particularly the United States and United Kingdom, the labor market was largely deregulated and the wages of these workers fell greatly. But if the lowest-wage Scottish or American workers are now close to Korean wages, they are far from having fallen to a level as low as in all the countries "in the process of industrializing." A wage of $600 per month in the Midwest of the United States is still thirty times more than $20 a month in Vietnam or China. Relocations are continuing to occur, and even if the fall in wages in some developed countries has already contributed to retaining or attracting some investments, unemployment has not disappeared as a result.

## Deregulation, Speculation, and Austerity

So-called neoliberal policies, which have been dominant in the world since the end of the 1970s, only worsen the general crisis. In the developing countries, these policies are expressed by the abandonment of autocentered development strategies, based on public investments aimed at satisfying the internal market by the production of import-substitution goods, and by the adoption of eternal-oriented development strategies, based on private investments from many sources, attracted by low wages and focused on exports. These new strategies, encouraged by the World Bank and other development institutions, widen the field for relocations, although they do not increase world demand.

In the developed countries, policies of full employment and maintenance of a high level of public and private demand were abandoned and replaced by policies of generalized deregulation and denationalizaton, which widen the possibilities for capital to move and profitably invest, without proportionately expanding global demand for consumer goods. These neoliberal policies favored the explosion of speculation in financial markets, derivative markets, and currency exchange markets. In the middle of the 1990s, it is estimated that 90 percent of the transactions on the exchange markets were speculative in nature. And, since there no longer exists an international system for regulating exchange rates and financial flows, each country is constrained to adopt policies that aim to maintain the parity of its currency and attract or retain capital, which is henceforth mobile and sensitive to the least risk and the least variation in the rates of return on capital. To do that, it is advisable to limit inflation and reduce deficits, both the public deficit and the deficit in the balance of payments current account. That explains the convergence of economic policies in the developed countries and the conformism of the thinking that justifies those policies.

## The Failure of Austerity Policies in the Developed Countries

Certainly, in theory, austerity policies have the effect of improving the competitiveness of companies in the countries that carry them out. By reducing public expenditures (reduction in investments, elimination of jobs, and freezing of wages for officials, reduction of social welfare, restriction of military programs) and by exercising generalized downward pressure on wages, an attempt is made to reduce social, fiscal, and wage obligations for companies, and thus increase their profits. But since that is obtained at the price of a reduction in demand from households and the administration, in a world economy where productive investments and job creation are precisely limited by the insufficiency of demand, austerity policies can only worsen the general crisis of the world economy. Moreover, the additional profits that investors retain from austerity policies are used more for speculation or to accelerate and accentuate movements of modernization and relocation, which reduce employment and income, than to create new jobs in the developed countries.

These policies, which pretend to be "virtuous" on a national scale in the countries that carry them out, are intrinsically "perverse" for the whole world. Further, it should be noted that if these policies have generally succeeded in reducing inflation and slowing down the increase of deficits, they have not led to a long-lasting return of growth or to the restoration of full employment.

## The Failure of National Policies of Stimulation in a Globalized Economy in Crisis

In this context, countries that attempted, in isolation, to carry out policies of stimulation through public investments and household and administrative consumption (as in France in 1981–1983) also failed. In a country where the production capacity is underutilized because of insufficient outlets, the increase of solvent demand leads to a certain stimulation of internal production. But if this country is open to competition from countries more competitive than it is, the increase in demand is expressed above all by a "stimulation" of imports. Then, if this increase in demand is obtained by increasing wage and fiscal obligations of companies, their competitiveness diminishes, causing a new increase in imports, another, more significant slowdown, indeed even a decline in national production, an accelerated relocation of investments, and an increase in unemployment.

In other words, a national policy of stimulation through demand contributes much, relatively speaking, to expanding the world market and to stimulating production, which is eminently "virtuous." But insofar as this country is open to competition and is less competitive, this policy in the end worsens its own crisis and benefits the more competitive countries. It cannot be sustained for very long.

In any case, the contemporary crisis cannot be treated as a particular crisis in the economy of any specific country, be it developed or developing, nor as the sum of particular crises of this type. Rather it is the global crisis of this completely new "world economy," which has been constructed in the last thirty years by means of reduction in the cost of transportation and communications and the liberalization of the movement of commodities and capital.[12] This is a global crisis that is worsened by the destabilization of the international monetary and financial system, by speculation, by the abandonment of policies of full employment and support for demand, and by the nearly general adoption of deflationist policies, which reduce employment and income.

In these conditions, the national policies of stimulation (through demand) or austerity (stimulation through investment), which only attack the symptoms of the crisis in each country, cannot bring the contemporary general crisis to an end. There can only be a solution to the global crisis of the world economy in a *global policy devised for every country*, which attacks the profound cause of this crisis. As we have seen, the profound cause of this crisis, which has lasted for more than a quarter-century, lies essentially in the massive fall in incomes and purchasing power in the poor agricultural countries, a fall that results both from the tendency toward a unified world market in basic agricultural commodities, beginning with grains, and the tendency of real agricultural prices to fall.

## 4. FOR A WORLD ANTI-CRISIS STRATEGY FOUNDED ON SAFEGUARDING AND DEVELOPING THE POOR PEASANT ECONOMY

To resolve this crisis of the completely new, increasingly decompartmentalized world economy that lacks solvent outlets for investments, it is necessary to raise purchasing power in the poor countries, where the largest sphere of unsatisfied social needs resides and therefore the largest possibilities for increasing world solvent demand.

In 1993, while barely a billion persons living in 24 countries "with high incomes" had an average income per capita of $63 per day, more than 3 billion persons living in 45 countries "with low incomes" had an average income of $1 per day, that is, 60 times less than the former. Moreover, 1.6 billion people living in 63 countries "with middle incomes" had an average income of $7 per day, or 7 times less than the high-income countries.[13] In addition, the poor peasantry, a majority in the developing countries, has incomes much lower still than the average. A simple doubling of its income would then have a limited effect on the increase of world demand. In order for this peasantry to get out of poverty and stimulate the world economy, it is necessary to envisage at least a tripling or quadrupling of its income. Perhaps this increase in income would be sufficient to reduce the pockets of extreme rural poverty, slow down the agricultural exodus, make possible a real return of productive investment for the poor peasantry (purchases of tools and inputs, improvement of fertility, etc.), and lead to a significant increase in agricultural production, thereby creating the conditions of an expanded development of the peasant economy that is both self-supporting and cumulative. In that way, in the medium term, the increase in the incomes of the poor peasantry will go much further than the tripling or quadrupling initially envisaged and would likely lead gradually to an increase in the incomes of other parts of the poor population, rural and urban. In the long term, beginning with a certain threshold of development, peasant agriculture, having become clearly more productive, will be able to support the costs of modernization and industrialization in the poor countries. Then, and only then, will these countries have sufficient purchasing power to contribute effectively to the stimulation of the world economy.

## For a Significant Increase in Agricultural Prices in the Poor Countries

If this analysis is correct, the most appropriate and powerful lever to reduce the immense sphere of poverty that slows down the development of the world economy resides in a *gradual, significant, and prolonged increase in prices of agricultural commodities*, beginning with basic food commodities, in the poor countries.

*Reduce Taxation and Protect Poor Agriculture.* In order to raise agricultural prices in these countries, it is first necessary to roll back policies of direct or indirect "taxation" of agriculture where it is still carried out: taxes on agricultural exports, subsidies on food imports, compulsory deliveries at low prices, overvaluation of the national currency, excessive protection of industry, etc. Policies of "removing taxation" on agriculture have already been adopted by numerous countries, but are not even close to being sufficient to move the mass of the peasantry to above the threshold of capitalization, which is the necessary condition of its development.

In order to increase the incomes of the poorest farmers in the world in a significant way, it is not sufficient, as proclaimed for fifteen years, to abolish the "taxation" and the "pillage" to which they have been subjected in the course of previous decades. It is still necessary above all to *protect* them, that is, *to tax the imports of basic agricultural commodities, beginning with grains*. The total absence of intervention in agricultural prices, i.e., the pure and simple free exchange of agricultural products, will not suffice to increase the purchasing power of the peasantry and other social classes in the poor agricultural countries significantly and persistently and get them out of the crisis.

*A Significant but Progressive Protection.* In order to put an end to rural poverty, the increase in the prices of basic food commodities must be *significant*. Naturally, such an increase in prices should not be established suddenly, because its positive effects on the production of foodstuffs, on wages, and on other types of incomes will not be rapid, while, conversely, the increase in the prices of food commodities and the negative effects that will result for buyers will be immediate. The increase in the prices of basic agricultural commodities should, then, be *gradual*, enough so that at any moment of the process the negative effects on buyers will not prevail over the positive effects for the producers. In other words, it should be gradual enough so that the economic agents have time to adapt and possibly move into a new type of employment. It takes time for the peasants who wander in search of a job and extra monetary income to rejoin the newly viable family farm. And it will take still more time for a part of the capital and the population exiled in the cities to be redirected to agriculture. The increase in agricultural prices, in order to be both large and gradual, must persist for a long time— ten years or twenty years if necessary. It is not possible to reverse in a few years the disastrous consequences of a half-century of low agricultural prices.

## For a Significant Increase in Wages in the Poor Countries

It is also indeed necessary to assess whether this policy of protecting the peasant economy in the poor countries will have the desirable consequence of raising all wages, today ridiculously low, and therefore raising the costs of production and the prices of products exported by these countries.

Of course, the raising of the prices of agricultural products and raw materials exported by the developing countries will have an effect on the economy of importing countries, in the first place, the developed countries. But in view of the fact that imports from the 86 poorest countries (45 countries with low incomes and 41 countries from the lower bracket of the countries with medium incomes) represent *less than 2.5 percent* of the gross domestic product of the high-income countries, this effect will be limited.[14] It is no less the case that, here too, a gradual approach will be necessary. The explosion in the prices of agricultural products, other raw materials, and above all petroleum in the 1970s demonstrated at what point a significant and sudden increase of these prices could worsen the general crisis.

The raising of prices of manufactured products exported by the developing countries will reduce the competitive pressure that the relocated industries in these countries exercise on those of the developed countries. But the outlets for the companies established in the developing countries will not be reduced since the expected result, and by far the most significant one, resides precisely in a strong increase in demand in these countries.

In other words, as opposed to national policies of stimulation through demand, which rapidly turn back against those who carry them out in isolation, and as opposed to austerity policies, which reduce employment and income, a global anti-crisis strategy, based on the expansion of the world market due to a significant and gradual increase in prices and incomes in the developing countries, will benefit simultaneously the poor agricultural countries, the new industrialized countries and the developed countries, because such a strategy attacks the true root of the crisis, to wit, the mass poverty in the developing countries and the resulting narrowness of world demand.

## The Necessity of a Hierarchical World Organization of Markets

The global strategy of raising prices and incomes in the developing countries should not be uniform: *the level of agricultural prices and thus the degree of protection of agriculture should be established in inverse ratio to its productivity*. It should, for example, be higher for the countries of intertropical Africa than for the countries of Southeast Asia and some Latin American countries.

With this intention, it is advisable, then, to select regional subgroups of countries whose agricultural productivities are of the same scale. Each of these main regions of the world would then form a customs union enjoying a degree of protection and a sufficiently high level of agricultural prices to safeguard the poor peasantry and allow it to develop. The selection of these main regions and the determination of the most pertinent level of prices for each of them could be within the competence of a new organization of the United Nations, charged with regulating international trade and rates of exchange. This is not the place to propose such groupings. However, beyond western Europe and North America, there are possible groups of countries in intertropical Africa, continental Asia, Southeast Asia, eastern Europe, the Near East and North Africa, etc. In each of these regional unions, raising the prices of basic food commodities will lead subsequently to an increase in the prices of all exported raw materials, which will then have to be raised and differentiated according to their region of origin.

This proposition is similar to the recommendations formulated by Maurice Allais, winner of the Nobel Prize for Economics (1988), in his speech to the first European food summit in 1993. After having underlined the dangers of generalized free trade under a system of floating rates of exchange, he asserts: "Total liberalization of trade is only possible, only desirable, in the context of regional groups that bring together economically and politically associated countries, of comparable economic and social development, and mutually agree not to make any unilateral decisions, yet still ensure a market that is large enough for effective competition to take place."[15]

Such a world organization of trade, with hierarchical prices, fixed in inverse ratio to the levels of agricultural productivity of each region of the world, presupposes negotiation and international agreement per product, bearing on the prices and quantities to produce in each region, as well as the organization of one or several international funds for balancing out prices for purchasers and funds for stabilization.

In order to have some chance of success, a world strategy based on a new organization of trade of this type should be devised and accepted by all countries, beginning with the developed countries. It would be enough for one group of developing and developed countries to agree on a policy of low prices and low wages to ruin such a strategy. In this respect, one should be suspicious of the current tendency to form large regional economic blocks that group developed and developing countries, such as the North American Free Trade Agreement (NAFTA), or the enlargement of the European Common Market to eastern Europe and North Africa or the association of Japan with Asian and Pacific countries. In fact, if these vast groupings were to lead to a new division of the world between a few large competing groups, focused more on economic war than cooperation, it would go precisely against the anti-crisis strategy proposed here.

## Increasing Prices and Incomes Rather than Financial Assistance

The anti-crisis strategy we propose comes down to improving the terms of trade to the benefit of the poor countries, in order to increase their income and their purchasing power. This raising of prices could be considered as a sort of income transfer from the rich countries to the poor countries, as a form of assistance. The great advantage of assistance by means of prices comes from the fact that it has a much better chance of directly benefiting the agricultural producers and indirectly the rest of the population than classic financial assistance. Of course, this is on condition that the price increases are not massively captured by the state under the form of taxes or by other economic agents under the form of unjustified margins.

Undoubtedly, such a policy of development assistance, by way of differentiated prices in an organized world market, will be difficult to negotiate and manage. But will it be more difficult than the current aid policies, which work via gifts and loans between institutions? Moreover, these forms of financial aid are quite often lost in unproductive expenditures or private savings that return to sustain banks and financial markets. They are increasingly discredited in the eyes of public opinion,

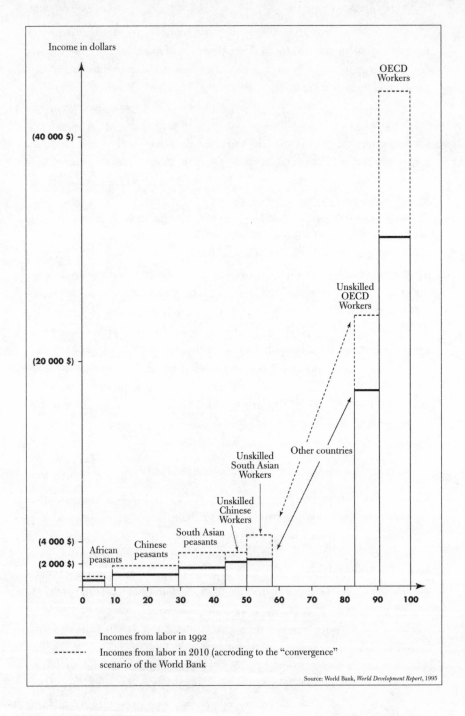

Figure 11.4 *Level of Annual Incomes from Labor in the World in 1992 and According to the World Bank "Convergence" Scenario For 2010*

as much in the developed countries as in the developing countries. In any case, financial assistance today comes up against increasingly limited public budgets in the developed countries and weaker repayment capabilities in the developing countries. It also encounters the limits of management capabilities of national and international development assistance institutions and national institutions in recipient countries. Experience has shown that it has no success in leading the poorest countries into a true process of development.

If, contrary to the scenario proposed here, liberalization of trade continues into the future, then there will be neither a strong reduction in unemployment or poverty in the developing countries nor the creation of adequate purchasing power to stimulate the world economy in a persistent manner nor, therefore, a reduction in unemployment and poverty in the developed countries.

Moreover, the projections established for 2010 by the World Bank show that, in the event of increasing liberalization of world trade, the gap between the highest wages (those of skilled workers in the developed countries of the OECD) and the lowest ones (those of African peasants) could still increase enormously.[16] Under this assumption, this gap would go from around $31,000 to $43,000 between 1992 and 2010. In other words, according to this scenario, the gap would again be increased by close to 1.4. This is far from a scenario of convergence. To realize this, it is sufficient to represent the large range of agricultural incomes and wages in different regions of the world by adopting an ordinary scale and not a logarithmic scale, which hides the gaps (*Figure 11.4*). This graph shows at what point, under the assumption of an increasing liberalism, these immense gaps in incomes will only grow larger. It also aims to show that the level of wages in each region of the world is indeed established, as we maintain here, as a function of the level of income, and thus of the productivity, of peasant agriculture.

## Necessity for National Policies of Safeguarding and Developing the Poor Peasant Economy

But if this international strategy of reorganizing trade to the benefit of the poor countries is necessary, it will not suffice alone to save the most underequipped peasant economy and provide vigorous stimulation for its development. Again it will be necessary that this peasantry actually has access to land, to credit, to adequate hydraulic installations in good repair, and to the results of research appropriate to its needs. Again it will be necessary that this peasantry benefits from a stability of prices and a security in land tenure sufficient to ensure that it collects the fruits of its labor and its investments, and that it benefits, in the end, from good maintenance of and improvements in the fertility of the lands it farms. Again it will be necessary that the incomes of this peasantry not be eroded by exorbitant processing and marketing costs or by excessive land rents, taxes, or duties. That is, to support this international strategy of development for the poor peasant economy, national economic and agricultural policies will have much to do.

### Agrarian Reform and Development Policies for the Poor Peasant Economy

In countries where the proliferation of minifundia and mass peasant poverty come from unequal distribution of land, agrarian reform is the first of the policies to implement. Naturally, agrarian reform is a domestic policy decision that is difficult to take and apply because it provokes violent opposition. It cannot, then, be decreed from the outside, but it could nevertheless be supported more vigorously than it is presently by international development institutions such as the United Nations Food and Agriculture Organization (FAO), the International Fund for Agricultural Development (IFAD), the United Nations Development Program (UNDP), the World Bank, and the regional development banks. In the context of the international strategy proposed here, these institutions would have to monitor the price increases for agricultural commodities to ensure they only benefited those countries that had actually carried out an agrarian reform where it was necessary and that practice a policy of development favorable to the poor peasantry.

Indeed, to have a lasting impact, an agrarian reform should be followed by an enlarged and not costly credit policy, making it possible for deprived peasants to stock and sell their harvests at opportune times, buy necessary inputs, and gradually equip themselves (installment plans for equipment). Without these measures, the beneficiaries of the reform are deprived of capital for farming, go into debt, mortgage, and soon lose their lands. An agrarian reform should also be followed by a consistent land policy, which aims to prevent the process of land concentration and unequal development from getting the upper hand again: an anti-accumulation law, priority allocation of freed or newly organized lands to the most deprived peasants, specific aid for setting up young poor farmers, etc. These land and credit policies would certainly not be necessary only in regions having recently benefited from agrarian reforms. They would be necessary in all countries with a peasant economy, to prevent the obstruction of the small peasantry's development and the reappearance of mass poverty that would result from continual unequal development between regions and farms.

Policies for managing the infrastructure (service roads, terraces and other anti-erosion works, irrigation, drainage, etc.) will have to be revised by avoiding disproportionate and nonprofitable projects dear to large institutions and by favoring more appropriate projects, worked out and managed in cooperation with the population, which rely greatly on off-season agricultural labor, experience of the peasantry, and other local resources. Moreover, the organization of markets will have to facilitate the selling and best increase in valorization of the peasantry's products.

## Redirection of Research Policies

Lastly, in order that the international strategy and national policies for stimulating the poor peasant economy have all the necessary assets, agricultural research policies must be massively directed, or more precisely redirected, to the benefit of the peasantry and disadvantaged regions. To contribute effectively to the reduction of poverty, research for agricultural development will have to be much more attentive than in the past to the needs and possibilities, but also and above all, to the knowledges and know-how of the peasantry. It will have to take advantage of all the diversity of local species, varieties and breeds of plants and domestic animals that it has previously neglected, in order to improve them to the benefit of the greatest number. Such research will have to focus on complex systems of production that the peasantry gradually developed, which combine crop growing, stockbreeding, and arboriculture, in order better to renew fertility and reduce the biological and economic risks with which that peasantry is confronted, risks that are more serious the greater the level of poverty.

Densely populated regions of the world today where the peasantry develops complex, sustainable forms of agriculture with high yields that are not costly in nonrenewable resources are undoubtedly the laboratories in which the most invaluable forms of agriculture for the future of humanity are elaborated. Only the deep ecological and economic study and accurate comprehension of these agrarian systems, which are the fruit of many centuries of continually renewed peasant experience, can make it possible for researchers to identify and propose appropriate improvements, and transfer, through adaptations, benefits from one agriculture to another.

In this regard, it seems that the Consultative Group on International Agricultural Research (CGIAR), a body that groups the research centers of the Green Revolution and is today essentially financed by thirty mostly developed or newly industrialized countries, and some national research services have begun to orient their work in this direction, indeed more than in the past. But for such a reversal in perspective to materialize, it will require much effort and many changes in attitude, not only in research but also in education and administration.

In a competitive economy, capital and knowledge are generally attracted to the most immediately profitable activities, regions, or types of farm. Policies and projects that aim at a balanced and long-lasting planetary development certainly do not consist in reinforcing this spontaneous tendency by allocating additional public means, both financial and human, to those who can develop without that. On the contrary, it involves struggling continually, across a large front, against stagnation and the impoverishment of the most disadvantaged by devoting, as a priority, the necessary public means to that struggle.

## 5. CONCLUSION

The experience of twenty-five years of rapid world economic growth, followed by more than twenty years of slow growth and latent crisis, shows that the lowering of transportation costs, the opening of national economies, and the growing liberalization of international trade have not reduced the disparity in equipment, productivity, and income between the poorest countries and the richest countries, nor resolved the problem of unemployment and poverty in the world. On the contrary, during the latest period, poverty, unemployment, and the inequalities between the most deprived and the most affluent have increased. Scenarios of the future resting on assumptions of continuing world economic integration and increased liberalization of trade show that these disparities are going to increase more. Is that surprising?

Among the developing countries that were still essentially agricultural at the end of World War Two, only a small group of them, which had inherited a relatively productive agriculture and, moreover, carried out a policy favorable to the latter, succeeded in extracting an agricultural surplus sufficient to develop the other sectors of the economy significantly and create the conditions for a high level of profitability for investments. These newly industrialized countries saw a portion of their population escape poverty.

But, in most of the developing countries, the dominant underequipped and relatively unproductive peasant agriculture, often taxed and in all cases insufficiently protected, did not have the means of equipping itself and developing. It was subjected to a North-South competition and a South-South competition above its resources and sustained a fall in prices that led hundreds of millions of peasants to ruin, rural exodus, unemployment, and extreme poverty.

Even if a small fraction of the ruined peasants, unemployed and poor from the developing countries, were able to emigrate to the industrialized countries and attain poorly paid employment, the great majority of them had neither the means nor permission to do so. And they still had less access to agricultural lands, however overabundant and to some extent lying fallow, or to agricultural credit from the developed countries. In other words, even if, in the world of today, the free circulation of commodities and capital is increasingly effective, there is no free circulation of people and even less free access to land and credit. It is capital that moves toward the immense reserves of low-priced labor formed by the most accommodating developing countries.

If, in the future, agricultural prices, incomes, and, following from them, wages in the developing countries should remain as low as they are today, then the industrialization of a handful of these countries would be principally directed toward exports to the countries still having a significant purchasing power (developed countries, petroleum-exporting countries, and some newly industrialized countries) and it would be to the detriment of already industrialized countries with higher wages. According to this scenario, industrialization of countries with low wages would, in the developed countries, lead to a large expansion of unemployment and lowering of wages that would, in turn, lead to a gradual strangulation of world demand and, in the end, to a reduction of global possibilities for productive investments and job creation, an upsurge of speculation, and generalized recession.

If the industrialization of the developing countries is to participate in a true stimulation of the world economy, it must be based on a significant and lasting growth of purchasing power in these countries. In order to form in this "two-thirds of the world" an effective demand equal to human needs and the possibilities of lasting growth in the world economy, it is necessary, as we have tried to demonstrate, to begin by raising the prices of basic agricultural commodities in a significant, gradual, and sustained manner. Such an increase in agricultural prices is the best means of increasing the incomes of the underequipped peasantry, favoring its development and consequently curbing the agricultural exodus, limiting the rise in unemployment and urban poverty, raising the general level of wages and other incomes, significantly increasing the potential for tax revenues and foreign exchange earnings in the developing countries, and, finally, releasing investment capabilities, thereby making it possible for these countries to modernize and industrialize.

To promote this anti-crisis scenario of stimulating the world economy, both through vigorous development on a broad front of the poor peasant economy and a huge enlargement of solvent demand in the developing countries, there is no other way than a world organization of trade, based on regional customs unions grouping countries having comparable levels of agricultural equipment and productivity. Each of these regional unions would benefit from a sufficiently high level of prices for agricultural commodities and raw materials, negotiated internationally, making it possible for peasant agriculture to develop and get the majority of the population out of poverty.

But for this strategy of stimulating the world economy to succeed, it is still necessary for it to be followed up by a policy of balanced agricultural development in each country, strongly directed—or rather redirected—toward the peasantry and disadvantaged regions. Further, it will be necessary that the world give itself a new international monetary and financial system that assures the preservation of relatively stable rates of exchange that vary within reasonable limits around fundamental equilibrium exchange rates, penalizes speculation, and favors concerted policies of development in all countries. These policies aim at full employment and an increase in solvent demand proportional to production and investment capacities existing in the world, rather than deflationist policies that reduce jobs and income.

# Conclusion

Today's world agricultural and food economy is less disorganized and chaotic than the price jolts, surpluses, shortages, famines, and hostile international trade negotiations lead one to believe. Just as the waves and tides of an ocean reflect the organization and functioning of the solar system, movements of the atmosphere, and marine currents, the surface agitations of markets and agricultural policies reflect the organization, functioning, and dynamics of the world agricultural and food system. This system has been formed only in the course of the last few decades by bringing together the diverse agricultures produced by 10,000 years of an agrarian history which is extraordinarily differentiated throughout the world.

This world agricultural and food system, composed of relatively specialized, competing, and unequally productive regional subsystems, develops in a contradictory and divergent manner. On one side, a reduced number of farms and regions of the world continually accumulate more capital, concentrate on the most productive crops and animal breeding operations, and continually conquer new market shares. On the other side, extensive regions and the majority of the world's peasantry are plunged into crisis and poverty, up to the point of being excluded from the world economy altogether. On one side, an agriculture that is able to be profligate through an excess of means. On the other side, an agriculture that, lacking the means, no longer renews the fertility of the environments it exploits.

This colossal distortion of the world agricultural and food system is the basis of the enormous inequalities of income and development that exist among countries. If, unfortunately, the world were allowed to drift according to such a violently contradictory law of development, it must be feared that the world will come to resemble more an archipelago of well-guarded islands of prosperity dispersed in an ocean of poverty than a universe of prosperity that conquers and absorbs, one after another, the residual pockets of poverty.

The crisis of the underequipped and relatively unproductive peasantry, by far the most numerous, is the cause of the rising tide of rural and urban poverty that makes the development of the poor agricultural countries impossible. This mass poverty—in other words, the unmet needs of more than half of humanity—is the cause of the manifest inadequacy of world effective demand, the slowdown in economic growth, and the rising unemployment and poverty that reach right into the developed countries themselves. As a result, capital, in search of profitable investments, is oriented more and more toward profuse speculation, a modernization that reduces employment and relocations that reduce income, all of which only worsen the general crisis, with its trail of poverty, despair, criminality, corruption, and wars.

If one really wants to get out of the contemporary general crisis and build a world of full employment and lasting, extensive, and equitably distributed prosperity, a world to which the great majority of the world's inhabitants aspire and in which everybody would be able to take advantage of opportunities, both material and moral, it is necessary to create the conditions for a real development of the underequipped peasant economy and an accumulation of long-term productive capital in the poor countries. In order to accomplish this, it is necessary to attack the roots of the problem—which is to say, the enormous inequalities in income that result from the thoughtless competition among highly unequal agrarian heritages.

These inequalities of income and development will remain insurmountable so long as a much more equitable world monetary, financial, and trade system has not yet been brought into being, to correct the huge inequalities of yield inherited from history and geography. These inequalities will remain insurmountable as long as the policies, the projects, and the pursuit of development in each country are not principally directed toward the most disadvantaged regions and toward safeguarding and developing the poorest peasant economy. In order to move in this direction and be both legitimate and effective, these policies and projects will have to be conceived and carried out in a democratic manner, with the effective participation of the populations concerned.

\*

By placing agriculture at the center of our analysis of the contemporary crisis and by crediting it with a primary role in the solution of that crisis, it was not our intent, certainly, to reduce the problem to this essential aspect. Moreover, we have taken into account many other aspects of this many-sided crisis, even if we have not treated them in a detailed manner. But insofar as agriculture generally forms the blind spot in analyses of the crisis, we have tried to emphasize and share what our personal origins and our professions allow us to understand better—to wit, that it is not possible to explain the contemporary world crisis without taking into account the immense and contradictory transformations that drive today's agricultures and without evaluating the part they play in the formation of planetary poverty and unemployment. Further, it is not possible to remedy a crisis of such magnitude without protecting the impoverished peasant economy and without turning to the immense possibilities this peasant economy, representing nearly half of the world's population and the majority of the poor, presents for creating employment and income.

This book is too global in perspective to present the daily life of the peasants who, for 10,000 years, have continually built and rebuilt the agrarian base upon which we live. Others with more talent than we have are doing that. For our part, we wanted to honor the work of yesterday's peasants and defend the work of today's peasants.

We have viewed from below, over a very long period of time, the world's self-construction based on agriculture. Thus we have given a quite different representation of this process than that obtained by viewing it from the heights of timeless theory or the heights of the financial and political conjuncture of the moment. As a result, the ideas we have presented of the causes and solutions for the contemporary crisis are quite different from the ideas that are politically dominant today. According to the latter, the ills of this world come essentially from insufficient competition, and the best economic policies are those that continually facilitate this competition, limiting and mitigating the most negative effects, which are viewed as transitory. But our position is close to the increasingly numerous analyses that point to the necessity for a coordinated world policy that focuses on the equitable reorganization of international trade, as well as of the international monetary and financial system, and on a balanced development of every country as the remedy for a crisis that is more global than ever.

In truth, this world, which is crumbling today from the bottom much more quickly than it is being built from above, has become a colossus with clay feet, a cracked colossus whose foundation must be reconstructed in all urgency. As the beautiful motto inscribed on the front of the house of the storyteller Charles, close to Lekana on the Koukouya plateau in Congo, declares: "When one wants to climb a tree, one does not begin at the top."

# Notes

PREFACE

1   Around 1.5 billion individuals lack iron, 740 million lack iodine, and 200 million lack vitamin A, according to the United Nations Food and Agriculture Organization, commonly designated by its English abbreviation, FAO.

2   According to the FAO, there are around 800 million undernourished people in the developing countries, or close to one individual in five, 30 million in the transitional countries (formerly the planned economies), and 10 million in the developed countries. These figures, which are uncertain, should be considered as orders of magnitude.

3   A cereal equivalent is the quantity of cereals having the same caloric value as the agricultural product under consideration.

CHAPTER 1

1   François Ramade, *Le Peuple des fourmis* (Paris: Presses Universitaires de France, 1965).

2   Y. Coppens, *Le Singe, l'Afrique et l'homme* (Paris: Fayard, 1983).

3   Göran Burenhult, ed., *The First Humans* (St. Lucia, Queensland: University of Queensland Press, 1994).

4   Coppens, *Le Singe, l'Afrique et l'Homme*

5   Burenhult, *Les Premiers Hommes.*

6   I. C. Glover, "Outils et cultures du paléolithique tardif en Asie du Sud–Est," in *Les Premiers Hommes* (Paris: Bordas, 1994), pp. 128–29.

7   M. Daumas, ed., *The Origins of Technological Civilization*, vol. 1 of *A History of Technology and Invention*, trans. Eileen B. Hennessy (New York: Crown, 1970).

8   Ibid.

9   V. G. Child, *Man Makes Himself*, p. 18.

10  R. Thom, *La Rencontre théorie-expérience,*

11  Ph. Duchafour and B. Souchier, eds., *Pédologie*, 2nd ed., 2 vols. (Paris: Masson, 1983–1994); Ph. Duchaufour, *Pédologie—sol, végétation, environnement*, 4th ed. (Paris: Masson, 1994).

12  The French is *une friche boisée*, a "wooded fallow period." In this context, "fallow" implies allowing the land to return to its original wooded state. —Trans.

13  The French is *une friche herbeuse*, a "grassy fallow period." In this context, "fallow" implies allowing the land to return to its original grassy state. —Trans.

14  The French is *la jachère*. This type of fallowing should not be confused with the different types of *friche* mentioned above.—Trans.

15  J.-N. Biraben, "Essai sur l'évolution du nombre des hommes,"*Population*, no. 1 (1979): 13–25; R. Krengel, *Die Weltbevölkerung von den Anfängen des anatomisch modernen Menschen bis zu den Problemen seiner Überlebensfähigkeit im 21 Jahrhundert* (Berlin: Duncker & Humboldt, 1994).

16  René Dumont and Lester Brown, Worldwatch Institute.

17  FAO, *World Agriculture: Toward 2010: An FAO Study* (1995).


CHAPTER 2

1  J. R. Harlan, *Les Plantes cultivées et l'Homme* (1987).

2  Jacques Cauvin, *The Birth of the Gods and the Origins of Agriculture*, trans. Trevor Watkins (New York: Cambridge University Press, 2000).

3  Harlan, *Les Plantes cultivées et l'Homme*.

4  Achilles Gautier, *La Domestication et l'homme créa ses animaux* (Paris: Errance, 1990).

5  G. O. Rollefson, *Le Néolithique de la vallée du Jourdain*, 1994; North American center, D. H. Thomas, *Agriculteurs due Nouveau Monde* (1994); Chinese center, Harlan, *Les Plantes cultivées et l'Homme*.

6  Cauvin, *Birth of the Gods and the Origins of Agriculture*.

7  Marshall Sahlins, *Stone Age Economics* (Chicago: Aldine–Atherton, 1972);and following him Cauvin, *Birth of the Gods and the Origins of Agriculture*, and Harlan, *Les Plantes cultivées et l'Homme*.

8  Cauvin, *Birth of the Gods and the Origins of Agriculture*.

9  Ibid.

10  Gerald Mendel, *La Chasse structurale: une interprétation du devenir humain* (Paris: Payot, 1977).

11  P. Bellwood, "Les origines des familles de langues," in *L'Âge de pierre* (Paris: Bordas, 1994), pp. 138–39.

12  C. Renfrew, *L'Énigme indo-européene*.

13  Arab, Hebrew, and Nilotic languages: Louis Hjelmslev, *Language: An Introduction*, trans. Francis J. Whitfield (Madison: University of Wisconsin Press, 1970); languages of African farmers: Cheikh Anta Diop, *Nations nègres et Culture*, 3rd ed. (Paris: Présence africaine, 1979).

14  Harlan, *Les Plantes cultivées et l'Homme*.

15  Gautier, *La Domestication*.

16  Harlan, *Les Plantes cultivées et l'Homme*.

17  Gautier, *La Domestication*.

18  J. P. White, "Peuples du Pacifique," in *L'Âge de pierre* (Paris: Bordas, 1994), pp. 145–61.

19  D. H. Thomas, *Agriculteurs du Nouveau monde* (1994).

20  P. Bellwood and G. Barnes, "Les agriculteurs d'Asie du Sud et de l'Est," in *L'Âge de pierre* (Paris: Bordas, 1994), pp. 123–43.

21  G. O. Rollefson, *Le Néolithique de la vallée du Jourdain* (1994).

22  Re the Blue River: Bellwood, "Les Origines des familles de langues."

23  J. Barrau, "Histoire et préhistoire horticoles de l'Océanie tropicale," *Journal de la Société des océanistes* (1965): 55–78.

24  P. Rolley-Conwy, "Chasseurs-Cueilleurs at Agriculteurs de l'âge de pierre en Europe," 1994.

25  G. Barnes and P. Bellwood, *Les Agriculteur d'Asie du Sud et de l'Est* (1994).

26  D. H. Thomas, "Agriculteurs du nouveau Monde," in *L'Âge de pierre* (Paris: Bordas, 1994), pp. 163–85.

27  Jack R. Harlan, "Les origines de l'agriculture," *La Recherche* 3, no. 29 (1972): 1035–43.

28  Gautier, *La Domestication.*

29  Cauvin, *Birth of the Gods and the Origins of Agriculture.*

30  Jack R. Harlan, *Crops and Man*, 2nd ed. (Madison: American Society of Agronomy: Crop Science Society of America, 1992).

31  Gautier, *La Domestication.*

32  J. Pernès, "La gégétique de la domestication des céréales," *La Recherche* 14, no. 146 (1983): 910–19.

33  Ibid.

34  Gautier, *La Domestication.*

## CHAPTER 3

1   Pierre Pétrequin and Christian Jeunesse, eds., *La Hache de pierre. Carrières vosgiennes et échanges de lames polies pendant le néolithique (5400–2100 avant J.-C.)* (Paris: Éditions Errance, 1996).

2   Marcel Mazoyer, "Le développement de la production marchande et la dégradation des systèmes agraires traditionnels en Afrique de l'Ouest," Communication au Colloque IDEP–IEDES–IDS–CLASCO sur Les stratégies du développement économique Afrique et Amérique latine comparées (Dakar, 1972).

3   Pierre Gourou, *Riz et Civilisation* (Paris: Fayard, 1984).

4   Gourou, citing R. Champsolaix in *Riz et Civilisation.*

5   Gourou, *Riz et Civilisation.*

6   Phillipe Duchaufour et al., *Pédologie–sol, végétation, environnement,* 4th ed. (Paris: Masson, 1994).

7   Plato, *Timaeus and Critias*, trans. Desmond Lee (New York : Penguin, 1977), pp. 133–34.

8   François Sigaut, *L'Agriculture et le Feu* (Paris: Mouton, 1975); Pierre Gourou, *L'Afrique* (Paris: Hachette, 1970); *Riz et Civilisation.*

9   Cl. Serre-Duhem, "Essai d'interprétation d'une famine. Les transformations d'un système agraire au Congo: le plateau Kukuya," Ph.D. diss., Institut national agronomique Paris–Grignon, 1995.

10  André Angladette, *Le Riz* (Paris: Maisonneuve et Larose, 1966).

11  Marcel Mazoyer et al., *Esquisse d'une nouvelle politique agricole au Congo* (Brazzaville: Ministére du Développement rural, ministère de l'Économie et du Plan, 1986).

## CHAPTER 4

1   A *shadouf* is a long pole with a bucket tied to a rope on one end and a counterweight on the other. This, in turn, is attached to a post in the ground which is free to swivel. The bucket is dipped into a source of water and then the whole unit is swiveled around and the bucket is emptied onto the field. — Trans.

2   Achilles Gautier, *La Domestication et l'homme créa ses animaux* (Paris: Errance, 1990).

3   R. G. Klein, "Chasse, cueillette et agriculture en Afrique," in *L'Âge de pierre* (Paris: Bordas, 1994), pp. 39–55.

4   F. Wendorf et al., *L'Utilisation des plantes au Sahara* (1994).

5   Göran Burenhult, *L'Art rupestre dans le Sahara central* (1994).

6   G. Hamdan, "Évolution de l'agriculture irriguée en Égypte," in *Histoire de l'utilisation des terres des régions arides* (Paris:UNESCO, 1961), pp. 133–59

7   G. Alleaume, *Les Systèmes hydrauliques de l'Égepte prémoderne.*

8   Ibid.

9   T. Ruf, Histoire contemporaine de l'agriculture égyptienne (Paris: Éditions de l'ORSTOM, 1988).

10  J. Vercoutter, *L'Égypte jusqu'à la fin du Nouvel Empire.*

11  S. Sauneron, *Histoire générale du travail*, vol. 1.

12  A. Aymard and J. Auboyer, *L'Orient et la Grèce antique,* 7th ed. (Paris: Presses universitaires de France, 1979)..

13  G. Alleaume, "L'évolution du paysage à l'époque arabe," in *Égyptes. Histoires et cultures,* no. 4 (1994): 35–41; "Les systèmes hydrauliques de l'Égypte pré–moderne. Essai d'histoire du paysage," in *Itinéraires d'Égypte. Mélanges offerts au père Maurice Martin, SJ* (Cairo: Institut français d'archéologie orientale du Caire, 1992), pp. 301–22

14  Hamdan, "Évolution de l'agriculture irriguée en Égypte."

15  Julien Hipployte Eugène Barois, *Irrigation in Egypt,* trans. A. M. Miller (Washington, D.C.: Government Printing Office, 1889).

16  Ruf, *Histoire contemporaine de l'agriculture égyptienne.*

## CHAPTER 5

1   Karl Wittfogel, *Oriental Despotism: A Comparative Study of Total Power* (New Haven: Yale University Press, 1957),

2   O. Dollfus, "Les sociétés paysannes andines: autonomie et dépendance," in *Sociétés paysannes du tiers–monde* (Lille, France: Presses universitaires de Lille, 1980_, pp. 13–24..

3   Rafael Karsten, *A Totalitarian State of the Past: the Civilization of the Inca Empire in Ancient Peru* (Port Washington, N.Y.: Kennikat Press, 1969).

4   Nathan Wachtel, *The Vision of the Vanquished: The Spanish Conquest of Peru through Indian Eyes, 1530–1570*, trans. Ben and Siân Reynolds (Hassocks, Eng.: Harvester Press, 1977).

5   Phillip Wolff, *Histoire générale du travail.*

6   Wachtel, *The Vision of the Vanquished.*

7   Ibid.

8   This policy was a form of *composition*, a term from ancient and medieval law that referred to a sum of money paid by a guilty party to the family of an injured or killed person and/or to the state. —Trans.

## CHAPTER 6

1   Epigraph: *The Republic of Plato*, rtrans. Francis MacDonald Cornford (London: Oxford University Press, 1941), pp. 60–62 —

2   The variety of idling of long and medium duration referred to here is *la friche* in French, i.e., formerly uncultivated land allowed to return to a natural state. The short-term fallowing is *la jachère*, i.e., that type of formerly cultivated land subject to human labor during the uncultivated part of its rotation. —Trans.

3    The *saltus* is of the *friche* variety of fallow land, i.e., it is not plowed.

4    Auguste Jardé, *Les Céréales dans l'Antiquité grecque* (Paris: Boccard, 1979).

5    In the French, the authors are discussing the proper use of the term *jachère* versus use
     of the term *friche*. I have changed the sentence so that the emphasis is on the conceptual
     distinction and not the term used to denote that distinction, since in English there is just
     the one term, i.e., *fallow*. The original reads: "*Il est donc impropre, soit dit en passant,
     d'employer le terme de jachère pour designer une friche boisée de moyenne ou de longue
     durée défrichée par abattis-brûlis, ou pour parler d'un pâturage naturel en rotation non
     encore labouré.*" —Trans.

6    F. Sigaut, *L'Agriculture et le Feu.*

## CHAPTER 7

1    Epigraph: Georges Duby, *The Age of the Cathedrals: Art and Society, 980–1420,* trans.
     Eleanor Levieux and Barbara Thompson. (Chicago: University of Chicago Press, 1981),
     p. 93.

2    G. and C. Bertrand in *Histoire de la France rurale,* vol. 1, ed. Georges Duby (Paris:
     Seuil, 1975).

3    Hesiod, *Works and Days*

4    Lefebvre des Noëttes, *L'Attelage: le cheval de selle à travers les âges: contribution
     à l'histoire de l'esclavage* (Paris: Picard, 1931).

5    Jean Gimpel, *The Medieval Machine: The Industrial Revolution of the Middle Ages,*
     2nd ed. (Aldershot, Hants, England: Wildwood House, 1988).

6    At this point in the French text, reference is made to the name given to this crop, for
     which there is no English equivalent. The parenthetical text continues: "*d'où le nom
     'trémois' qui lui est parfois donné.*" This may be translated as "hence the name 'trémois'
     that is sometimes given to it." In other words, this cereal crop is sown in March, say,
     and harvested three months later. —Trans.

7    The French text refers to the names often given to these kitchen gardens in France:
     *potager* from *potage,* meaning "soup" —Trans.

8    A coppice is a forest of shrubs and small trees that has grown from shoots and suckers
     on stumps left behind from previous cuttings, rather than from seedlings. A coppice
     with standards, then, is a forest in which some of these smaller trees and shoots are
     protected in order to encourage the growth of a new forest of mature trees. —Trans.

9    Perrine Mane, *Calendriers et Techniques agricoles (France-Italie, XIIe–XIIIe siècle)*
     (Paris: Le Sycomore, 1983).

10   Georges Duby, *Rural Economy and Country Life in the Medieval West,* trans. Cynthia
     Postan (Columbia, S.C.: University of South Carolina Press, 1968).

11   P. Goubert in , *Histoire économique et sociale de la France,* vol. 2, ed. Fernand Braudel
     and Ernest Labrousse (Paris: Presses Universitaires de France, 1977).

12   Emilio Sereni, *History of the Italian Agricultural Landscape,* trans. R. Burr Litchfield
     (Princeton, N.J.: Princeton University Press, 1997).

13   E. Perroy, *Le Moyen Âge,* (Paris: Presses Universitaires de France, 1993).

14   G. Fourquin, in *Histoire de la France rurale,* vol. 1 (Paris: Seuil, 1975), pp. 439–44,

15  Jacques Le Goff, *Medieval Civilization, 400–1500*, trans. Julia Barrow (Oxford: Blackwell, 1990).

16  Perroy, *Le Moyen Âge*.

17  Gimpel, *The Medieval Machine*.

18  Marie-Claire Amouretti, "La diffusion du moulin à eau dans l'Antiquité," in *L'Eau et les Hommes en Méditerranée* (Paris: CNRS, 1987).

19  Perroy, *Le Moyen Âge*.

20  Gimpel, *The Medieval Machine*.

21  The French expression is *une charte de franchise.* —Trans.

22  Jean Gimpel, *Les Bâtisseurs de cathédrales*.

23  A. Guerreau, *Le Féodalisme: un horizon théorique* (Paris: Le Sycomore, 1980).

24  E. Fournial, *Les Villes et l'Économie d'échange en Forez aux XIIIe et XIVe siècles*.

25  Emmanuel Le Roy Ladurie, *Histoire économique et sociale de la France*, vol. 1, pt. 2, *Histoire économique et sociale de la France*, ed. Fernand Braudel and Ernest Labrousse (Paris: Presses Universitaires de France, 1977),

26  Duby, *Rural Economy and Country Life in the Medieval West*.

27  Le Roy Ladurie , *Histoire économique et sociale de la France* .28 Ester Boserup, *The Conditions of Agricultural Growth: The Economics of Agrarian Change Under Population Pressure* (London: Routledge, 2003).

## CHAPTER 8

1  Lucie Bolens, *Agronomes andalous du Moyen Âge*.

2  Georges Duby, *Rural Economy and Country Life in the Medieval West*, trans. Cynthia Postan (Columbia, S.C.: University of South Carolina Press, 1968).

3  M. Postan and Christopher. Hill, *Histoire économique et sociale de la Grande-bretagne*, vol. 1.

4  Emmanuel Le Roy Ladurie, *Histoire économique et sociale de la France*, vol. 1, pt. 2, *Histoire économique et sociale de la France*, ed. Fernand Braudel and Ernest Labrousse (Paris: Presses Universitaires de France, 1977),

5  L. de Lavergne, *Essai sur l'économie rurale de l'Angleterre, de l'Écosse et de l'Irlande* (Paris: Librairie agricole, Guillaumin et Cie, 1882).

6  Naomi Riches, *The Agricultural Revolution in Norfolk*, 2nd ed. (London: Frank Cass, 1967).

7  J.-C. Toutain, *Cahiers de l'ISEA*.

8  *Histoire de la France rurale*, vol. 3 (Paris: Seuil, 1975), p. 400.

9  J.-N. Biraben, "Essai sur l'évolution du nombre des hommes," *Population* 1 (1979): 13–25.

10  Paul Bairoch, *Third World at an Impasse.*, trans. Don Fillinger (New York: Prentice-Hall, 1996).

11  P. Giraud, *La Propriété foncière en Grèce* (Paris, 1893).

12  Marc Bloch, *French Rural History: An Essay on Its Basic Characteristics,* trans. Janet Sondheimer (Berkeley: University of California Press, 1966).

13  Postan and Hill, *Histoire économique et social de la Grande Bretagne.*

14  Michael Tracy, *Government and Agriculture in Western Europe, 1880–1988*, 3rd ed. (New York; London: Harvester Wheatsheaf, 1989).

15   Cited by Roger Barny, in *L'Éclatement révolutionnaire du rousseauisme*.

16   Jethro Tull, *The New Horse-Hoeing Husbandry* (1731).

17   Duhamel du Monceau, *Traité de la culture des terres,* 1750–1760.

18   M. Augé-Laribé, *La Révolution agricole* (Paris: Albin Michel, 1955).

19   Vauban, *La Dîme royale* (1707).


CHAPTER 9

1    Michael Tracy, *Government and Agriculture in Western Europe, 1880–1988*, 3rd ed.
     (New York; London: Harvester Wheatsheaf, 1989), pp. 36–37.


CHAPTER 10

1    Marcel Mazoyer, "Les modalités d'applications de la recherche opérationnelle en
     agriculture," *Revue française de recherche opérationnelle* 27 (1963): 107–29.

2    The change in price of one commodity over several years is not easy to estimate. The
     prices recorded each year are expressed in current monetary units (franc, dollar, etc.)
     and the value of this monetary unit is not constant. In general, it diminishes from one
     year to another because of the tendency for the prices of all commodities to increase,
     i.e., inflation. In order to assess the "real" change in price of a particular commodity, it is
     necessary to reevaluate the recorded prices over the years (current prices) in a constant
     monetary unit, which is that of the chosen reference year. This amounts to "deflating"
     the current prices, that is, tcorrecting them for inflation by means of an appropriate price
     index, based on the reference year.

3    David Ricardo, *Principles of Political Economy and Taxation*.

4    René Dumont, *Le Problème agricole français* (Paris: Les Éditions Nouvelles, 1946).

5    René Dumont, *Les Leçons de l'agriculture américaine* (Paris: Flammarion, 1949).

6    Marcel Mazoyer et al., "Essai d'appréciation des conditions d'application et des résultats
     d'une politique de réforme de l'agriculture dans les régions difficiles,"
     *Informations internes sur l'agriculture* 138 (1974).


CHAPTER 11

1    World Bank, *World Development Report 1986*

2    Ibid.

3    Ibid.

4    Marcel Mazoyer et al., *Esquisse d'une nouvelle politique agricole au Congo* (Brazzaville:
     Ministére du Développement rural, ministère de l'Économie et du Plan, 1986).

5    A. Krueger, M. Schiff, and A. Valdès, *The Political Economy of Agricultural Pricing Policy*

6    Ibid.

7    D. Diakosavvas, P.-L. Scandizzo, *Trends in the Terms of Trade of Primary
     Commodities, 1900-1982: The Controversy and its Origins*).

8    World Bank, *World Development Report*, 1986

9    A. Singh and A. Zammit, "Employment and Unemployment, North and South,"
     in *Managing the Global Economy*, ed. Jonathan Michie and John Grieve Smith
     (Oxford: Oxford University Press, 1995), pp. 93–110.

10  M. Kitson and J. Michie, "Trade and Growth: A Historical Perspective,"
    in *Managing the Global Economy*, ed. Jonathan Michie and John Grieve Smith (Oxford:
    Oxford University Press, 1995), pp. 3–36.

11  Ibid.

12  World Bank, *World Development Report 1995: Workers in an Integrating World* (1995).

13  Ibid.

14  Ibid.

15  Maurice Allais, European Food Summit, 1993.

16  World Bank, *World Development Report 1995*.

# Metric Units of Measurement Converted into Imperial Units

1 centimetre (cm) . . . . . . . . . . 10 mm . . . . . . . . . . . . . . 0.3937 inches

1 meter (m) . . . . . . . . . . . . . . . 100 cm . . . . . . . . . . 1.0936 yards (yd)

1 kilometer (km) . . . . . . . . . . 1,000 m . . . . . . . . . . . . . . . 0.6214 mile

AREA

1 hectare (ha) . . . . . . . . . . . . 10,000 m$^2$ . . . . . . . . . . . . . . 2.4711 acres

1 sq km (km$^2$) . . . . . . . . . . . . . 100 ha . . . . . . . . . . . . . . . . 0.3861 mile2

WEIGHT

1 gram (g) . . . . . . . . . . . . . . . . 1,000 mg . . . . . . . . . . . . . 0.0353 ounce

1 kilogram (kg) . . . . . . . . . . . 1,000 g . . . . . . . . . . . . 2.2046 pounds

1 tonne (t) . . . . . . . . . . . . . . . 1,000 kg . . . . . . . . . . . . . . . 0.9842 ton

TEMPERATURE

0 degrees Celsius . . . . . . . . . . . . . . . . . . . . . . . . 32 degrees Fahrenheit

100 degrees Celsius . . . . . . . . . . . . . . . . . . . . . 212 degrees Fahrenheit

# Index

A History of World Agriculture